AF394264

Dynamics of Mechanical Systems

Written by experts in the field, this text provides a modern introduction to three-dimensional dynamics for multibody systems. It covers rotation matrices, the twist-wrench formalism for multibody dynamics, and Lagrangian dynamics, an approach that is often overlooked at the undergraduate level. The only prerequisites are differential equations and linear algebra as covered in a first-year engineering mathematics course.

The text focuses on obtaining and understanding the equations of motion, featuring a rich set of examples and exercises that are drawn from real-world scenarios. Readers develop a reliable physical intuition that can then be used to apply dynamic analysis software tools, and to develop simplified approximate models. With this foundation, they will be able to confidently use the equations of motion in a variety of applications, ranging from simulation and design to motion planning and control.

Frank C. Park is Professor of Mechanical Engineering at Seoul National University. He is author (with Kevin Lynch) of the book *Modern Robotics* (2017) and developer of the EdX course "Robot Mechanics and Control Parts I-II." He is a fellow of the IEEE, former editor-in-chief of the *IEEE Transactions on Robotics*, and past president of the IEEE Robotics and Automation Society.

Kyu Min Park is Assistant Professor with the Department of Artificial Intelligence and Robotics, Sejong University, Seoul, where he teaches courses related to robotics, dynamics, mathematics, artificial intelligence, and programming. He coauthored the book *Collision Detection for Robot Manipulators: Methods and Algorithms* (2023).

Dynamics of Mechanical Systems

A Modern Introduction to Three-Dimensional Rigid Body Dynamics

FRANK C. PARK
Seoul National University

KYU MIN PARK
Sejong University

Shaftesbury Road, Cambridge CB2 8EA, United Kingdom

One Liberty Plaza, 20th Floor, New York, NY 10006, USA

477 Williamstown Road, Port Melbourne, VIC 3207, Australia

314–321, 3rd Floor, Plot 3, Splendor Forum, Jasola District Centre,
New Delhi – 110025, India

103 Penang Road, #05–06/07, Visioncrest Commercial, Singapore 238467

Cambridge University Press is part of Cambridge University Press & Assessment,
a department of the University of Cambridge.

We share the University's mission to contribute to society through the pursuit of
education, learning and research at the highest international levels of excellence.

www.cambridge.org
Information on this title: www.cambridge.org/9781009609333

DOI: 10.1017/9781009609371

When citing this work, please include a reference to the DOI 10.1017/9781009609371

First published 2026

Cover image: mycola/iStock/Getty Images Plus

A catalogue record for this publication is available from the British Library

A Cataloging-in-Publication data record for this book is available from the Library of Congress

ISBN 978-1-009-60933-3 Hardback

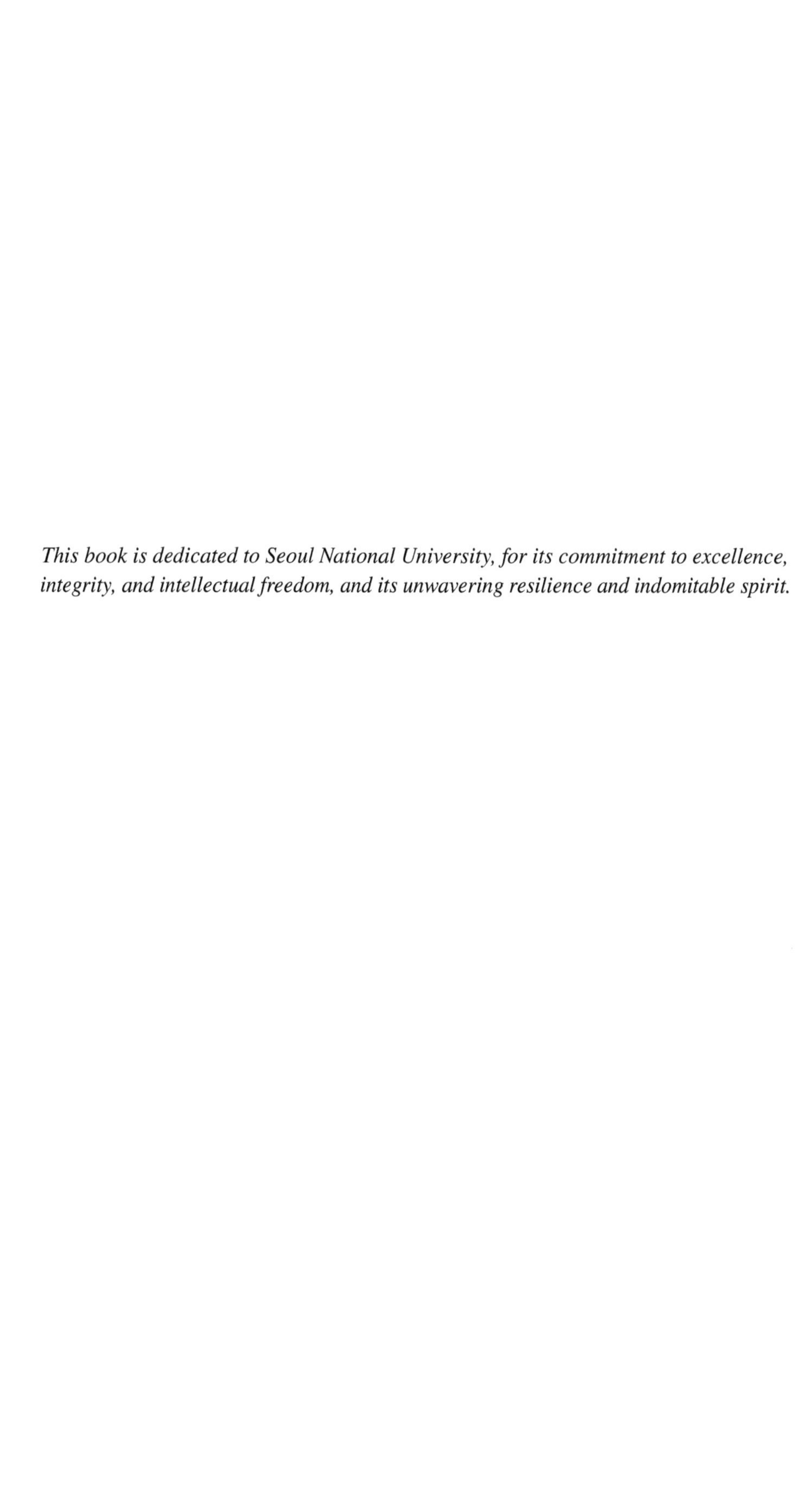

This book is dedicated to Seoul National University, for its commitment to excellence, integrity, and intellectual freedom, and its unwavering resilience and indomitable spirit.

Contents

Preface

This book is intended to be used as a textbook for a one-semester undergraduate introductory course in dynamics. The book is written for second- or third-year undergraduate engineering students for whom dynamics is a core course, although anyone who is studying dynamics for the first time and has the needed background – differential equations, linear algebra, and classical physics at a first-year undergraduate level – should find the contents accessible and suitable for self-study.

The motivation for this book was initially borne out of frustration at the lack of what we felt were appropriate undergraduate textbooks on dynamics. Most of the widely used introductory dynamics textbooks ignore the Lagrangian approach, and never make it to the analysis of multibody systems in three-dimensional space, most likely because of a fear that students will be overwhelmed by the mathematics. As a consequence, the skills that students acquire in a typical dynamics course are an ability to solve mostly simple toy problems, learning how to calculate, for example, how far an object has traveled or what its velocity is at some specific time, what is the tension in a cable connecting a system of pulleys, or other similarly contrived problems. Later, when students are asked to analyze the dynamics of a real mechanical system like a simple robot or vehicle suspension system, they usually have no idea where to begin, because they do not know how to go about modeling the system, let alone formulating the equations of motion.

We believe this is a tremendous lost opportunity. The reality is that most engineering students will complete their undergraduate education having taken at most one course in dynamics, if that. Given this narrow window of learning opportunity, the needed skills cannot be developed by studying a large repository of simple problems that have been crafted to reduce to a set of algebraic equations solvable by hand. The analysis of, for example, roller coasters, pulley systems, and objects rolling or sliding down inclines, especially using work-energy or energy-momentum methods, covers only a very small subset of the skills needed to design, analyze, and control the types of mechanical systems that students will encounter later in their professional careers.

Given the wide availability today of numerical software tools for dynamic analysis, we believe the most important skill to be acquired in a dynamics course is to be able to formulate, for realistic mechanical systems that the students will likely encounter later, the equations of motion. Students can then study the ways in which these equations can be transformed for different application contexts as well as techniques for their solution. Dynamics, particularly the Lagrangian perspective, is an excellent example of the power of mathematical abstraction when it comes to solving problems in the physical world.

A second important skill that students should acquire in a dynamics course is the ability to simplify complex mechanical systems, to be able to intuitively reason about their qualitative behavior, and to be able to construct approximate dynamic models that are more amenable to analysis. Linearization of a nonlinear mechanical system about a solution trajectory is a case in point. While it may seem paradoxical, we believe the first step to developing a reliable intuition about mechanical system dynamics is the ability to derive and analyze the equations of motion.

We also believe that three-dimensional dynamics for multibody systems (at least of the holonomic variety), and also Lagrangian dynamics, can be made accessible to undergraduate students who have studied differential equations and linear algebra at a basic level, for example, introductory undergraduate courses on these two subjects, or an introductory engineering mathematics course. Once the student is able to derive the equations of motion, it is then a simple matter to use an appropriate software tool to numerically integrate these equations of motion, and to simulate and analyze the resulting motion of the system.

If students can also develop the reasoning and intuition to construct, for example, simplified point mass models that approximate more complex rigid multibody systems, then the student will have acquired a set of skills that can be put to immediate practical use in solving real engineering problems. More importantly, with this foundation the student can learn how to use the equations of motion in a wide range of practical applications ranging from simulation and design to motion planning and control.

The book assumes that students have studied differential equations and linear algebra at the introductory college level, and are familiar with solution methods for scalar second-order linear differential equations as well as with basic matrix concepts like the inverse, determinant, eigenvalues and eigenvectors (a review of the needed concepts is provided in an Appendix). With this assumed background, concepts like rotation matrices, the Lagrangian dynamics formalism, and twist–wrench multibody dynamics can be introduced in an accessible manner. A rich set of examples and exercises drawn from real-world scenarios can also be presented. It is for these reasons that we believe our book offers a more modern introduction to mechanical system dynamics that is quite distinct from the classical textbooks.

For those intending to use this textbook in an introductory undergraduate dynamics course, the contents should be just about appropriate for a one-semester course. The chapters are best covered in sequence, although some of the problem solution techniques based on work-energy methods can be skipped if one wishes to spend more time on Lagrangian methods. The chapter on the unified twist–wrench formulation of rigid body dynamics can be skipped in a first pass of the book, but the methods of analysis, particularly with respect to reference frame transformations, are so powerful and practically useful that students are highly urged to come back to it later. The material on vibration analysis can also be selectively covered depending on the interests and background of the instructor and students.

It is a pleasure to acknowledge the many students, colleagues, and friends at Seoul National University who have contributed to the development of this book throughout the years, especially members of the SNU Robotics Laboratory who have helped create the many examples and exercises. F. C. Park wishes to thank his family (Hyunmee, Shiyeon, Soonkyu) for their unwavering support and encouragement throughout this entire endeavor. Kyu Min Park wishes to thank members of the Sejong University Interactive Robotics Laboratory, his parents, younger sister Suhyun, and most importantly, his wife Seoyoung for their support and encouragement.

1   Introduction

1.1   Mechanical Systems and Their Dynamics

Dynamics is the branch of mechanics that studies the physics of motion, particularly the relationship between the motion of a physical object and the forces that cause it. This book focuses on the dynamics of **mechanical systems**, which can be informally regarded as a collection of interconnected, interacting bodies. Several examples are shown in Figure 1.1.

The mechanical systems that we consider in this book are assumed to consist of **rigid bodies**, which are solid objects that do not undergo any elastic deformations. Rigid bodies can be connected to each other by a variety of devices: mechanical joints, gears, cables, tendons, cams, springs, dampers, and various other mechanical elements. Many of these connecting elements, such as springs, can exhibit some elastic behavior.

Forces that act on a mechanical system can include gravity and various other attractive or repulsive forces caused by aerodynamic, electromagnetic, or other physical field-like

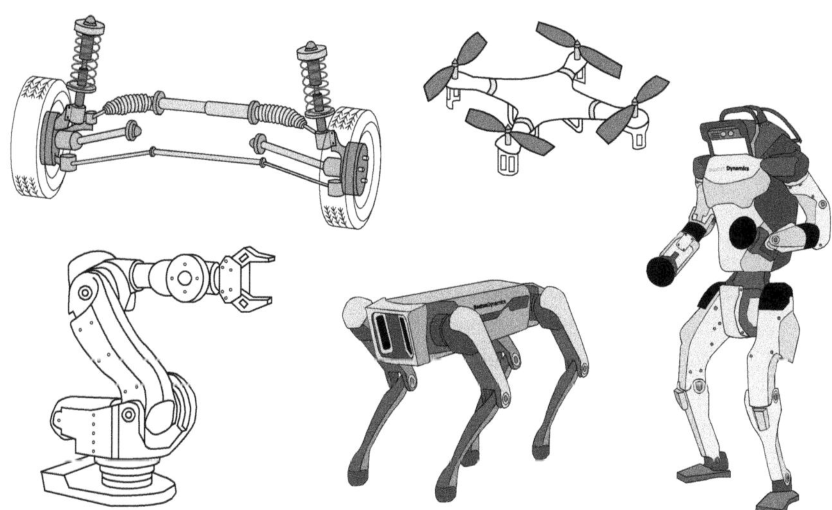

Figure 1.1 Examples of mechanical systems.

forces. Forces can result from contact with external objects or the environment. Forces can also be generated by actuators – for example, electric motors – that are attached to elements like joints and gears. The forces generated by the actuators then cause relative motion between the connected bodies of the system, and eventually motion of the overall mechanical system.

1.2   Learning Objectives

1.2.1   Deriving and Applying the Equations of Motion

One of the primary goals of this book will be to develop systematic methods for deriving the **equations of motion** of a mechanical system; these are a set of ordinary differential equations that relate the forces acting on the system to its overall motion. Sometimes the differential equations are augmented by algebraic equations that capture constraints on how the bodies can move relative to each other.

The equations of motion serve as a model of the dynamics of a mechanical system. As a general rule, mathematical models are only an approximation of reality, and only as good as the accuracy of the model parameters. In the case of the equations of motion, while the simplifying assumptions that we make about mechanical systems may not always hold – for example, rigid bodies may not be perfectly rigid, cables may not be perfectly inextensible, and springs may not behave in a perfectly linear way – these assumptions considerably simplify the structure and derivation of the equations of motion without sacrificing too much in the way of accuracy.

Along with their derivation, solutions to the equations of motion are also examined, where the notion of a 'solution' depends on the application – for example, determining the motion of the system from the applied forces or, conversely, determining the forces required to produce a desired motion. For many practical systems the equations of motion are nonlinear and quite complex, and usually do not admit closed-form analytical solutions. In such cases solutions must be obtained by numerically integrating the differential equations.

Practical applications of dynamic models are numerous. The most obvious one is simulation: If one wishes to observe the behavior of a mechanical system subjected to a given set of input and external forces, the equations of motion can be numerically integrated to predict the system's response. Of course, the accuracy of the simulation depends on the accuracy of the model parameters. Other applications include design (for example, determining a mechanical system's shapes, dimensions, masses, spring stiffnesses, damping coefficients, etc., to maximize a performance criterion), planning (for example, determining the motion that optimizes a criterion such as maximum payload lifting), and control (for example, using the dynamic model to compute the input required to achieve a desired motion; this *feedforward* control is typically combined with a *feedback control law* to correct for execution errors in real time).

1.2.2 Linearization and Its Uses

Suppose one has not only derived the equations of motion for a particular mechanical system but also obtained a specific solution in the form of a motion trajectory corresponding to a given input force profile. If one is interested in how the motion changes when the input is slightly perturbed, then, as it happens, there is no need to re-integrate the equations of motion; a good first-order approximation of the solution can be obtained with much less effort.

Imagine, for example, a satellite in stable geosynchronous orbit around the earth, and one of its thrusters accidentally misfires for a brief moment; will the satellite still remain in a stable orbit (perhaps at a different altitude), or will it fall out of orbit? For these and other related questions about the "local" behavior of the system, one can in fact obtain analytical solutions that provide answers about the qualitative behavior, as well as analytical solutions of motion trajectories that closely approximate the actual trajectory.[1]

Another practical phenomenon observed in many real-life mechanical systems is **vibration**, especially in systems characterized by periodic motions – such as washing machines or robots performing repetitive tasks. It is important to understand at what frequencies such vibrations occur, so that they can be suppressed (for example, to prevent imbalance in washing machines), or in some cases to exploit them (for example, adjusting a robot's gait for more energy-efficient locomotion). These and other vibration-related questions can also be answered by investigating the linearized version of the equations of motion.

1.2.3 Developing Intuition

While it may sound paradoxical, being able to think qualitatively about a mechanical system's dynamics, and having a rigorous and reliable physical intuition about its behavior, depends to a large extent on knowing how to derive and analyze its equations of motion. This becomes even more true for more complex mechanical systems. The first step in deriving the equations of motion is to determine the **degrees of freedom** of the system together with an appropriate set of coordinates, and these notions are vital to developing a physical intuition about the system.

All mechanical systems – those involving not only rigid bodies but also elastic elements like springs, dampers, and different types of actuators and energy storage and delivery devices – share a common structure in their equations of motion. Specifically, each of the terms can be given a certain physical interpretation, and the relative importance of these terms can differ significantly depending on the motion context. For relatively slow motions, for example, some of the terms in the equations of motion can be safely ignored without significantly influencing the resulting motion, while for high speed motions other terms may dominate.

---

[1] In brief, given a solution to a nonlinear differential equation, it is possible to obtain a "linearized" version of these nonlinear equations that admits closed-form analytic solutions; questions about stability can then be answered directly via analysis of these linearized equations.

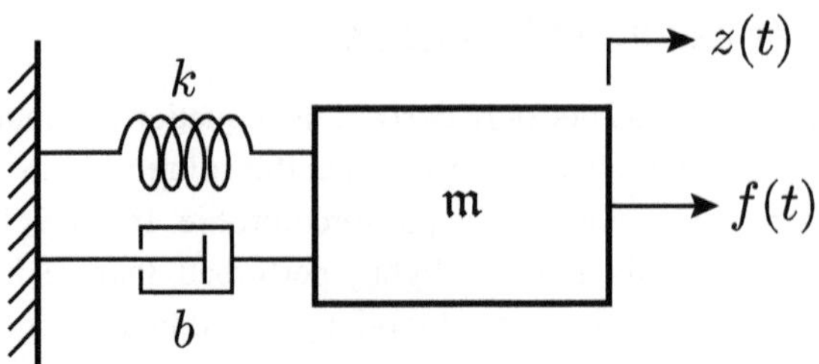

Figure 1.2 Linear mass-spring-damper system.

Mechanical systems can also be approximately modeled using simplified models, for example, replacing a rigid body by a small number of rigidly connected point masses. In such cases the equations of motion and their corresponding solutions can be more easily obtained, providing useful insight and intuition about the actual system. Having such an intuition is important not only for understanding the solutions produced by a numerical simulation[2] but also in the system design. For example, a typical design objective is to determine the shapes and masses of a robot's links, or to specify the motors, transmissions, and gears that deliver power to the joints, so that the robot can execute the desired motions as efficiently and accurately as possible. A reliable physical intuition about the system's dynamics is crucial for effective design.

1.3   Some Examples

1.3.1   Linear Mass-Spring-Damper System

Perhaps the simplest and most ubiquitous mechanical system, and one whose underlying physical intuition most readily generalizes to more complex mechanical systems, is the linear mass-spring-damper system illustrated in Figure 1.2. Here a mass m is attached to a linear spring with stiffness k, and at the same time a damper applies a resistive force against the direction of translation (or equivalently, in the opposite direction as the velocity). The mass is restricted to move along a line – denote the displacement from the equilibrium position by z – and subject to an external force f. Modeling the mass as a particle and applying Newton's second law of motion (i.e., force = mass × acceleration), the equations of motion take the form

$$f = m\ddot{z} + b\dot{z} + kz, \tag{1.1}$$

where $\dot{z}$ and $\ddot{z}$ respectively denote the velocity and acceleration of the mass, $b\dot{z}$ denotes the resistive force generated by the damper, and kz denotes the spring force.

---

[2] There are still cases where simulators produce results that are spectacularly wrong, particularly when contact and friction are involved.

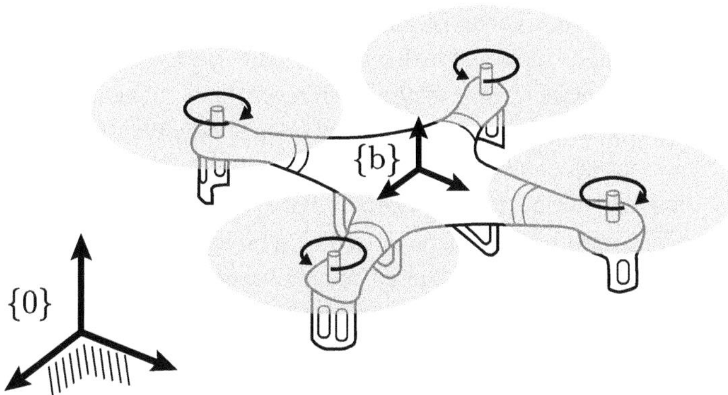

Figure 1.3 A quadrotor aerial vehicle.

Equation (1.1) is a second-order linear differential equation. Details of its solution can be found in any introductory text on differential equations or systems modeling, but intuitively it should be clear that depending on the relative values of the mass m, damping coefficient b, and spring stiffness k, the system can exhibit a range of behaviors. For example, assuming $f = 0$ and the mass starts at some non-equilibrium position, the mass can oscillate indefinitely ($b = 0$), oscillate with decreasing amplitude (b is nonzero but the spring force dominates), or simply decay slowly to the equilibrium in the limit (the damping force dominates).

Generalizations of the linear mass-spring-damper system are widely used in mechanical systems modeling – for example, in vehicle suspension systems, in modeling contact between a soft finger and a rigid object, or in modeling foot-ground contact. Deformable bodies are also sometimes modeled as a collection of rigid bodies connected to each other by springs and dampers. More complex mechanical systems naturally have equations of motion that are more complex than (1.1), but in fact there are structural similarities in the equations of all mechanical systems. For example, terms that correspond to the mass, damping, and spring forces can be identified in the equations of motion of general mechanical systems. The techniques for mechanical vibration analysis also originate from the basic analysis of the linear mass-spring-damper system.

1.3.2   Quadrotor Aerial Vehicle

Figure 1.3 illustrates a quadrotor aerial vehicle. In its most basic form, a quadrotor consists of a rigid body with four rotors attached. Typically the rotor axes are parallel to each other and attached symmetrically about the body. The rotors generate thrust, and when properly controlled allow the quadrotor to fly both horizontally and vertically as well as hover.

The quadrotor can basically be modeled as a single rigid body subject to external forces and moments, with the external forces and moments resulting from the thrust generated by the rotors as well as gravity, aerodynamic, and other forces. Referring again to Figure 1.3, the four rotors are parallel and symmetrically located about the quadrotor's center of mass. Opposing pairs of rotors typically rotate in opposite directions as indicated by the arrows. The *inertial reference frame* (or *fixed frame*) is indicated by {0}, while the *body reference frame* (or *moving frame*), indicated by {b}, is rigidly attached to some fixed point on the body. We will assume that the rotor masses and inertias can be ignored, and that the rotor axes are all perfectly parallel to each other and normal to the quadrotor plane.

The rigid body is characterized by its shape and material properties. Expressing everything in terms of the chosen body frame coordinates, for the purposes of constructing a dynamic model the needed information is captured by the rigid body's mass $\mathfrak{m} > 0$, its first mass moment $h = \mathfrak{m} \cdot p \in \mathbb{R}^3$, where $p \in \mathbb{R}^3$ is the vector from the body frame origin to the body's center of mass, and its *inertia matrix* $\mathcal{I} \in \mathbb{R}^{3\times 3}$. $\mathcal{I}$ must further satisfy the following additional properties: (i) $\mathcal{I}$ is symmetric; (ii) $\mathcal{I}$ is positive-definite (or equivalently, $\det \mathcal{I} > 0$); (iii) the three eigenvalues of $\mathcal{I}$, denoted $\lambda_1, \lambda_2, \lambda_3$, must satisfy $\lambda_i + \lambda_j > \lambda_k$ for any $\{i, j, k\} = \{1, 2, 3\}$. Denoting the elements of $\mathcal{I}$ by

$$\mathcal{I} = \begin{bmatrix} I_{xx} & I_{xy} & I_{xz} \\ I_{xy} & I_{yy} & I_{yz} \\ I_{xz} & I_{yz} & I_{zz} \end{bmatrix}, \tag{1.2}$$

the mass-inertial properties of the rigid body are then completely characterized by a set of ten parameters:

$$\left[\mathfrak{m}, h, I_{xx}, I_{yy}, I_{zz}, I_{xy}, I_{yz}, I_{xz} \right] \in \mathbb{R}^{10}, \tag{1.3}$$

subject to the constraints described above. If the body frame {b} is attached to the center of mass, then $h = 0$, and the equations of motion can be further simplified. Assume in what follows that {b} is attached to the center of mass. With the above specification of the mass-inertial properties of the rigid body, let $v \in \mathbb{R}^3$ and $\omega \in \mathbb{R}^3$ be the linear and angular velocity vectors of the body frame, respectively. Further let $f_i \in \mathbb{R}^3$ be the force generated by rotor i and $r_i \in \mathbb{R}^3$ be the vector from the body frame origin to rotor i (more accurately, the point on the rigid body at which the force generated by rotor i is applied), $i = 1, \ldots, 4$. We emphasize again that all vectors are expressed in coordinates of the body frame.

With the above definitions, the equations of motion for the quadrotor are then given by the pair of equations

$$\sum_i f_i = \mathfrak{m} \left(\dot{v} + \omega \times v \right), \tag{1.4}$$

$$\sum_i m_i = \mathcal{I} \dot{\omega} + \omega \times \mathcal{I} \omega, \tag{1.5}$$

where $\sum_i f_i$ denotes the (vector) sum of all forces applied to the body, $\sum_i m_i$ denotes the (vector) sum of all moments applied to the body, and $\dot{\omega}$ and $\dot{v}$, respectively, denote the time derivatives of ω and v. For the quadrotor, the resultant forces and moments include the thrust generated by the four rotors as well as gravity and aerodynamic forces. Denoting the angular speed of rotor i by $\dot{\theta}_i$, the vertical force f_i produced by rotor i has magnitude $k_f \dot{\theta}_i^2$, where k_f is a scalar constant. Rotor i also produces a moment about the vertical axis according to $m_i = k_m \dot{\theta}_i^2$ for some scalar constant k_m. More complex models of the rotor actuator and propeller dynamics, drag, and other aerodynamic forces are also possible.

The rotors may also not be parallel to each other, or for that matter may not even be arranged in a symmetric fashion. All of these require further modifications to the model, likely with the introduction of additional reference frames to describe the placement of the rotors. In fact, different numbers of rotors can be attached in a wide variety of arrangements, in which case the system is referred to as a *multirotor aerial vehicle*. Other possible variations include the addition of wings, and actuators for the active tilting of rotors, all of which lead to even more complex dynamic models.

1.3.3   Multi-Link Robot

A typical multi-link robot is shown in Figure 1.4. This robot consists of six rigid bodies (called *links*) connected in serial fashion by revolute joints. Actuators such as electric motors deliver torques to the joints, causing relative motion between adjacent links. By carefully designing these input joint torques, the robot can be controlled to move in desired ways. In most practical robots the actuators are not attached directly to the joints, but placed at different locations; transmission elements such as gears, cables, and other machine elements then deliver the motor torques to the joints.

If a dynamic model relating the joint torques to the motion of the robot is available, the model can then be used to design input torques that realize a given desired motion. Whereas the previous quadrotor example can be modeled as a single rigid body, the robot involves multiple rigid bodies, each of which may have different mass-inertial properties. For the robot of Figure 1.4, the equations of motion assume the following form:

$$\tau = M(\theta)\ddot{\theta} + c(\theta, \dot{\theta}) + g(\theta), \qquad (1.6)$$

where $\theta \in \mathbb{R}^6$ is the vector of joint angle positions, and $\dot{\theta}$ and $\ddot{\theta}$ are the respective joint angle velocities and accelerations, $\tau \in \mathbb{R}^6$ is the vector of joint torques, $M(\theta)$ is the 6×6 mass matrix, $c(\theta, \dot{\theta}) \in \mathbb{R}^6$ is the vector of Coriolis and centripetal torques, and $g(\theta) \in \mathbb{R}^6$ is the vector reflecting gravitational forces. The mass-inertial parameters of the rigid bodies are all embedded in each of these terms.

Despite the seeming simplicity of (1.6), in reality each of the terms in this equation is immensely complex. If one were to further take into account the dynamics of the actuators, gears, and other transmission elements, the dynamic model becomes even more complicated. Sometimes the robot may be subject to a set of constraints on its motion, for example, when the robot makes contact with the environment; in such cases there exist constraints on possible motions of the robot that need to be accounted for in the dynamic model. In this book we will examine the derivation of the dynamic equations of motion for serially connected rigid bodies like this robot.

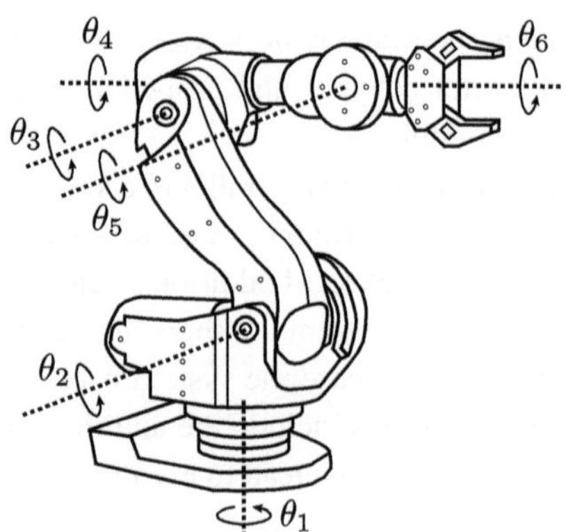

Figure 1.4 A multi-link robot.

1.4   Summary

- **Dynamics** is the branch of mechanics that studies the physics of motion, particularly the relationship between the motion of a physical object and the forces that cause it.
- **Mechanical systems** are a collection of interconnected interacting bodies, connected by mechanical devices such as joints, gears, cables, and tendons. **Actuators** such as motors deliver forces and torques that cause motion of the bodies. Other forces acting on the body can include gravity and forces arising from contact. The bodies will be assumed to be **rigid** – that is, non-deformable.
- The **equations of motion** of a mechanical system are a set of ordinary differential equations that relate the forces acting on the system to its overall motion. Sometimes the differential equations are augmented by a set of algebraic equations that capture constraints on how the bodies can move relative to each other.
- The **linearized** equations of motion are a set of linear differential equations that provide a first-order approximation of the behavior of the mechanical system. They are useful for analyzing how small perturbations affect the system's motion, as well as for studying its vibration and oscillation characteristics.
- **Examples of mechanical systems** include the linear mass-spring-damper system, systems like the quadrotor aerial vehicle that can be modeled as a single rigid body subject to external forces, and multibody systems like multi-link robots.

2 Kinematics

Classical physics tells us that the motion of an object is determined by the forces applied to the object. The subject of dynamics is concerned with the relationship between these applied forces and the motion of the object, together with the mass-shape properties of the object. **Kinematics** is the study of motion without regard to the forces applied to the object or the object's mass-shape properties. We begin our study of dynamics by examining the kinematics of moving objects, starting with single objects that can be modeled as a point or rigid body, to more complex systems consisting of multiple interconnected bodies.

The motion of an object can be described in a number of different ways – for example, by marking a set of reference points on the object and tracking their positions throughout its motion. Our preferred method of choice will be to attach a reference frame to the moving object – we call this the **moving frame** – and to describe the position and orientation of the moving frame with respect to some **fixed frame**. **Rotation matrices** are introduced to describe the orientation of a reference frame. The **linear** and **angular velocity** of the moving frame then, respectively describe instantaneous changes in position and orientation. Later for dynamic analysis, we will also need the **linear** and **angular accelerations** of the moving frame.

One of the key concepts introduced in this chapter is the use of moving frames for kinematic analysis. Arguably one of the most complicated and difficult steps in dynamic analysis is the calculation of velocities and accelerations of various points on rigid bodies. By choosing appropriate moving frames – a skill that requires some practice – and expressing the position of the point in question in terms of these moving frames, the velocity and acceleration can then be obtained simply by taking repeated derivatives of the position in a systematic manner. By understanding the rules for taking derivatives with respect to moving frames, the kinematic analysis of even the most complex mechanical systems can be reduced to a straightforward procedure, once moving frames have been assigned.

2.1 Points, Vectors, Frames

The mechanical systems that we study all live in the familiar three-dimensional physical space of classical physics and Euclidean geometry. While it is tempting to immediately equate physical space with three-dimensional Euclidean space $\mathbb{R}^3$, so that a point in

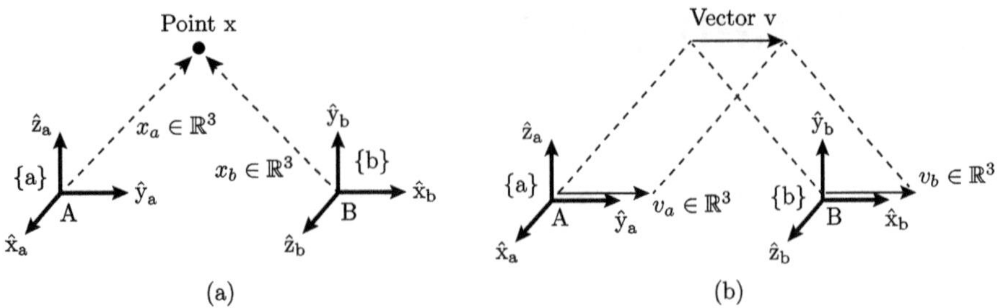

Figure 2.1 (a) A point x in physical space and its coordinates with respect to reference frames {a} and {b}. (b) A vector v and its representation with respect to reference frames {a} and {b}.

physical space can be represented by a three-tuple of real numbers $x = (x_1, x_2, x_3)^\top \in \mathbb{R}^3$, this identification is possible only after certain choices and assumptions are made.

First, choosing a **length scale** for physical space allows one to measure the distance between two points x and y in physical space – this is done by drawing a line segment between x and y and measuring its length in terms of the chosen length scale. For most of the examples and problems in this book the metric unit of length is used where needed, but it should be noted that there is no natural choice of length scale for physical space.

Once a length scale has been chosen, the next step is to designate a point in physical space to be the **origin**, and to attach an **orthonormal** and **right-handed reference frame** at the origin. In Figure 2.1(a), suppose point A is the designated origin, and denote the axes of the reference frame by $\{\hat{x}_a, \hat{y}_a, \hat{z}_a\}$; we will also denote this frame using the notation {a}. The frame being orthonormal implies that $\hat{x}_a$, $\hat{y}_a$, $\hat{z}_a$ are all of unit length and normal (perpendicular) to each other, while the frame being right-handed implies that $\hat{x}_a \times \hat{y}_a = \hat{z}_a$ (if the frame were **left-handed**, then $\hat{x}_a \times \hat{y}_a = -\hat{z}_a$). Coordinates for point x with respect to reference frame {a} can now be expressed by the three-dimensional column vector

$$x_a = \begin{bmatrix} x_{a,x} \\ x_{a,y} \\ x_{a,z} \end{bmatrix} \in \mathbb{R}^3, \tag{2.1}$$

which we will also sometimes write more compactly as $x_a = (x_{a,x}, x_{a,y}, x_{a,z})^\top$. If instead another reference frame {b} with origin at B were chosen, then the coordinates for the same point x are given by another three-dimensional column vector

$$x_b = \begin{bmatrix} x_{b,x} \\ x_{b,y} \\ x_{b,z} \end{bmatrix} \in \mathbb{R}^3. \tag{2.2}$$

For the specific case of Figure 2.1(a), we have $x_a = (0, 1, 1)^\top$ while $x_b = (-1, 1, 0)^\top$.

As the above example illustrates, once a length scale and reference frame for physical space have been chosen, a point x in physical space can be represented uniquely by a three-dimensional vector $x \in \mathbb{R}^3$. One can even think of $x \in \mathbb{R}^3$ as the point x itself.

However, a different choice of reference frame and length scale leads to a different vector representation $x' \in \mathbb{R}^3$ for x. The point x can be regarded as an **intrinsic** element of physical space (i.e., independent of any choice of coordinates for physical space), whereas $x \in \mathbb{R}^3$ is a coordinate representation of x that depends on the choice of reference frame and length scale.

A **vector** v is another intrinsic quantity in physical space, defined by a **direction** and **length** and visualized as a directed line segment as illustrated in Figure 2.1(b) (the fact that v is characterized by its length implies that a length scale has a priori been chosen). In some textbooks a distinction is made between a **bound vector** whose base point is assumed attached to some fixed point in physical space, and a **free vector** whose base point is unconstrained. Since the vectors we consider in this book are for the most part free vectors, this distinction will not be important.

Figure 2.1(b) illustrates two coordinate representations for v, in which $v_a \in \mathbb{R}^3$ is with respect to frame {a} and $v_b \in \mathbb{R}^3$ is with respect to frame {b}:

$$v_a = \begin{bmatrix} v_{a,x} \\ v_{a,y} \\ v_{a,z} \end{bmatrix} \in \mathbb{R}^3, \quad v_b = \begin{bmatrix} v_{b,x} \\ v_{b,y} \\ v_{b,z} \end{bmatrix} \in \mathbb{R}^3. \tag{2.3}$$

In terms of the axes of frames {a} and {b}, v can be written

$$\mathrm{v} = v_{a,x}\hat{x}_a + v_{a,y}\hat{y}_a + v_{a,z}\hat{z}_a \tag{2.4}$$
$$= v_{b,x}\hat{x}_b + v_{b,y}\hat{y}_b + v_{b,z}\hat{z}_b. \tag{2.5}$$

For those familiar with the linear algebraic notion of a basis for a vector space, the $\{\hat{x}_a, \hat{y}_a, \hat{z}_a\}$ and $\{\hat{x}_b, \hat{y}_b, \hat{z}_b\}$ serve as two different bases for the vector space in which v resides, and (2.4)–(2.5) express v in terms of these two bases. For the specific case of Figure 2.1(b) we have $v_a = (0, 1, 0)^\top$ while $v_b = (1, 0, 0)^\top$.

So far we have made a distinction between two types of intrinsic quantities: a point x and a vector v. Choosing a reference frame $\{\hat{x}, \hat{y}, \hat{z}\}$ and length scale for physical space allows us to specify coordinates $x \in \mathbb{R}^3$ for x. At the same time, an intrinsic vector v can be expressed in terms of the basis $\{\hat{x}, \hat{y}, \hat{z}\}$ as $\mathrm{v} = v_x\hat{x} + v_y\hat{y} + v_z\hat{z}$, where $v = (v_x, v_y, v_z)^\top \in \mathbb{R}^3$. Although both x and v are elements of $\mathbb{R}^3$, they represent different intrinsic quantities, and also transform differently under a change of reference frame. It is important to remember that two vectors $x, y \in \mathbb{R}^n$ that represent different intrinsic quantities, with different transformation rules, cannot be added or subtracted together in a meaningful way.

It is possible to represent a point x in physical space by an intrinsic vector once an origin O (but not necessarily a reference frame at O) for physical space has been chosen. Referring to Figure 2.2, the intrinsic vector p is directed from O to x. If however a different origin O′ is chosen, then the intrinsic vector representing x also changes, from p to p′ = d + p, where d is directed from O′ to O. Clearly intrinsic vectors can be added and subtracted (and also multiplied by scalars), but again, intrinsic vectors representing different physical quantities – for example, linear velocities, angular velocities, forces, and moments (forces and moments are introduced later) – cannot be added together in any meaningful way.

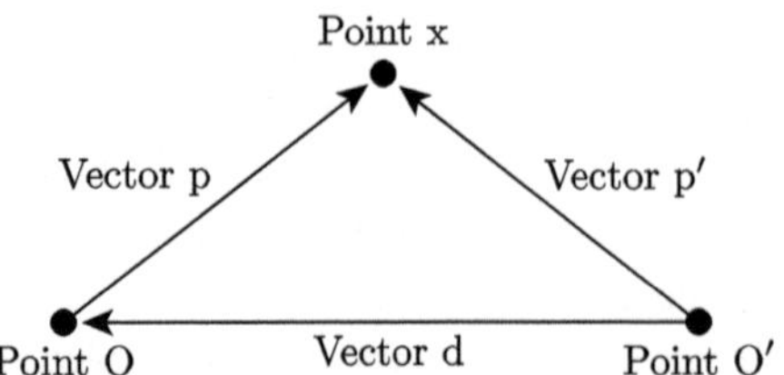

Figure 2.2 Representing a point in physical space by an intrinsic vector. The intrinsic vector depends on the choice of origin.

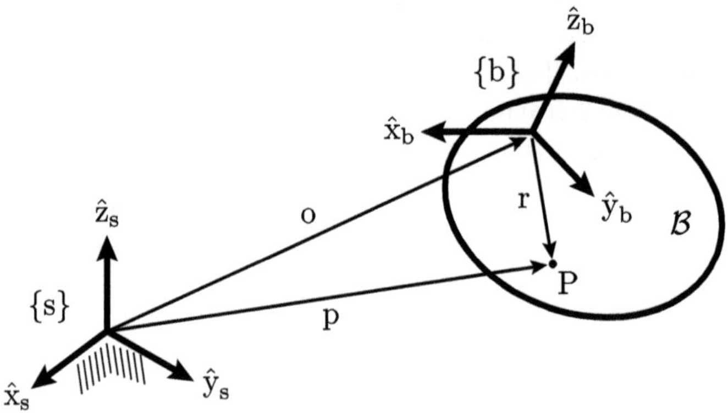

Figure 2.3 A rigid body moving in space.

Remarks on Notation

We will adopt the following notational convention throughout the book. Intrinsic quantities like points in physical space, or vectors represented by a directed line segment with a length and direction, will be written in roman letters, for example, x, p, v, ω. The coordinate representations of these intrinsic quantities will be written in italics, for example, $x, p, v, \omega \in \mathbb{R}^3$. All reference frames are right-handed and orthonormal, and will be denoted by notation of the form {a}, {b}, {s}. The axes of a reference frame will be written in roman letters with a hat on top, for example, $\{\hat{x}_b, \hat{y}_b, \hat{z}_b\}$, where the subscript indicates the reference frame label.

2.2   Particle Kinematics Using Moving Frames

Referring to Figure 2.3, a fixed frame of reference, denoted {s} and also called the **space frame**, is placed in three-dimensional physical space as shown. For the present purposes the space frame can be regarded to be stationary (later in Chapter 3 we will introduce the notion of an **inertial reference frame**; the fixed frames used in our later dynamic analysis will all be inertial reference frames). Denote the axes of frame {s} by

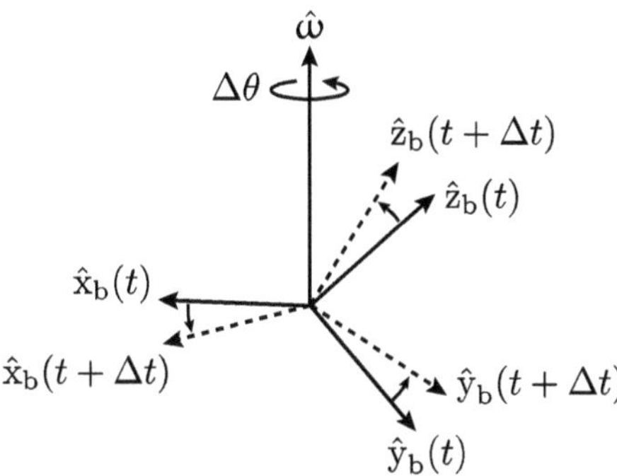

Figure 2.4 Illustration of the angular velocity vector.

$\{\hat{x}_s, \hat{y}_s, \hat{z}_s\}$. This frame, and all reference frames described in this book, are orthonormal and right-handed.

Let $\mathcal{B}$ be a rigid body in physical space. Attach to $\mathcal{B}$ a reference frame {b}, called the **body frame** or **moving frame** since $\mathcal{B}$ is assumed to move in space. We emphasize that {b} need not be attached to a specific physical point on $\mathcal{B}$; rather, what matters is that {b} moves as if rigidly attached to $\mathcal{B}$ – that is, the displacement between $\mathcal{B}$ and {b} remains constant at all times. Denote the unit axes of {b} by $\{\hat{x}_b, \hat{y}_b, \hat{z}_b\}$.

Given some arbitrary point P on $\mathcal{B}$, let p be the vector from the fixed frame origin to this point. Denoting by o the vector from the fixed frame origin to the origin of the body frame {b}, and by r the vector from the body frame origin to the point P, the vector equation

$$p = o + r \tag{2.6}$$

then always holds wherever $\mathcal{B}$ may be situated. Once a length scale for physical space has been chosen, o and r (and therefore p) can be expressed in terms of coordinates for frames {s} and {b} as follows:

$$\begin{aligned}
o(t) &= x_o(t)\hat{x}_s + y_o(t)\hat{y}_s + z_o(t)\hat{z}_s, \\
r(t) &= x_r(t)\hat{x}_b + y_r(t)\hat{y}_b + z_r(t)\hat{z}_b, \\
p(t) &= x_o(t)\hat{x}_s + y_o(t)\hat{y}_s + z_o(t)\hat{z}_s + x_r(t)\hat{x}_b + y_r(t)\hat{y}_b + z_r(t)\hat{z}_b,
\end{aligned} \tag{2.7}$$

where t denotes time. Observe that o is written in terms of the fixed frame coordinates, while r is written in terms of the body frame coordinates. If P corresponds to some specific point on $\mathcal{B}$, then the values of x_r, y_r, z_r are constant. On the other hand, if P corresponds to, for example, the location of a bug crawling on the surface of $\mathcal{B}$, then x_r, y_r, z_r can be time-varying.

Differentiating $p(t)$ with respect to t to obtain the velocity vector $v(t)$,

$$\begin{aligned}
v(t) = \dot{p}(t) &= \dot{x}_o\hat{x}_s + \dot{y}_o\hat{y}_s + \dot{z}_o\hat{z}_s + \dot{x}_r\hat{x}_b + \dot{y}_r\hat{y}_b + \dot{z}_r\hat{z}_b \\
&\quad + x_r\frac{d}{dt}\hat{x}_b + y_r\frac{d}{dt}\hat{y}_b + z_r\frac{d}{dt}\hat{z}_b.
\end{aligned} \tag{2.8}$$

In the above we have made use of the fact that since the fixed frame $\{\hat{x}_s, \hat{y}_s, \hat{z}_s\}$ is stationary, its time derivatives are all zero and can be ignored.

If the orientation of the body frame changes with time, then the time derivatives of $\hat{x}_b$, $\hat{y}_b$, and $\hat{z}_b$ are nonzero. To determine these derivatives, consider Figure 2.4, which shows the body frame $\{\hat{x}_b, \hat{y}_b, \hat{z}_b\}$ at times t and $t + \Delta t$, where Δt is a small increment of time (the two frames are drawn with their origins overlapping). We now make the following claim: There exists some unit vector $\hat{\omega}$ and angle $\Delta\theta$ such that frame $\{\hat{x}_b(t),$ $\hat{y}_b(t), \hat{z}_b(t)\}$ can be rotated by $\Delta\theta$ radians about $\hat{\omega}$ into frame $\{\hat{x}_b(t + \Delta t), \hat{y}_b(t + \Delta t),$ $\hat{z}_b(t + \Delta t)\}$ (see Figure 2.4). Here $\hat{\omega}$ is the axis of rotation, passing through the frame origin, with positive rotation defined to be counterclockwise. While intuitively it should make sense that such an $\hat{\omega}$ and $\Delta\theta$ always exists – observe that it is not unique, since rotating about $-\hat{\omega}$ by angle $-\Delta\theta$ also achieves the same effect – the proof of their existence, together with the formula for extracting $\hat{\omega}$ and $\Delta\theta$ from the two frames, requires additional concepts involving rotation matrices covered later in this chapter.

The **angular velocity** of moving frame $\{\hat{x}_b, \hat{y}_b, \hat{z}_b\}$ is then defined via the following limit:

$$\omega = \lim_{\Delta t \to 0} \hat{\omega} \frac{\Delta\theta}{\Delta t}. \tag{2.9}$$

The direction of ω corresponds to the **instantaneous axis of rotation**, while the length of ω corresponds to the rate of rotation (i.e., radians per second). Note that both the direction and length of ω can vary with time as the frame undergoes rotation.

With the angular velocity vector ω of frame $\{b\}$ defined in this way, the velocities of the unit axes are then given by the following cross-products:

$$\frac{d}{dt}\hat{x}_b = \omega \times \hat{x}_b,$$

$$\frac{d}{dt}\hat{y}_b = \omega \times \hat{y}_b, \tag{2.10}$$

$$\frac{d}{dt}\hat{z}_b = \omega \times \hat{z}_b.$$

The reader should try using the right-hand rule to verify that the cross-product gives the correct direction for the velocity vectors. For this purpose it may be helpful to recall that for two vectors u, v, the length of their cross-product $u \times v$ satisfies $\|u \times v\| = \|u\| \cdot \|v\| \sin\theta$, where θ is the angle between u and v.

Returning to the formula (2.8) for the velocity, replacing $\frac{d}{dt}\hat{x}_b$ by $\omega \times \hat{x}_b$, $\frac{d}{dt}\hat{y}_b$ by $\omega \times \hat{y}_b$, and $\frac{d}{dt}\hat{z}_b$ by $\omega \times \hat{z}_b$, $v(t)$ can be rewritten

$$v(t) = \dot{x}_o\hat{x}_s + \dot{y}_o\hat{y}_s + \dot{z}_o\hat{z}_s + \dot{x}_r\hat{x}_b + \dot{y}_r\hat{y}_b + \dot{z}_r\hat{z}_b + \omega \times (x_r\hat{x}_b + y_r\hat{y}_b + z_r\hat{z}_b)$$

$$= v_o + \omega \times r + \dot{x}_r\hat{x}_b + \dot{y}_r\hat{y}_b + \dot{z}_r\hat{z}_b, \tag{2.11}$$

where v_o denotes the velocity of point o corresponding to the frame $\{b\}$ origin. Observe that if P is a fixed point on $\mathcal{B}$, then x_r, y_r, z_r are all constant and the velocity simplifies to $v = v_o + \omega \times r$.

Differentiating $v(t)$ once more to get the acceleration vector $a(t)$,

$$a(t) = \ddot{x}_o\hat{x}_s + \ddot{y}_o\hat{y}_s + \ddot{z}_o\hat{z}_s + \ddot{x}_r\hat{x}_b + \ddot{y}_r\hat{y}_b + \ddot{z}_r\hat{z}_b$$
$$+ \dot{\omega} \times (x_r\hat{x}_b + y_r\hat{y}_b + z_r\hat{z}_b) + 2\omega \times (\dot{x}_r\hat{x}_b + \dot{y}_r\hat{y}_b + \dot{z}_r\hat{z}_b)$$
$$+ \omega \times (\omega \times (x_r\hat{x}_b + y_r\hat{y}_b + z_r\hat{z}_b))$$
$$= a_o + \dot{\omega} \times r + \omega \times (\omega \times r)$$
$$+ (\ddot{x}_r\hat{x}_b + \ddot{y}_r\hat{y}_b + \ddot{z}_r\hat{z}_b) + 2\omega \times (\dot{x}_r\hat{x}_b + \dot{y}_r\hat{y}_b + \dot{z}_r\hat{z}_b), \tag{2.12}$$

where a_o is the acceleration of the point o corresponding to the frame {b} origin. The above equation makes use of the relation $\frac{d}{dt}(\omega \times r) = (\dot{\omega} \times r) + (\omega \times \dot{r})$. Again, observe that if P is a fixed point on $\mathcal{B}$, then x_r, y_r, z_r are all constant and the acceleration simplifies to $a = a_o + \dot{\omega} \times r + \omega \times (\omega \times r)$.

Example 2.1   General velocity and acceleration formula. Many dynamics textbooks provide general formulas for the velocity and acceleration in terms of moving frames. Referring again to Figure 2.3, frame {s} is fixed, frame {b} is the moving frame, and the position of a moving particle P is expressed by the vector equation

$$p = o + r. \tag{2.13}$$

Note that P here is not necessarily assumed to be a fixed point on the rigid body $\mathcal{B}$; P can correspond to, for example, the position of a bug crawling on the surface of $\mathcal{B}$. The moving frame {b} rotates with angular velocity ω. The velocity of p is then written

$$v = v_o + \omega \times r + (\dot{r})_b, \tag{2.14}$$

while the acceleration is expressed as

$$a = a_o + \dot{\omega} \times r + \omega \times (\omega \times r) + 2\omega \times (\dot{r})_b + (\ddot{r})_b. \tag{2.15}$$

The interpretation of the $(\dot{r})_b$ and $(\ddot{r})_b$ terms in (2.15) may seem somewhat mysterious. Comparing (2.15) with (2.12), it can be seen that these terms correspond to the following:

$$(\dot{r})_b = \dot{x}_r\hat{x}_b + \dot{y}_r\hat{y}_b + \dot{z}_r\hat{z}_b, \tag{2.16}$$
$$(\ddot{r})_b = \ddot{x}_r\hat{x}_b + \ddot{y}_r\hat{y}_b + \ddot{z}_r\hat{z}_b. \tag{2.17}$$

The reader is strongly encouraged **not** to attempt to memorize the general formulas (2.14) and (2.15) for the velocity and acceleration. Rather, expressing the position vector in terms of a combination of fixed and moving frame unit axes, and applying (2.10) when derivatives of the moving frame unit axes are needed, is a more straightforward procedure that is much less prone to error. $\qquad\qquad\square$

Example 2.2   Velocity and acceleration analysis of a rotating telescopic arm. Consider the mechanism of Figure 2.5, in which an extensible arm rotates about a fixed vertical axis at the base, and at the same time the arm extends horizontally. Denote by θ the angle of rotation about the vertical axis, and by s the length of the arm. The fixed frame {s} is attached at the rotating base.

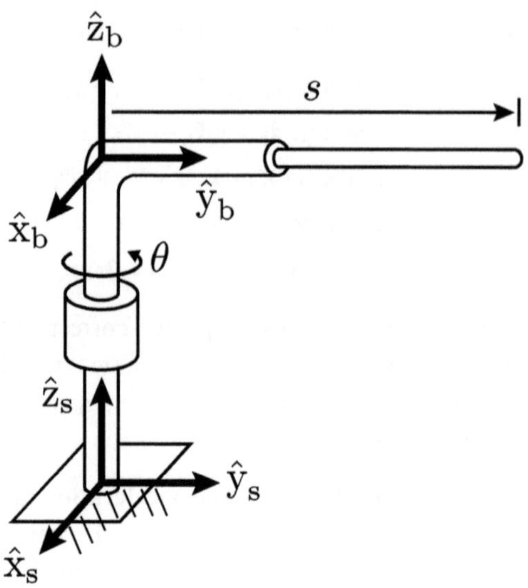

Figure 2.5 Rotating telescopic arm for Example 2.2.

To find the velocity and acceleration of the tip, attach a moving frame {b} at the intersection of the arm and the vertical axis of rotation, so that the $\hat{y}_b$ axis is always directed along the length of the arm. The angular velocity of frame {b} is

$$\omega = \dot{\theta}\hat{z}_b \tag{2.18}$$

(observe that $\hat{z}_b = \hat{z}_s$). With the moving frame defined in this way, the tip position is simply

$$p = s\hat{y}_b + L\hat{z}_s, \tag{2.19}$$

where L is a constant denoting the distance between the origins of {b} and {s}. The velocity v is then given by

$$v = \dot{s}\hat{y}_b + s\dot{\hat{y}}_b = \dot{s}\hat{y}_b + \omega \times s\hat{y}_b, \tag{2.20}$$

while the acceleration is given by

$$a = \ddot{s}\hat{y}_b + \dot{s}\dot{\hat{y}}_b + \dot{\omega} \times s\hat{y}_b + \omega \times (\dot{s}\hat{y}_b + s\dot{\hat{y}}_b) \tag{2.21}$$

$$= \ddot{s}\hat{y}_b + 2(\omega \times \dot{s}\hat{y}_b) + (\dot{\omega} \times s\hat{y}_b) + \omega \times (\omega \times s\hat{y}_b). \tag{2.22}$$

$\square$

Example 2.3   Velocity and acceleration in polar coordinates. In this example we derive the velocity and acceleration for a particle moving in the plane using polar coordinates. Referring to Figure 2.6, assign the fixed frame {s} as shown, and assume particle P moves in the $\{\hat{x}_s, \hat{y}_s\}$ plane. Define a moving frame {b} with axes $\{\hat{r}, \hat{\theta}\}$, such that its origin always coincides with the fixed frame origin, and $\hat{r}$ always points in the

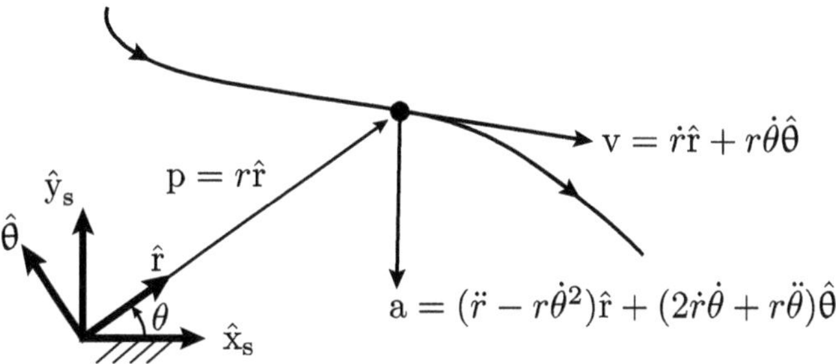

Figure 2.6 Velocity and acceleration in polar coordinates.

direction of point P, while $\hat{\theta}$ is orthogonal to $\hat{r}$ such that $\hat{r} \times \hat{\theta} = \hat{z}_s$ is always satisfied. The vector p from the fixed frame origin to P is then given by

$$p = r\hat{r}, \tag{2.23}$$

where r is the distance from the fixed frame origin to P. Let θ be the angle measured counterclockwise from $\hat{x}_s$ to $\hat{r}$ as indicated in the figure. The angular velocity of the moving frame is then

$$\omega = \dot{\theta}\hat{z}_s. \tag{2.24}$$

The velocity v can now be obtained by differentiating p with respect to t:

$$v = \dot{p} = \dot{r}\hat{r} + r\dot{\hat{r}}, \tag{2.25}$$

where

$$\dot{\hat{r}} = \omega \times \hat{r} = \dot{\theta}\hat{\theta}. \tag{2.26}$$

Substituting into (2.25),

$$v = \dot{r}\hat{r} + r\dot{\theta}\hat{\theta}. \tag{2.27}$$

The acceleration is obtained by differentiating v with respect to t:

$$a = \ddot{r}\hat{r} + \dot{r}\dot{\hat{r}} + \dot{r}\dot{\theta}\hat{\theta} + r\ddot{\theta}\hat{\theta} + r\dot{\theta}\dot{\hat{\theta}}, \tag{2.28}$$

where

$$\dot{\hat{\theta}} = \omega \times \hat{\theta} = -\dot{\theta}\hat{r}. \tag{2.29}$$

Substituting the above into (2.28),

$$a = (\ddot{r} - r\dot{\theta}^2)\hat{r} + (2\dot{r}\dot{\theta} + r\ddot{\theta})\hat{\theta}. \tag{2.30}$$

$$\square$$

Example 2.4   Velocity and acceleration in tangential-normal coordinates. In this example we derive the velocity and acceleration for a particle moving in the plane using **tangential-normal** coordinates. Referring to Figure 2.7, assign the fixed frame {s} as shown, and assume particle P moves in the $\{\hat{x}_s, \hat{y}_s\}$ plane. Define a moving frame {b}

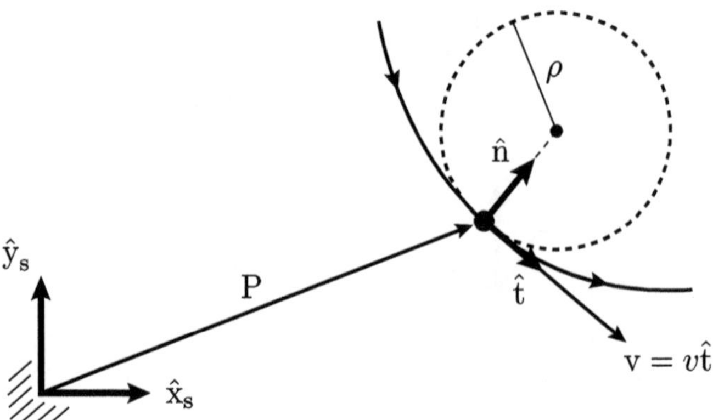

Figure 2.7 Velocity and acceleration in tangential-normal coordinates.

with axes $\{\hat{t}, \hat{n}\}$, such that its origin is attached to P, and $\hat{t}$ always points in the direction of the velocity of P:

$$v = v\hat{t}, \tag{2.31}$$

where v is the speed of the particle ($\hat{t}$ is therefore always tangent to the path of P). $\hat{n}$ is orthogonal to $\hat{t}$ and points toward the center of curvature of the curve traced by P; in other words, find the circle of radius ρ that is tangent to the path at P and such that its curvature $\kappa = 1/\rho$ matches the curvature of the path at P. $\hat{n}$ then points toward the center of this circle of radius ρ. The formula for ρ, also known as the **radius of curvature**, can be obtained from v as

$$\rho = \frac{\|v\|^3}{\|v \times \dot{v}\|}, \tag{2.32}$$

where $\| \cdot \|$ denotes the Euclidean vector norm.

Write the angular velocity of the moving frame {b} as

$$\omega = \omega\hat{z}_s \tag{2.33}$$

for some ω to be determined. Differentiate the velocity $v = v\hat{t}$ to obtain the acceleration:

$$a = \dot{v} = \dot{v}\hat{t} + v\dot{\hat{t}}, \tag{2.34}$$

where

$$\dot{\hat{t}} = \omega \times \hat{t} = \omega\hat{n}. \tag{2.35}$$

Substituting into (2.34),

$$a = \dot{v}\hat{t} + \omega v\hat{n}. \tag{2.36}$$

If the radius of curvature of the path at P is known, then $v = \rho\omega$, or $\omega = v/\rho$, from which it follows that

$$a = \dot{v}\hat{t} + \frac{v^2}{\rho}\hat{n}. \tag{2.37}$$

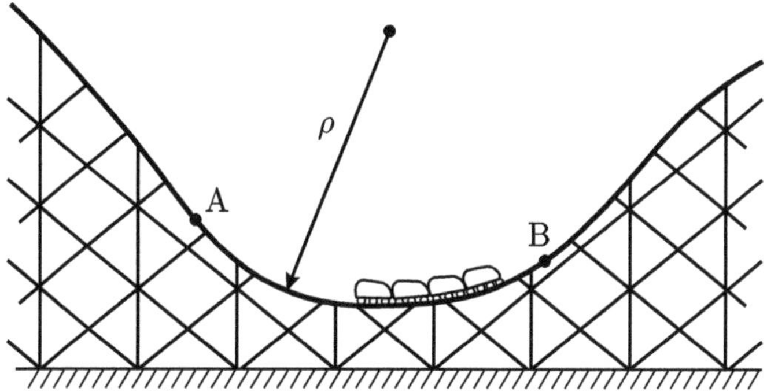

Figure 2.8 Measuring the velocity of a roller-coaster.

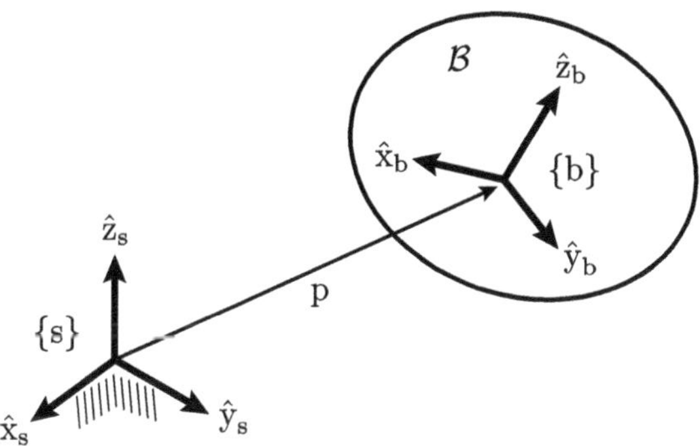

Figure 2.9 Representing a frame's orientation as a rotation matrix.

The above formula is useful when P undergoes a circular motion of known radius, in which case ρ is simply this radius. If P undergoes a purely circular motion of constant speed v, (i.e., $\dot{v} = 0$), then the above formula implies that the acceleration is directed only toward the center of the circle.

As an example of the use of tangential-normal coordinates, the roller-coaster shown in Figure 2.8 travels along a circular portion AB of the track with radius $\rho = 20$ m. Suppose the maximum acceleration experienced by the riders at the bottom of the track is three times the gravitational force. The maximum speed of the roller-coaster at this instant can then be computed from the normal component of (2.37):

$$\frac{v^2}{\rho} = a_n \rightarrow v = \sqrt{a_n \rho}, \tag{2.38}$$

where $a_n = 3 \times 9.81 \, \text{m/s}^2$ and $\rho = 20$ m. $\qquad\qquad\square$

2.3   Rotation Matrices

How many independent parameters are required to specify the position and orientation of a rigid body in space? To answer this question, consider the rigid body of Figure 2.9. Assigning a fixed frame {s} in space and a body frame {b} to the rigid body as shown, the position of the rigid body can be represented by the vector p from the fixed frame origin to the body frame origin:

$$p = p_1\hat{x}_s + p_2\hat{y}_s + p_3\hat{z}_s. \tag{2.39}$$

The orientation can be represented by the directions of the three orthonormal axes of the body frame:

$$\hat{x}_b = r_{11}\hat{x}_s + r_{21}\hat{y}_s + r_{31}\hat{z}_s, \tag{2.40}$$
$$\hat{y}_b = r_{12}\hat{x}_s + r_{22}\hat{y}_s + r_{32}\hat{z}_s, \tag{2.41}$$
$$\hat{z}_b = r_{13}\hat{x}_s + r_{23}\hat{y}_s + r_{33}\hat{z}_s. \tag{2.42}$$

Using this scheme requires a total of twelve parameters (p_i and $r_{ij}, i, j = 1, 2, 3$) to completely specify the position and orientation of the body-fixed frame. Note however that the r_{ij} are not independent. For example, if the components of $\hat{x}_b$ and $\hat{y}_b$ were known, then the components of $\hat{z}_b$ could be determined from the fact that $\hat{z}_b = \hat{x}_b \times \hat{y}_b$. This observation alone suggests that orientations can be specified with six, and likely fewer, parameters. What, then, is the smallest number of parameters required to describe a frame's orientation?

To answer this question, arrange the components of $\hat{x}_b$, $\hat{y}_b$, $\hat{z}_b$ into the 3×3 matrix R:

$$R = \begin{bmatrix} r_{11} & r_{12} & r_{13} \\ r_{21} & r_{22} & r_{23} \\ r_{31} & r_{32} & r_{33} \end{bmatrix}. \tag{2.43}$$

Note that the three columns of R are formed by the components of $\hat{x}_b$, $\hat{y}_b$, and $\hat{z}_b$, respectively. R satisfies the following conditions:

1. $\|\hat{x}_b\|^2 = \|\hat{y}_b\|^2 = \|\hat{z}_b\|^2 = 1$, or in components of R,

$$r_{11}^2 + r_{21}^2 + r_{31}^2 = 1, \tag{2.44}$$
$$r_{12}^2 + r_{22}^2 + r_{32}^2 = 1, \tag{2.45}$$
$$r_{13}^2 + r_{23}^2 + r_{33}^2 = 1. \tag{2.46}$$

2. $\hat{x}_b \cdot \hat{y}_b = \hat{x}_b \cdot \hat{z}_b = \hat{y}_b \cdot \hat{z}_b = 0$, or in components of R,

$$r_{11}r_{12} + r_{21}r_{22} + r_{31}r_{32} = 0, \tag{2.47}$$
$$r_{12}r_{13} + r_{22}r_{23} + r_{32}r_{33} = 0, \tag{2.48}$$
$$r_{11}r_{13} + r_{21}r_{23} + r_{31}r_{33} = 0. \tag{2.49}$$

A total of six constraints are imposed on the nine components of R, implying that only three of the nine components of R are independent. The above six constraints can be compactly expressed as a single constraint on the matrix R:

$$R^\top R = I, \tag{2.50}$$

where $R^\top$ denotes the matrix transpose of R, and I denotes the 3×3 identity matrix.

There is still the matter of accounting for the fact that the frame is right-handed (i.e., $\hat{x}_b \times \hat{y}_b = \hat{z}_b$) rather than left-handed (i.e., $\hat{x}_b \times \hat{y}_b = -\hat{z}_b$), which is not distinguished by the relation $R^\top R = I$. We make use of the following formula for evaluating the determinant of a 3×3 matrix M: Denoting its three columns by a, b, c, respectively, the determinant of M is then given by

$$\det M = a^\top(b \times c) = c^\top(a \times b) = b^\top(c \times a). \tag{2.51}$$

Now, if $R^\top R = I$, then from the determinant identity $\det AB = \det A \cdot \det B$ for any $A, B \in \mathbb{R}^{n \times n}$, it follows that $\det(R^\top R) = \det R^\top \cdot \det R = (\det R)^2 = 1$, or $\det R = \pm 1$. If R corresponds to a right-handed frame, then $\det R = 1$, while for a left-handed frame $\det R = -1$. The requirement $\det R = 1$ ensures that the chosen frame is right-handed; however, the number of independent continuous parameters needed to parameterize R still remains three.

Based on the above we make the following definition:

Definition 2.1 A **rotation matrix** is any 3×3 real matrix R that satisfies $R^\top R = I$ and $\det R = 1$.

Properties of Rotation Matrices

We now list some basic properties of rotation matrices:

1. The inverse of a rotation matrix R is its transpose $R^\top$. This follows from the definition $R^\top R = I$, implying that $R^{-1} = R^\top$. Another consequence is that $RR^\top = I$.
2. The product of two rotation matrices is also a rotation matrix. This can be verified by noting that given two rotation matrices R_1 and R_2, their product $R_1 R_2$ satisfies $(R_1 R_2)^\top(R_1 R_2) = I$ and $\det R_1 R_2 = \det R_1 \cdot \det R_2 = 1$.
3. The third property of rotation matrices provides a physical interpretation of the product of two rotation matrices. We first introduce some notation. Given three reference frames $\{a\}$, $\{b\}$, $\{c\}$, suppose the axes of frame $\{b\}$ can be expressed in terms of the axes of frame $\{a\}$ as

$$\begin{aligned}
\hat{x}_b &= r_{11}\hat{x}_a + r_{21}\hat{y}_a + r_{31}\hat{z}_a, \\
\hat{y}_b &= r_{12}\hat{x}_a + r_{22}\hat{y}_a + r_{32}\hat{z}_a, \\
\hat{z}_b &= r_{13}\hat{x}_a + r_{23}\hat{y}_a + r_{33}\hat{z}_a.
\end{aligned} \tag{2.52}$$

Similarly, suppose the axes of frame $\{c\}$ can be expressed in terms of the axes of frame $\{b\}$ as

$$\begin{aligned}
\hat{x}_c &= s_{11}\hat{x}_b + s_{21}\hat{y}_b + s_{31}\hat{z}_b, \\
\hat{y}_c &= s_{12}\hat{x}_b + s_{22}\hat{y}_b + s_{32}\hat{z}_b, \\
\hat{z}_c &= s_{13}\hat{x}_b + s_{23}\hat{y}_b + s_{33}\hat{z}_b.
\end{aligned} \tag{2.53}$$

The above equations can be expressed more compactly as

$$\begin{bmatrix} \hat{x}_b & \hat{y}_b & \hat{z}_b \end{bmatrix} = \begin{bmatrix} \hat{x}_a & \hat{y}_a & \hat{z}_a \end{bmatrix} \begin{bmatrix} r_{11} & r_{12} & r_{13} \\ r_{21} & r_{22} & r_{23} \\ r_{31} & r_{32} & r_{33} \end{bmatrix}, \tag{2.54}$$

$$\begin{bmatrix} \hat{x}_c & \hat{y}_c & \hat{z}_c \end{bmatrix} = \begin{bmatrix} \hat{x}_b & \hat{y}_b & \hat{z}_b \end{bmatrix} \begin{bmatrix} s_{11} & s_{12} & s_{13} \\ s_{21} & s_{22} & s_{23} \\ s_{31} & s_{32} & s_{33} \end{bmatrix}. \tag{2.55}$$

Denote the 3×3 rotation matrix of Equation (2.54) by R_{ab}, and that of Equation (2.55) by R_{bc}. Substitution of (2.54) into (2.55) yields

$$\begin{bmatrix} \hat{x}_c & \hat{y}_c & \hat{z}_c \end{bmatrix} = \begin{bmatrix} \hat{x}_a & \hat{y}_a & \hat{z}_a \end{bmatrix} \begin{bmatrix} r_{11} & r_{12} & r_{13} \\ r_{21} & r_{22} & r_{23} \\ r_{31} & r_{32} & r_{33} \end{bmatrix} \begin{bmatrix} s_{11} & s_{12} & s_{13} \\ s_{21} & s_{22} & s_{23} \\ s_{31} & s_{32} & s_{33} \end{bmatrix}. \tag{2.56}$$

From the above, it follows that

$$R_{ab} R_{bc} = R_{ac}. \tag{2.57}$$

Since $R_{aa} = I$, by choosing $c = a$ in (2.57), we get that $R_{ab}^{-1} = R_{ba}$.

4. Referring to Figure 2.10, given two reference frames {a} and {b} and a vector v, let

$$v = v_{a,x}\hat{x}_a + v_{a,y}\hat{y}_a + v_{a,z}\hat{z}_a \tag{2.58}$$

$$= v_{b,x}\hat{x}_b + v_{b,y}\hat{y}_b + v_{b,z}\hat{z}_b, \tag{2.59}$$

with

$$\begin{aligned} \hat{x}_b &= r_{11}\hat{x}_a + r_{21}\hat{y}_a + r_{31}\hat{z}_a, \\ \hat{y}_b &= r_{12}\hat{x}_a + r_{22}\hat{y}_a + r_{32}\hat{z}_a, \\ \hat{z}_b &= r_{13}\hat{x}_a + r_{23}\hat{y}_a + r_{33}\hat{z}_a. \end{aligned} \tag{2.60}$$

Using the notation

$$v_a = \begin{bmatrix} v_{a,x} \\ v_{a,y} \\ v_{a,z} \end{bmatrix}, \quad v_b = \begin{bmatrix} v_{b,x} \\ v_{b,y} \\ v_{b,z} \end{bmatrix}, \quad R_{ab} = \begin{bmatrix} r_{11} & r_{12} & r_{13} \\ r_{21} & r_{22} & r_{23} \\ r_{31} & r_{32} & r_{33} \end{bmatrix}, \tag{2.61}$$

a straightforward substitution establishes the following transformation rule between two different coordinate frame representations of the same vector:

$$v_a = R_{ab}v_b, \quad v_b = R_{ba}v_a. \tag{2.62}$$

5. Referring to Figure 2.11, let P be a point in space and {a} and {b} be two reference frames. Vectors u, v, w are defined as shown in the figure, with u = w + v. Let $p_a \in \mathbb{R}^3$ be the (x, y, z) coordinates for point P with respect to frame {a}; this is just the vector u expressed in frame {a} coordinates, which we also denote $u_a \in \mathbb{R}^3$. Similarly, let $p_b \in \mathbb{R}^3$ be the (x, y, z) coordinates for the same point P with respect to frame {b}; this is just v expressed in frame {b} coordinates, which we also denote $v_b \in \mathbb{R}^3$. Expressing u = w + v in frame {a} coordinates,

$$u_a = R_{ab}v_b + w_a, \tag{2.63}$$

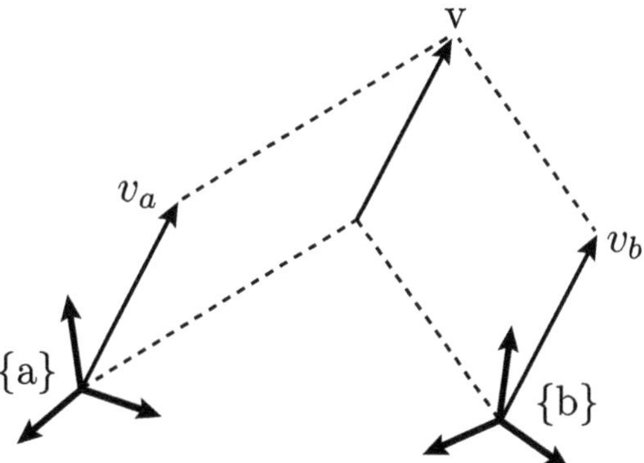

Figure 2.10 A vector v expressed in two different reference frames.

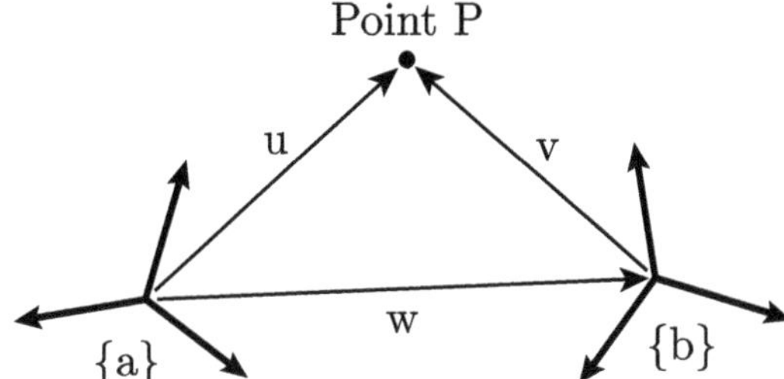

Figure 2.11 Coordinates for a point P expressed in two different reference frames.

where $w_a \in \mathbb{R}^3$ is the vector w expressed in frame {a} coordinates, and we make use of the previous property $v_a = R_{ab}v_b$. Since $u_a = p_a$ and $v_b = p_b$, we have the following relation between p_a and p_b:

$$p_a = R_{ab}p_b + d_{ab}, \tag{2.64}$$

where $d_{ab} = w_a$; that is, $d_{ab} \in \mathbb{R}^3$ is the vector from frame {a} to frame {b}, expressed in frame {a} coordinates.

Example 2.5   Orientation of a camera gimbal. The camera gimbal of Figure 2.12 consists of three rigid links serially connected by three revolute joints, in such a way that successive joint axes are always orthogonal to each other, and the three axes always intersect at a common point. The camera is rigidly attached to the final link, and placed at the intersection of the three joint axes. Figure 2.12 shows the gimbal when all of its joint angles are set to zero. The orientation of the camera is then determined by the angles of rotation about each of the three joint rotation axes.

To determine the orientation of the camera as a function of the three joint angles θ, ϕ, ψ, first place a fixed frame {0} at the stationary base, and attach moving reference frames {1}, {2}, {3} to each of the three links. The camera frame {c} is attached to the

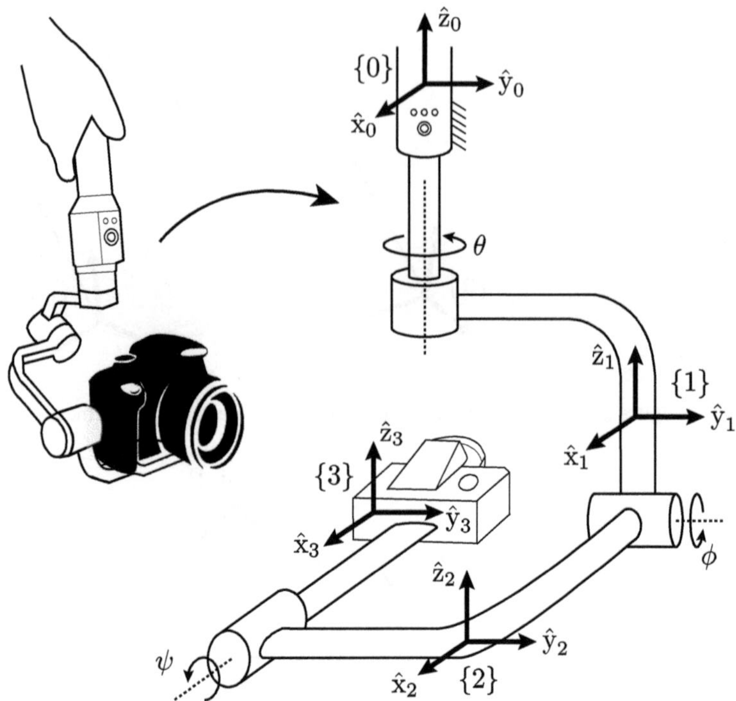

Figure 2.12 A camera gimbal shown with all joint angles set to zero.

camera in the same orientation as that of frame $\{3\}$. When the three joint angles are set to zero, all reference frames have the same orientation.

We now derive the relative frame orientations R_{01}, R_{12}, and R_{23} as a function of θ, ϕ, ψ. First, it can be seen that R_{01} depends only on the angle θ, and is given by

$$R_{01} = \begin{bmatrix} \cos\theta & -\sin\theta & 0 \\ \sin\theta & \cos\theta & 0 \\ 0 & 0 & 1 \end{bmatrix}. \tag{2.65}$$

Similarly, R_{12} depends only on ϕ and is given by

$$R_{12} = \begin{bmatrix} \cos\phi & 0 & \sin\phi \\ 0 & 1 & 0 \\ -\sin\phi & 0 & \cos\phi \end{bmatrix}. \tag{2.66}$$

Finally, R_{23} depends only on ψ, and is given by

$$R_{23} = \begin{bmatrix} 1 & 0 & 0 \\ 0 & \cos\psi & -\sin\psi \\ 0 & \sin\psi & \cos\psi \end{bmatrix}. \tag{2.67}$$

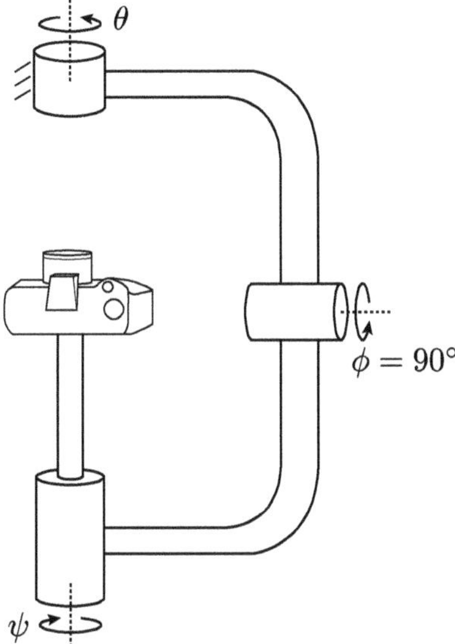

Figure 2.13 Configuration corresponding to $\phi = 90°$ for ZYX Euler angles.

$R_{03} = R_{01}R_{12}R_{23}$ is therefore given by

$$
R_{03} = \begin{bmatrix} c_\theta c_\phi & c_\theta s_\phi s_\psi - s_\theta c_\psi & c_\theta s_\phi c_\psi + s_\theta s_\psi \\ s_\theta c_\phi & s_\theta s_\phi s_\psi + c_\theta c_\psi & s_\theta s_\phi c_\psi - c_\theta s_\psi \\ -s_\phi & c_\phi s_\psi & c_\phi c_\psi \end{bmatrix},
\tag{2.68}
$$

where s_θ is shorthand for $\sin\theta$, c_θ for $\cos\theta$, and so on.

We can also ask what the angles (θ, ϕ, ψ) should be to place the camera in some desired orientation $R_{03} = R$. To answer this question, let r_{ij} be the ijth element of R. Then from Equation (2.68) we know that $r_{11}^2 + r_{21}^2 = \cos^2 \phi$; as long as $\cos\phi \neq 0$, or equivalently $\phi \neq \pm 90°$, we have

$$
\phi = \tan^{-1}\left(\frac{\sin\phi}{\cos\phi}\right) = \tan^{-1}\left(\frac{-r_{31}}{\pm\sqrt{r_{11}^2 + r_{21}^2}}\right).
\tag{2.69}
$$

Assuming the value for ϕ obtained from above is not $\pm 90°$, θ and ψ can then be determined from the following relations:

$$
\theta = \tan^{-1}\frac{r_{21}}{r_{11}},
\tag{2.70}
$$

$$
\psi = \tan^{-1}\frac{r_{32}}{r_{33}}.
\tag{2.71}
$$

In the event that $\phi = \pm\pi/2$, there exists a one-parameter family of solutions for θ and ψ. This is most easily seen from Figure 2.13: If for example, $\phi = \pi/2$, θ and ψ then represent rotations (in the opposite direction) about the same vertical axis. Hence, if

$(\theta, \phi, \psi) = (\bar{\theta}, \pi/2, \bar{\psi})$ is a solution for a given rotation R, then any triple $(\bar{\theta}', \pi/2, \bar{\psi}')$ where $\bar{\theta}' - \bar{\psi}' = \bar{\theta} - \bar{\psi}$, is also a solution.

The angles (θ, ϕ, ψ) as defined above correspond to the **ZYX Euler angles** for rotation matrices. Any rotation matrix R can be parameterized by these three angles. $\square$

2.4   Angular Velocity and Rotations

For a moving frame {b} with axes $\{\hat{x}_b, \hat{y}_b, \hat{z}_b\}$ that rotates with angular velocity ω, earlier we showed that

$$\dot{\hat{x}}_b = \omega \times \hat{x}_b, \tag{2.72}$$

$$\dot{\hat{y}}_b = \omega \times \hat{y}_b, \tag{2.73}$$

$$\dot{\hat{z}}_b = \omega \times \hat{z}_b. \tag{2.74}$$

We now write the above explicitly in terms of coordinates of the fixed frame {s}. To do so we will make use of the following formula for calculating the cross-product of two vectors:

Proposition 2.1   *For $u, v \in \mathbb{R}^3$, their cross-product $u \times v$ is given by*

$$u \times v = \begin{bmatrix} 0 & -u_3 & u_2 \\ u_3 & 0 & -u_1 \\ -u_2 & u_1 & 0 \end{bmatrix} \begin{bmatrix} v_1 \\ v_2 \\ v_3 \end{bmatrix}. \tag{2.75}$$

Defining the notation

$$[u] = \begin{bmatrix} 0 & -u_3 & u_2 \\ u_3 & 0 & -u_1 \\ -u_2 & u_1 & 0 \end{bmatrix} \tag{2.76}$$

for any $u \in \mathbb{R}^3$, Equation (2.75) can then be written $u \times v = [u]v$.

We will refer to $[u]$ as the 3×3 **skew-symmetric matrix representation** of the vector $u \in \mathbb{R}^3$. We will also need the following result:

Proposition 2.2   *Given a rotation matrix R and vector $\omega \in \mathbb{R}^3$, the following equality always holds:*

$$R^\top [\omega] R = [R^\top \omega]. \tag{2.77}$$

To prove (2.77), denote the columns of R by $r_i \in \mathbb{R}^3$, $i = 1, 2, 3$. Since the columns of R are orthonormal it follows that $r_1 \times r_2 = r_3$, $r_2 \times r_3 = r_1$, $r_3 \times r_1 = r_2$. Expanding $[\omega]R$,

$$[\omega]R = \begin{bmatrix} \omega \times r_1 & \omega \times r_2 & \omega \times r_3 \end{bmatrix}. \tag{2.78}$$

Using the above to expand $R^\top[\omega]R$,

$$R^\top[\omega]R = \begin{bmatrix} r_1^\top(\omega \times r_1) & r_1^\top(\omega \times r_2) & r_1^\top(\omega \times r_3) \\ r_2^\top(\omega \times r_1) & r_2^\top(\omega \times r_2) & r_2^\top(\omega \times r_3) \\ r_3^\top(\omega \times r_1) & r_3^\top(\omega \times r_2) & r_3^\top(\omega \times r_3) \end{bmatrix}$$

$$= \begin{bmatrix} 0 & -r_3^\top\omega & r_2^\top\omega \\ r_3^\top\omega & 0 & -r_1^\top\omega \\ -r_2^\top\omega & r_1^\top\omega & 0 \end{bmatrix}$$

$$= [R^\top\omega], \tag{2.79}$$

where we have made use of (2.51) corresponding to the general vector identity $a^\top(b \times c) = b^\top(c \times a) = c^\top(a \times b)$ for any $a, b, c \in \mathbb{R}^3$. Since Equation (2.77) holds for all rotation matrices R (including its transpose $R^\top$), the following is also true:

$$R[\omega]R^\top = [R\omega]. \tag{2.80}$$

With the above, we now derive the relation between rotations and angular velocities. Let R_{sb} be the rotation matrix describing the orientation of $\{b\}$ as seen from $\{s\}$. The first column of R_{sb}, denoted r_1, is $\hat{x}_b$ expressed in fixed frame coordinates; similarly, r_2 and r_3, respectively, describe $\hat{y}_b$ and $\hat{z}_b$ in fixed frame coordinates. Let $\omega_s \in \mathbb{R}^3$ be the angular velocity ω expressed in fixed frame coordinates. The previous three velocity relations, expressed in fixed frame coordinates, then become

$$\dot{r}_i = \omega_s \times r_i = [\omega_s]r_i, \quad i = 1, 2, 3, \tag{2.81}$$

where $[\omega_s]$ is the 3×3 skew-symmetric matrix representation of ω_s. The above three equations can be rearranged into the following single matrix equation:

$$\dot{R}_{sb} = [\omega_s]R_{sb}. \tag{2.82}$$

Observe that since $R_{sb}^{-1} = R_{sb}^\top$, the above can also be written

$$\dot{R}_{sb}R_{sb}^\top = [\omega_s]. \tag{2.83}$$

This result shows that not only is the quantity $\dot{R}_{sb}R_{sb}^\top$ always skew-symmetric, but also admits the physical interpretation as the angular velocity of the moving frame expressed in fixed frame coordinates.

Now let ω_b be ω expressed in the moving frame $\{b\}$ coordinates. Since ω_s and ω_b are simply two different coordinate frame representations of the same vector ω, it follows from Equation (2.62) that

$$\omega_b = R_{bs}\omega_s = R_{sb}^\top\omega_s \tag{2.84}$$

and

$$\omega_s = R_{sb}\omega_b. \tag{2.85}$$

Using the general relation $R^\top[\omega]R = [R^\top\omega]$ derived earlier, we can write

$$
\begin{aligned}
[\omega_b] &= [R_{sb}^\top \omega_s] \\
&= R_{sb}^\top [\omega_s] R_{sb} \\
&= R_{sb}^\top (\dot{R}_{sb} R_{sb}^\top) R_{sb} \\
&= R_{sb}^\top \dot{R}_{sb}.
\end{aligned}
\tag{2.86}
$$

In summary, we have the following two fundamental equations that relate the rotation matrix R to the angular velocity ω:

Proposition 2.3 *Let the rotation matrix $R_{sb}(t)$ represent the orientation, at time t, of the moving frame $\{b\}$ as seen from the fixed frame $\{s\}$, $\omega_s(t) \in \mathbb{R}^3$ be the angular velocity of the moving frame expressed in the $\{s\}$ frame, and $\omega_b(t) \in \mathbb{R}^3$ be the angular velocity of the moving frame expressed in the $\{b\}$ frame. Then*

$$
\dot{R}_{sb} R_{sb}^\top = [\omega_s],
\tag{2.87}
$$

$$
R_{sb}^\top \dot{R}_{sb} = [\omega_b],
\tag{2.88}
$$

with $\omega_s = R_{sb}\omega_b$, $\omega_b = R_{bs}\omega_s = R_{sb}^\top \omega_s$.

Example 2.6 The angular velocity of a rigid body is independent of the choice of body frame. To show this, suppose two moving frames $\{c\}$ and $\{d\}$ are attached to different points on the same rigid body, in different orientations. With respect to the fixed frame $\{s\}$, let R_{sc} and R_{sd} be rotation matrices representing the orientations of $\{c\}$ and $\{d\}$ with respect to $\{s\}$, respectively. The rotation R_{cd}, which is known in advance, is constant. Then

$$
[\omega_c] = \dot{R}_{sc} R_{sc}^\top,
\tag{2.89}
$$

$$
[\omega_d] = \dot{R}_{sd} R_{sd}^\top
\tag{2.90}
$$

are, respectively, the angular velocities of frames $\{c\}$ and $\{d\}$ expressed in the $\{s\}$ frame. Since $R_{sd} = R_{sc} R_{cd}$,

$$
\begin{aligned}
[\omega_d] &= \dot{R}_{sd} R_{sd}^\top \\
&= (\dot{R}_{sc} R_{cd} + R_{sc} \dot{R}_{cd}) R_{cd}^\top R_{sc}^\top \\
&= \dot{R}_{sc} R_{sc}^\top \\
&= [\omega_c],
\end{aligned}
\tag{2.91}
$$

where we use the fact that $\dot{R}_{cd} = 0$. Therefore $\omega_c = \omega_d$ as claimed. $\square$

2.5   Multibody Kinematics

We begin this section with the example of Figure 2.14. The fire truck is parked, and the fixed frame is indicated by $\{0\}$. The ladder can rotate about the vertical and horizontal axes indicated by the angles θ_1 and θ_2 as shown in the figure. For this problem we shall

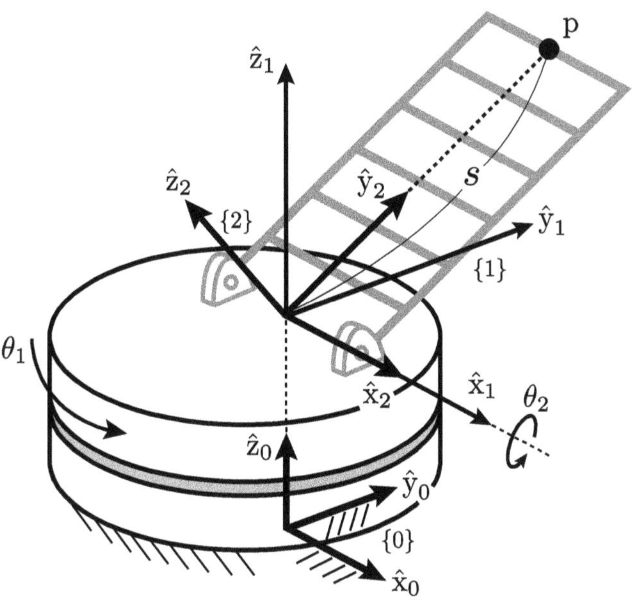

Figure 2.14 A fire truck ladder.

use two moving frames. Moving frame {1} is attached to the rotating base, and rotates about the vertical axis with angle θ_1 such that

$$R_{01} = \begin{bmatrix} \cos\theta_1 & -\sin\theta_1 & 0 \\ \sin\theta_1 & \cos\theta_1 & 0 \\ 0 & 0 & 1 \end{bmatrix}. \tag{2.92}$$

Moving frame {2} is attached to the ladder, and rotates about the horizontal axis with angle θ_2 such that

$$R_{12} = \begin{bmatrix} 1 & 0 & 0 \\ 0 & \cos\theta_2 & -\sin\theta_2 \\ 0 & \sin\theta_2 & \cos\theta_2 \end{bmatrix}. \tag{2.93}$$

Using moving frame {2}, the tip position p of the ladder can be written simply as $p = s\hat{y}_2$. To determine the velocity and acceleration of p requires knowledge of the angular velocity of frame {2}, which we now derive.

Let $\omega_1 \in \mathbb{R}^3$ be the angular velocity vector of frame {1} expressed in fixed frame {0} coordinates:

$$[\omega_1] = \dot{R}_{01} R_{01}^\top. \tag{2.94}$$

A straightforward calculation shows that $\omega_1 = (0, 0, \dot{\theta}_1)^\top$. Similarly, let $\omega_2 \in \mathbb{R}^3$ be the angular velocity vector of frame {2} expressed in fixed frame {0} coordinates:

$$[\omega_2] = \dot{R}_{02} R_{02}^\top. \tag{2.95}$$

Since $R_{02} = R_{01}R_{12}$, the above can also be written

$$
\begin{aligned}
[\omega_2] &= \frac{d}{dt}(R_{01}R_{12}) \cdot (R_{01}R_{12})^\top \\
&= (\dot{R}_{01}R_{12} + R_{01}\dot{R}_{12})R_{12}^\top R_{01}^\top \\
&= \dot{R}_{01}R_{01}^\top + R_{01}(\dot{R}_{12}R_{12}^\top)R_{01}^\top \\
&= [\omega_1] + R_{01}[\omega_{12}]R_{01}^\top,
\end{aligned}
\tag{2.96}
$$

where we have made use of the identity $RR^\top = I$ for any rotation matrix R, and we use the notation $[\omega_{12}]$ to denote $\dot{R}_{12}R_{12}^\top$. The vector $\omega_{12} \in \mathbb{R}^3$ can be interpreted as the relative angular velocity of frame $\{2\}$ as seen from frame $\{1\}$; i.e., imagine frame $\{1\}$ to be the fixed frame, and express the angular velocity of frame $\{2\}$ as observed from frame $\{1\}$. For this particular example ω_{12} evaluates to $\omega_{12} = (\dot{\theta}_2, 0, 0)^\top$.

Using the general rotation matrix identity $R[\omega]R^\top = [R\omega]$ from Equation (2.80), Equation (2.96) can be expressed in the form

$$
\omega_2 = \omega_1 + R_{01}\omega_{12}.
\tag{2.97}
$$

What is $R_{01}\omega_{12}$? To answer this question, recall the physical meaning behind the relation $v_a = R_{ab}v_b$ from (2.62): If $v_b \in \mathbb{R}^3$ is some vector v expressed in frame $\{b\}$ coordinates, then $v_a = R_{ab}v_b$ is the same vector v but this time expressed in frame $\{a\}$ coordinates.

With this interpretation, $R_{01}\omega_{12}$ can then be understood to be the representation of the same relative angular velocity vector, but this time in fixed frame $\{0\}$ coordinates. Indeed, calculation reveals that

$$
R_{01}\omega_{12} = \begin{bmatrix} \cos\theta_1 & -\sin\theta_1 & 0 \\ \sin\theta_1 & \cos\theta_1 & 0 \\ 0 & 0 & 1 \end{bmatrix} \begin{bmatrix} \dot{\theta}_2 \\ 0 \\ 0 \end{bmatrix} = \begin{bmatrix} \cos\theta_1 \\ \sin\theta_1 \\ 0 \end{bmatrix} \dot{\theta}_2,
\tag{2.98}
$$

which is the angular velocity vector of joint axis θ_2 expressed in fixed frame $\{0\}$ coordinates.

In summary, Equation (2.97) implies that the angular velocity vector of frame $\{2\}$ is just the sum of the angular velocity vectors of the two rotating axes:

$$
\begin{aligned}
\omega_2 &= \dot{\theta}_1\hat{z}_0 + \dot{\theta}_2\hat{x}_1 \\
&= \dot{\theta}_1\hat{z}_0 + \dot{\theta}_2(\cos\theta_1\hat{x}_0 + \sin\theta_1\hat{y}_0).
\end{aligned}
\tag{2.99}
$$

In short, the overall angular velocity vector can be obtained by taking the vector sum of the axes of rotation. The angular acceleration α_2 can then be obtained by differentiating ω_2 in the usual fashion:

$$
\begin{aligned}
\alpha_2 &= \dot{\omega}_2 \\
&= \ddot{\theta}_1\hat{z}_0 + \ddot{\theta}_2\hat{x}_1 + \dot{\theta}_2\frac{d}{dt}\hat{x}_1 \\
&= \ddot{\theta}_1\hat{z}_0 + \ddot{\theta}_2\hat{x}_1 + \dot{\theta}_2(\omega_1 \times \hat{x}_1),
\end{aligned}
\tag{2.100}
$$

which, by replacing ω_1 and $\hat{x}_1$ by their appropriate fixed frame $\{0\}$ representations, can be expressed entirely in terms of fixed frame coordinates.

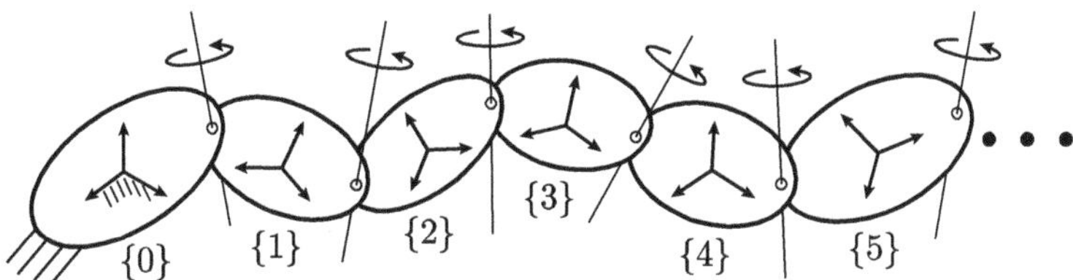

Figure 2.15 A serial chain.

The above procedure in fact extends straightforwardly to more general systems consisting of serially connected rigid bodies of the kind shown in Figure 2.15. For the sake of completeness we will now show this formally; the reasoning and key steps in the derivation are exactly identical to that of our earlier example.

For the n-link serial chain of Figure 2.15, attach reference frames to each link, numbered from $\{1\}$ to $\{n\}$, with $\{0\}$ the fixed frame. The orientation of the last link frame is then given by the product of rotation matrices

$$R_{0n} = R_{01}R_{12}\cdots R_{n-1,n}. \tag{2.101}$$

The angular velocity of frame $\{n\}$ expressed in the fixed frame $\{0\}$, $\omega_n \in \mathbb{R}^3$, is derived as

$$\begin{aligned}
[\omega_n] &= \dot{R}_{0n}R_{0n}^\top \\
&= \dot{R}_{01}R_{01}^\top + R_{01}(\dot{R}_{12}R_{12}^\top)R_{01}^\top + R_{02}(\dot{R}_{23}R_{23}^\top)R_{02}^\top \\
&\quad + \cdots + R_{0,n-1}(\dot{R}_{n-1,n}R_{n-1,n}^\top)R_{0,n-1}^\top.
\end{aligned} \tag{2.102}$$

The term $\dot{R}_{i-1,i}R_{i-1,i}^\top$ can be interpreted as the relative angular velocity of frame $\{i\}$ with respect to frame $\{i-1\}$: imagine frame $\{i-1\}$ to be the fixed frame, and express the angular velocity of frame $\{i\}$ in frame $\{i-1\}$ coordinates. The term $R_{0,i-1}(\dot{R}_{i-1,i}R_{i-1,i}^\top)R_{0,i-1}^\top$ then simply transforms $\dot{R}_{i-1,i}R_{i-1,i}^\top$ into fixed frame $\{0\}$ coordinates (this follows from the identity (2.80) and the transformation rule (2.62) for vectors). Using the notation

$$[\omega_{i-1,i}] = \dot{R}_{i-1,i}R_{i-1,i}^\top, \quad i = 1, \ldots, n, \tag{2.103}$$

where $\omega_{i-1,i} \in \mathbb{R}^3$ denotes the relative angular velocity of frame $\{i\}$ as seen from frame $\{i-1\}$, Equation (2.102) can be written in vector form as

$$\omega_n = \omega_{01} + R_{01}\omega_{12} + \cdots + R_{0,n-1}\omega_{n-1,n}. \tag{2.104}$$

Equation (2.104) expresses the angular velocity $\omega_n \in \mathbb{R}^3$ as a column vector in fixed frame $\{0\}$ coordinates. Alternatively, each of the terms in Equation (2.104) can be expressed in terms of the corresponding moving frame coordinates, similar to what was done in our earlier example.

Example 2.7 Angular velocity and angular acceleration of the camera gimbal. In this example we calculate the angular velocity and angular acceleration of the camera gimbal of Figure 2.12. Since the camera gimbal is a serial chain structure, the angular velocity ω_3 of the tip frame $\{3\}$ is

$$\omega_3 = \dot{\theta}\hat{z}_0 + \dot{\phi}\hat{y}_1 + \dot{\psi}\hat{x}_2. \tag{2.105}$$

We now express ω_3 as a column vector $\omega_3 \in \mathbb{R}^3$ in fixed frame $\{0\}$ coordinates, also deriving the explicit dependence of ω_3 on the rotation angles θ, ϕ, ψ. From the link frame assignments as indicated in the figure,

$$[\omega_3] = \dot{R}_{03} R_{03}^\top. \tag{2.106}$$

From the relation $R_{03} = R_{01} R_{12} R_{23}$, it can be verified that

$$[\omega_3] = \dot{R}_{01} R_{01}^\top + R_{01}(\dot{R}_{12} R_{12}^\top) R_{01}^\top + R_{02}(\dot{R}_{23} R_{23}^\top) R_{02}^\top. \tag{2.107}$$

1. The first term $\dot{R}_{01} R_{01}^\top$ is the angular velocity of frame $\{1\}$ expressed in frame $\{0\}$:

$$\dot{R}_{01} R_{01}^\top = \begin{bmatrix} 0 & -\dot{\theta} & 0 \\ \dot{\theta} & 0 & 0 \\ 0 & 0 & 0 \end{bmatrix}. \tag{2.108}$$

The above is the 3×3 skew-symmetric matrix representation of the vector

$$\begin{bmatrix} 0 \\ 0 \\ 1 \end{bmatrix} \dot{\theta}, \tag{2.109}$$

which in turn is just $\dot{\theta}\hat{z}_0$ expressed as a column vector in frame $\{0\}$ coordinates.

2. For the second term $R_{01}(\dot{R}_{12} R_{12}^\top) R_{01}^\top$, note that $\dot{R}_{12} R_{12}^\top$ is the relative angular velocity of frame $\{2\}$ as seen from frame $\{1\}$; i.e., imagine frame $\{1\}$ to be the fixed frame, and express the angular velocity of frame $\{2\}$ in frame $\{1\}$ coordinates:

$$\dot{R}_{12} R_{12}^\top = \begin{bmatrix} 0 & 0 & \dot{\phi} \\ 0 & 0 & 0 \\ -\dot{\phi} & 0 & 0 \end{bmatrix}. \tag{2.110}$$

The above is just the 3×3 skew-symmetric matrix representation of

$$\begin{bmatrix} 0 \\ 1 \\ 0 \end{bmatrix} \dot{\phi}, \tag{2.111}$$

which in turn is just $\dot{\phi}\hat{y}_1$ expressed as a column vector in frame $\{1\}$ coordinates. Using the general formula (2.80) to evaluate $R_{01}(\dot{R}_{12} R_{12}^\top) R_{01}^\top$, the second term simplifies to

$$R_{01}(\dot{R}_{12} R_{12}^\top) R_{01}^\top = \begin{bmatrix} 0 & 0 & \dot{\phi}\cos\theta \\ 0 & 0 & \dot{\phi}\sin\theta \\ -\dot{\phi}\cos\theta & -\dot{\phi}\sin\theta & 0 \end{bmatrix}. \tag{2.112}$$

The above is just the 3×3 skew-symmetric matrix representation of the vector

$$
\begin{bmatrix} -\sin\theta \\ \cos\theta \\ 0 \end{bmatrix} \dot\phi,
\tag{2.113}
$$

which in turn is just $\dot\phi \hat{y}_1$ expressed in frame $\{0\}$ coordinates:

$$
\dot\phi \hat{y}_1 = -\dot\phi \sin\theta \hat{x}_0 + \dot\phi \cos\theta \hat{y}_0.
\tag{2.114}
$$

3. For the third term, $R_{02}(\dot R_{23} R_{23}^\top) R_{02}^\top$, first evaluate $\dot R_{23} R_{23}^\top$:

$$
\dot R_{23} R_{23}^\top = \begin{bmatrix} 0 & 0 & 0 \\ 0 & 0 & -\dot\psi \\ 0 & \dot\psi & 0 \end{bmatrix}.
\tag{2.115}
$$

The above is just the 3×3 skew-symmetric matrix representation of the vector

$$
\begin{bmatrix} 1 \\ 0 \\ 0 \end{bmatrix} \dot\psi.
\tag{2.116}
$$

Now using (2.80) to evaluate $R_{02}(\dot R_{23} R_{23}^\top) R_{02}^\top$, the third term simplifies to

$$
R_{02}(\dot R_{23} R_{23}^\top) R_{02}^\top = \begin{bmatrix} 0 & \sin\phi & \sin\theta\cos\phi \\ -\sin\phi & 0 & -\cos\theta\cos\phi \\ -\sin\theta\cos\phi & \cos\theta\cos\phi & 0 \end{bmatrix} \dot\psi.
\tag{2.117}
$$

The above is just the 3×3 skew-symmetric matrix representation of the vector

$$
\begin{bmatrix} \cos\theta\cos\phi \\ \sin\theta\cos\phi \\ -\sin\phi \end{bmatrix} \dot\psi,
\tag{2.118}
$$

which in turn is just $\dot\psi \hat{x}_2$ expressed in frame $\{0\}$ coordinates:

$$
\dot\psi \hat{x}_2 = \dot\psi \cos\theta\cos\phi \hat{x}_0 + \dot\psi \sin\theta\cos\phi \hat{y}_0 - \dot\psi \sin\phi \hat{z}_0.
\tag{2.119}
$$

In summary, the expression for the angular velocity of Equation (2.107) is

$$
\begin{aligned}
\omega_3 &= \dot\theta \hat{z}_0 + \dot\phi \hat{y}_1 + \dot\psi \hat{x}_2 \\
&= (-\dot\phi \sin\theta + \dot\psi \cos\theta\cos\phi)\hat{x}_0 + (\dot\phi \cos\theta + \dot\psi \sin\theta\cos\phi)\hat{y}_0 \\
&\quad + (\dot\theta - \dot\psi \sin\phi)\hat{z}_0,
\end{aligned}
\tag{2.120}
$$

which can also be expressed as the column vector $\omega_3 \in \mathbb{R}^3$ in frame $\{0\}$ coordinates.

The angular acceleration is obtained by differentiating the angular velocity. Differentiating (2.105) directly,

$$
\alpha_3 = \dot\omega_3 = \ddot\theta \hat{z}_0 + \ddot\phi \hat{y}_1 + \dot\phi \dot{\hat{y}}_1 + \ddot\psi \hat{x}_2 + \dot\psi \dot{\hat{x}}_2,
\tag{2.121}
$$

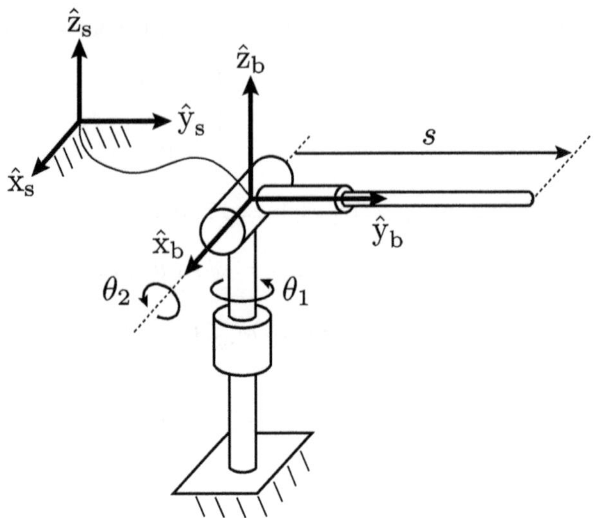

Figure 2.16 A spherical telescopic arm.

where $\dot{\hat{y}}_1 = \omega_1 \times \hat{y}_1$, and $\dot{\hat{x}}_2 = \omega_2 \times \hat{x}_2$, with ω_1 and ω_2 the angular velocities of frames $\{1\}$ and $\{2\}$, respectively:

$$\omega_1 = \dot{\theta}\hat{z}_0, \tag{2.122}$$

$$\omega_2 = \dot{\theta}\hat{z}_0 + \dot{\phi}\hat{y}_1. \tag{2.123}$$

Alternatively, if (2.120) is differentiated directly, the angular acceleration can be obtained explicitly in fixed frame coordinates (the resulting calculation is tedious). $\square$

Example 2.8   Velocity and acceleration analysis of a spherical telescopic arm. In this example we calculate the velocity and acceleration of the spherical telescopic arm of Figure 2.16. The fixed frame $\{s\}$ is placed at the intersection of the two revolute joint axes as shown. The moving frame $\{b\}$ is also attached at the intersection of the two revolute joint axes as shown in the figure, with its $\hat{y}_b$ axis always directed along the length of the arm; when the joints θ_1 and θ_2 are both set to zero, the $\{b\}$ frame exactly overlaps the fixed $\{s\}$ frame.

With the frames defined as above, the position of the tip can be written

$$p = s\hat{y}_b. \tag{2.124}$$

The velocity $v = \dot{p}$ is

$$\begin{aligned} v &= \dot{s}\hat{y}_b + s\dot{\hat{y}}_b \\ &= \dot{s}\hat{y}_b + \omega \times s\hat{y}_b, \end{aligned} \tag{2.125}$$

where we make use of the identity $\dot{\hat{y}}_b = \omega \times \hat{y}_b$, with the angular velocity ω given by

$$\omega = \dot{\theta}_1\hat{z}_s + \dot{\theta}_2\hat{x}_b. \tag{2.126}$$

The tip acceleration $a = \dot{v}$ is

$$a = \ddot{s}\hat{y}_b + \dot{\omega} \times s\hat{y}_b + \omega \times (2\dot{s}\hat{y}_b + (\omega \times s\hat{y}_b)), \qquad (2.127)$$

where

$$\dot{\omega} = \ddot{\theta}_1 \hat{z}_s + \ddot{\theta}_2 \hat{x}_b + \dot{\theta}_2(\omega \times \hat{x}_b). \qquad (2.128)$$

If the moving frame axes $\{\hat{x}_b, \hat{y}_b, \hat{z}_b\}$ can be determined explicitly as a function of the joint angles θ_1 and θ_2, then (2.128) can be used to determine the tip velocity and acceleration for any given values of $\theta_1, \theta_2, \dot{\theta}_1, \dot{\theta}_2$. $\qquad\qquad \Box$

2.6 Summary

- Once a length scale for physical space has been assigned, a **fixed reference frame** $\{s\}$, also called the **space frame**, is defined. The axes of $\{s\}$ are denoted $\{\hat{x}_s, \hat{y}_s, \hat{z}_s\}$. $\{s\}$ is stationary and **right-handed** (that is, $\hat{x}_s \times \hat{y}_s = \hat{z}_s$).
- Once a fixed frame has been defined, a **moving frame**, $\{b\}$, also called the **body frame**, can be attached to a moving rigid body $\mathcal{B}$ in physical space. All body frames are also right-handed.
- The **angular velocity** ω of a rotating moving frame is a three-dimensional vector whose direction represents the instantaneous axis of rotation, and whose length represents the rate of rotation (in radians per second, with positive rotations defined to be counterclockwise about the axis of rotation).
- Given a moving frame with axes $\{\hat{x}_b, \hat{y}_b, \hat{z}_b\}$ rotating with angular velocity ω, the derivatives of the moving frame axes are given by

$$\dot{\hat{x}}_b = \omega \times \hat{x}_b, \qquad (2.129)$$

$$\dot{\hat{y}}_b = \omega \times \hat{y}_b, \qquad (2.130)$$

$$\dot{\hat{z}}_b = \omega \times \hat{z}_b. \qquad (2.131)$$

  For any particle whose position is expressed in terms of the axes of the fixed frame and any number of moving frames, with the angular velocities of each of the moving frames known, the velocity of the particle can be obtained by differentiating the position with respect to time and using (2.129)–(2.131). The acceleration can be obtained by differentiating the velocity in a similar fashion.

- A **rotation matrix** is a 3×3 real matrix R satisfying $R^\top R = I$ and $\det R = 1$. From the definition, R also satisfies $R^{-1} = R^\top$. The product of two rotation matrices is also a rotation matrix. The orientation of a moving frame with respect to the fixed frame can be represented as a rotation matrix: The columns of the rotation matrix represent the axes of the moving frame, expressed as three-dimensional vectors in fixed frame coordinates.
- Given three reference frames $\{a\}$, $\{b\}$, $\{c\}$, let R_{ab} be the rotation matrix describing the orientation of $\{b\}$ as seen from $\{a\}$; R_{bc} and R_{ac} are analogously defined. Then $R_{ab}R_{bc} = R_{ac}$. Since by definition $R_{aa} = I$, it follows that $R_{ab}^{-1} = R_{ba}$.

- Given a vector v, let $v_a \in \mathbb{R}^3$ be its representation with respect to some frame {a}, and $v_b \in \mathbb{R}^3$ its representation with respect to frame {b}. Then $v_b = R_{ba}v_a$ and $v_a = R_{ab}v_b$.

- Given a point P in space, let $p_a \in \mathbb{R}^3$ be the (x, y, z) coordinates for P with respect to frame {a}, and $p_b \in \mathbb{R}^3$ be the (x, y, z) coordinates for P with respect to frame {b}. Then p_a and p_b are related by

$$p_a = R_{ab}p_b + d_{ab}, \tag{2.132}$$

where $d_{ab} \in \mathbb{R}^3$ is the vector from frame {a} to frame {b}, expressed in frame {a} coordinates.

- The **ZYX Euler angles** (θ, ϕ, ψ) parameterize the rotation matrices via the formula

$$R(\theta, \phi, \psi) = \mathrm{Rot}(\hat{z}, \theta) \cdot \mathrm{Rot}(\hat{y}, \phi) \cdot \mathrm{Rot}(\hat{x}, \psi) \tag{2.133}$$

$$= \begin{bmatrix} \cos\theta & -\sin\theta & 0 \\ \sin\theta & \cos\theta & 0 \\ 0 & 0 & 1 \end{bmatrix} \begin{bmatrix} \cos\phi & 0 & \sin\phi \\ 0 & 1 & 0 \\ -\sin\phi & 0 & \cos\phi \end{bmatrix} \begin{bmatrix} 1 & 0 & 0 \\ 0 & \cos\psi & -\sin\psi \\ 0 & \sin\psi & \cos\psi \end{bmatrix}. \tag{2.134}$$

For every rotation matrix R, there exists at least one triple (θ, ϕ, ψ) satisfying the above.

- The cross-product of two vectors $u, v \in \mathbb{R}^3$ can be calculated as

$$u \times v = \begin{bmatrix} 0 & -u_3 & u_2 \\ u_3 & 0 & -u_1 \\ -u_2 & u_1 & 0 \end{bmatrix} \begin{bmatrix} v_1 \\ v_2 \\ v_3 \end{bmatrix} = [u]v, \tag{2.135}$$

where $[u]$ is referred to as the 3×3 skew-symmetric matrix representation of $u \in \mathbb{R}^3$.

- Any rotation matrix R and vector $\omega \in \mathbb{R}^3$ always satisfies

$$R[\omega]R^\top = [R\omega], \quad R^\top[\omega]R = [R^\top\omega]. \tag{2.136}$$

- Let the rotation matrix R_{sb} represent the orientation of the moving frame {b} as seen from the fixed frame {s}. Let ω be the angular velocity of the moving frame, and $\omega_s \in \mathbb{R}^3$ and $\omega_b \in \mathbb{R}^3$ be ω expressed in the {s} and {b} frames, respectively. Then

$$[\omega_s] = \dot{R}_{sb}R_{sb}^\top, \tag{2.137}$$

$$[\omega_b] = R_{sb}^\top\dot{R}_{sb}. \tag{2.138}$$

Also $\omega_b = R_{bs}\omega_s$, $\omega_s = R_{sb}\omega_b$.

- Given n rigid bodies (links) attached serially to each other by joints, suppose moving frames labeled {1}, ..., {n} are attached to the links (let {0} be the fixed frame attached to ground). Let $\omega_{i-1,i} \in \mathbb{R}^3$ be the relative angular velocity of {i} with respect to {i–1}:

$$[\omega_{i-1,i}] = \dot{R}_{i-1,i}R_{i-1,i}^\top. \tag{2.139}$$

Then the angular velocity $\omega \in \mathbb{R}^3$ of the tip frame, expressed in fixed frame coordinates, is the sum of the relative angular velocities of the preceding frames, all transformed to fixed frame coordinates:

$$\omega = \omega_{01} + R_{01}\omega_{12} + \cdots + R_{0,n-1}\omega_{n-1,n}. \tag{2.140}$$

Alternatively, each of the terms in ω above can be written in terms of the corresponding moving frame axes.

- The angular acceleration α can be obtained by differentiating the angular velocity with respect to time in the usual fashion, making use of the relations $\dot{\hat{x}} = \omega \times \hat{x}$, $\dot{\hat{y}} = \omega \times \hat{y}$, $\dot{\hat{z}} = \omega \times \hat{z}$.

2.7 Exercises

Exercise 2.1 Figure 2.17 shows a design for an amusement park ride. The fixed frame $\{s\}$ is attached to point O as shown in the figure. The rider sits at point B, and the disk of radius $R = 1$ m rotates about the $\hat{z}_s$-axis at a constant rate $\omega_1 = \frac{\pi}{2}$ rad/s. Rod AC extends vertically at a constant speed $u = 1$ m/s out of tube CD, while at the same time rotating at a constant rate $\omega_2 = \frac{\pi}{2}$ rad/s. The distance from rod AC to point B is $L_2 = 0.5$ m, and at the instant shown in the figure, $L_1 = 1$ m.
(a) Express the velocity of point B at the instant shown in the figure in fixed frame coordinates.
(b) Express the acceleration of point B at the instant shown in the figure in fixed frame coordinates.
(c) After 1 s from the instant shown in the figure, express the position, velocity, and acceleration of point B in fixed frame coordinates.

Exercise 2.2 The wheel of a shopping cart of radius R rotates about two axes as shown in Figure 2.18 ($\theta_1 = \theta_2 = 0$ at the instant shown in the figure). θ_1 rotates at a constant rate $\dot{\theta}_1 = \omega_1$ rad/s. A moving frame $\{b\}$ is attached to the wheel as shown in the figure.
(a) Suppose θ_2 is set to some arbitrary nonzero constant. Express the velocity and acceleration of point P in moving frame coordinates.
(b) Now suppose θ_2 rotates at a constant rate $\dot{\theta}_2 = \omega_2$ rad/s. Express the velocity and acceleration of point P in moving frame coordinates at the instant shown in the figure.

Exercise 2.3 A robot with two revolute joints is shown in Figure 2.19. The fixed frame $\{s\}$ is attached to the robot base and a moving frame $\{b\}$ is attached to the robot end-effector as shown in the figure. The link lengths L_1, L_2, and L_3 are constants.
(a) Find the angular velocity of the robot end-effector expressed in fixed frame coordinates, where $\theta_1 = t$ and $\theta_2 = 2t^2$ for $t > 0$.
(b) For general θ_1 and θ_2, find the velocity of the end-effector expressed in fixed frame coordinates.

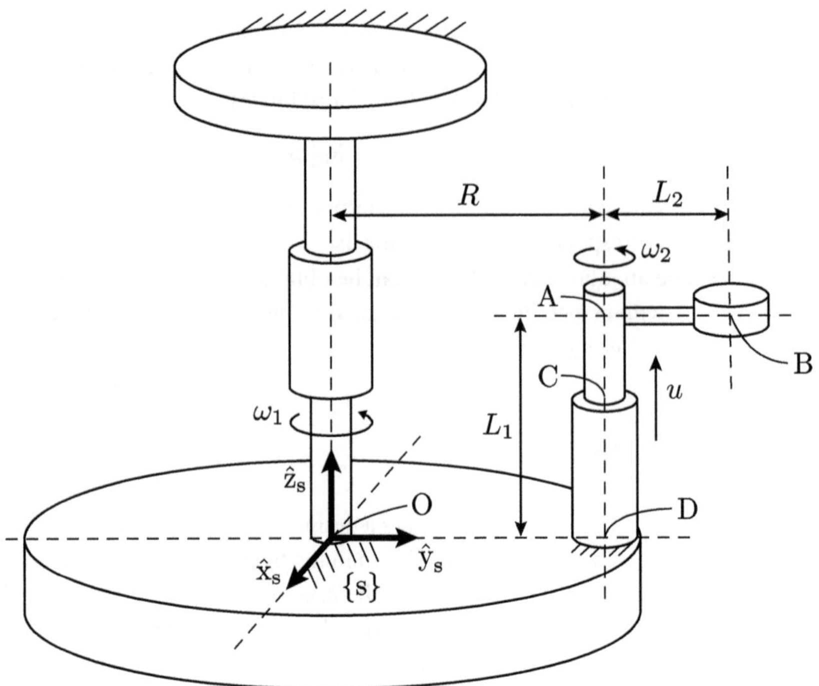

Figure 2.17 Amusement park ride of Exercise 2.1.

Exercise 2.4 Consider the three-dimensional compass shown in Figure 2.20. The fixed frame {s} is attached to the bottom of the compass. Point C and point P are the center and tip of the needle, respectively. The distance from the fixed frame origin to point C is L_1 and the distance from point C to point P is L_2. A unit vector $\hat{h}$ along the needle (from point C to point P) expressed in {s} frame coordinates is $\hat{h}_s = (\cos t, \sin t \cos 2t, \sin t \sin 2t)^\top$.

(a) Find the angular velocity ω of point P when $t = \pi$ in terms of {s} frame coordinates. Assume that ω and $\hat{h}$ are perpendicular to each other.

(b) Express the velocity and acceleration of point P in {s} frame coordinates.

Exercise 2.5 In the amusement park ride of Figure 2.21, shaft OC rotates about the $\hat{y}_s$-axis of the fixed frame {s} attached to point O at a constant rate $\omega_1 = \frac{\pi}{2}$ rad/s, and a disk rotates about rod OA at a constant rate $\omega_2 = \frac{\pi}{4}$ rad/s. The angle between shaft OC and rod OA is θ. The radius of the disk is $r = 4$ m and the length of rod OA is $R = 8$ m.

(a) Assume that θ is fixed to $\frac{\pi}{2}$. Express the velocity and acceleration of point P on the wheel at the instant shown in the figure in fixed frame coordinates.

(b) Now suppose rod OA swings about point O, with θ varying with time according to $\theta(t) = \pi - \frac{\pi}{2} \cos \frac{\pi}{4} t$. Express the velocity of point P when $t = 2$ (assume $t = 0$ at the instant shown in the figure) in fixed frame coordinates.

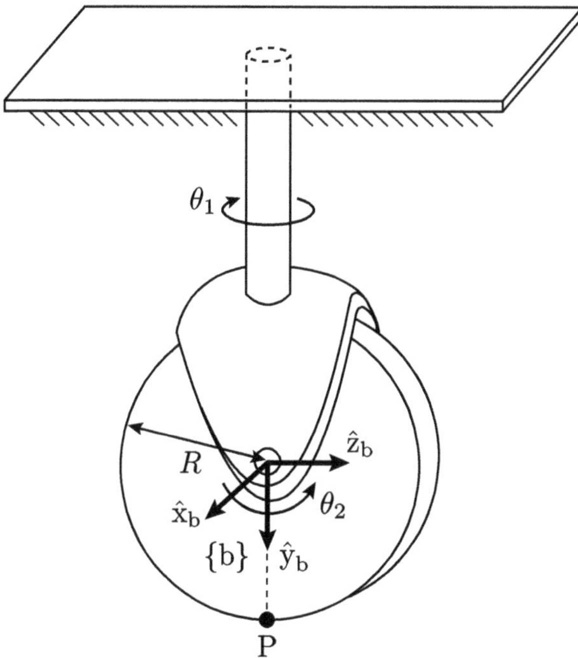

Figure 2.18 Wheel of a shopping cart of Exercise 2.2 shown when $\theta_1 = \theta_2 = 0$.

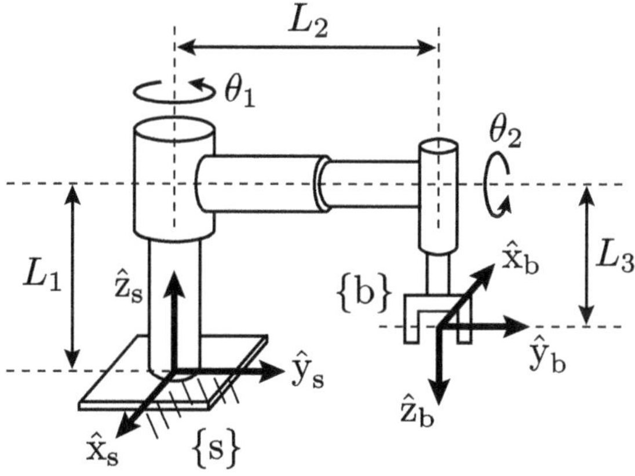

Figure 2.19 Robot with two revolute joints of Exercise 2.3.

Exercise 2.6   A particle is moving in the plane along the path parameterized in terms of polar coordinates as $r = r_0 e^{\beta t}$ and $\theta = \omega_0 t$, where r_0, β, and ω_0 are constants.

(a) Describe the position, velocity and acceleration of the particle in polar coordinates $\{\hat{r}, \hat{\theta}\}$.

(b) Find β that makes the radial ($\hat{r}$-direction) acceleration of the particle vanish. Is the radial velocity constant for such β? Explain your answer.

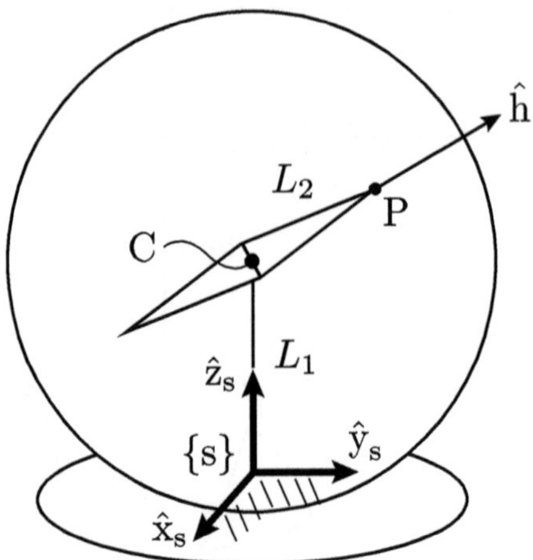

Figure 2.20 Three-dimensional compass of Exercise 2.4.

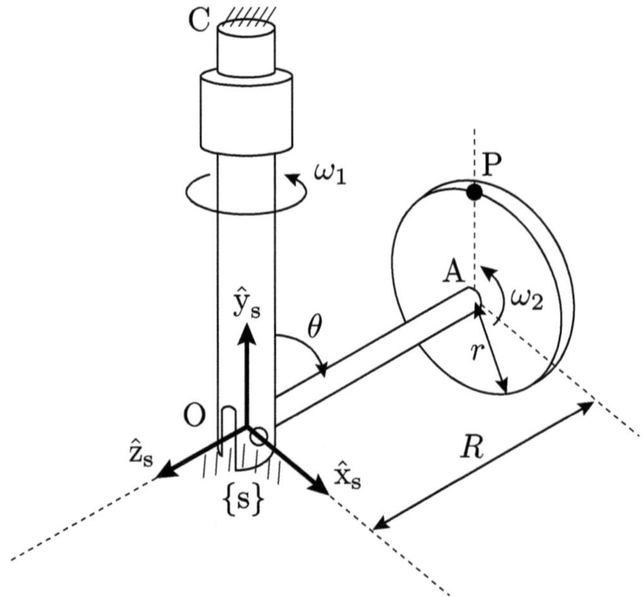

Figure 2.21 Amusement park ride of Exercise 2.5.

Exercise 2.7   A particle is moving in the plane along the path shown in Figure 2.22. The path is parameterized in terms of polar coordinates as $r = 1 - \cos\theta$, where $\dot\theta$ is constant. The fixed frame $\{s\}$ is attached to the origin as shown in the figure.

(a) Find the position at which the particle reaches its maximum speed. Express your answer in fixed frame coordinates.

(b) Express the acceleration of the particle when $\theta = \frac{\pi}{2}$ in fixed frame coordinates.

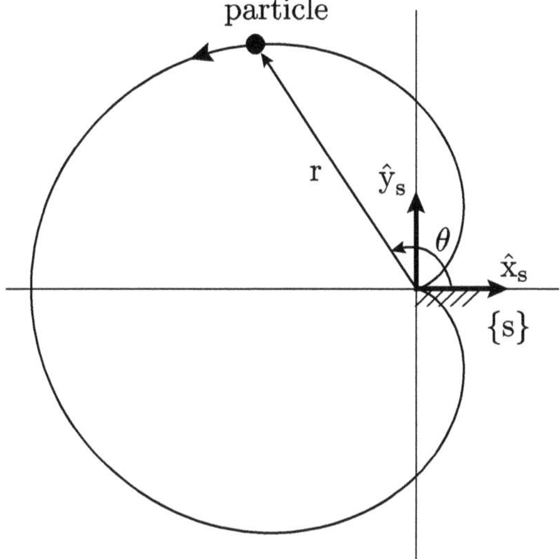

Figure 2.22 Figure for Exercise 2.7.

Exercise 2.8   A particle is moving in the plane along the path shown in Figure 2.23. The position of the particle is parameterized by

$$p(t) = (\cos t + t \sin t)\hat{x}_s + (\sin t - t \cos t)\hat{y}_s$$

for $t > 0$, where $\{\hat{x}_s, \hat{y}_s\}$ are the unit axes of the fixed frame $\{s\}$.
(a) Express the velocity and acceleration of the particle in tangential-normal coordinates $\{\hat{t}, \hat{n}\}$.
(b) Find the radius of curvature of the path when $t = 1$.

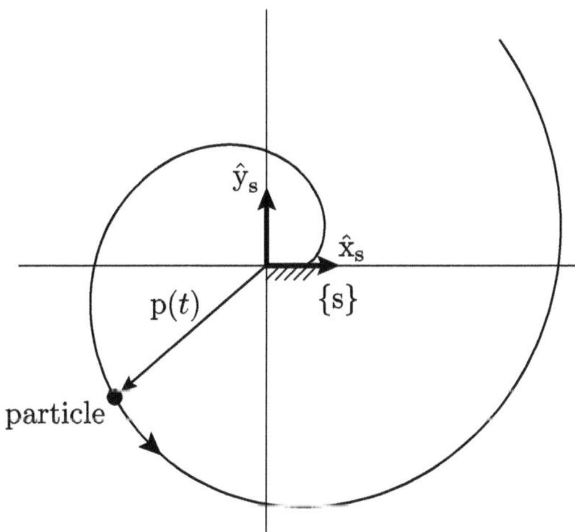

Figure 2.23 Figure for Exercise 2.8.

Exercise 2.9   Two reference frames $\{s\}$ and $\{b\}$ are given with the following relation:

$$\begin{bmatrix} \hat{x}_b & \hat{y}_b & \hat{z}_b \end{bmatrix} = \begin{bmatrix} \hat{x}_s & \hat{y}_s & \hat{z}_s \end{bmatrix} \begin{bmatrix} \frac{1}{\sqrt{2}} & -\frac{1}{\sqrt{2}} & 0 \\ \frac{1}{\sqrt{2}} & \frac{1}{\sqrt{2}} & 0 \\ 0 & 0 & 1 \end{bmatrix},$$

where $\{\hat{x}_s, \hat{y}_s, \hat{z}_s\}$ and $\{\hat{x}_b, \hat{y}_b, \hat{z}_b\}$ are the unit axes of frame $\{s\}$ and frame $\{b\}$, respectively. The vector from $\{s\}$ frame origin to $\{b\}$ frame origin expressed in $\{s\}$ frame coordinates is $(5, 10, 0)^\top$.

(a) Given a free vector $(2, 1, 3)^\top$ in terms of $\{b\}$ frame coordinates, find the corresponding vector in terms of $\{s\}$ frame coordinates.

(b) Now given a point Q in physical space, suppose its coordinates with respect to $\{s\}$ frame coordinates are $(1, 2, 3)^\top$. What are the coordinates of point Q with respect to $\{b\}$ frame coordinates?

(c) Generalize your answer to part (b); i.e., given two reference frames $\{a\}$ and $\{b\}$ in physical space, with the vector $p_{ab} \in \mathbb{R}^3$ and rotation matrix R_{ab} known (p_{ab} is the vector from $\{a\}$ frame origin to $\{b\}$ frame origin, expressed in $\{a\}$ frame coordinates), suppose the coordinates for a point Q in physical space are given in terms of $\{a\}$ frame coordinates as $q_a = (q_{a,x}, q_{a,y}, q_{a,z})^\top$. What are the coordinates of point Q in terms of $\{b\}$ frame coordinates?

Exercise 2.10   Consider a robot arm mounted on a spacecraft as shown in Figure 2.24, where frames are attached to the earth $\{e\}$, satellite $\{s\}$, the spacecraft $\{a\}$, and the robot arm $\{r\}$, respectively.

(a) Let R_{ea} be the rotation matrix describing the orientation of frame $\{a\}$ as seen from frame $\{e\}$, and p_{ea} be the vector from frame $\{e\}$ origin to frame $\{a\}$ origin, expressed in $\{e\}$ frame coordinates. Rotation matrices R_{ar}, R_{es}, and R_{rs}, and vectors p_{ar}, p_{es}, and p_{rs} are defined similarly. Given (R_{ea}, p_{ea}), (R_{ar}, p_{ar}), and (R_{es}, p_{es}), find (R_{rs}, p_{rs}).

(b) Suppose the coordinates of frame $\{s\}$ origin as seen from frame $\{e\}$ is $(1, 1, 1)^\top$. Suppose furthermore that

$$R_{er} = \begin{bmatrix} -1 & 0 & 0 \\ 0 & 1 & 0 \\ 0 & 0 & -1 \end{bmatrix}, p_{er} = \begin{bmatrix} 1 \\ 1 \\ 1 \end{bmatrix}.$$

Find the coordinates of frame $\{s\}$ origin as seen from frame $\{r\}$.

Exercise 2.11   (a) For a rotation matrix R and two vectors $u, v \in \mathbb{R}^3$, prove that

$$R(u \times v) = Ru \times Rv.$$

(b) For a rotation matrix R and a vector $u \in \mathbb{R}^3$, prove that the length of u and Ru are always the same, i.e., $\|Ru\| = \|u\|$.

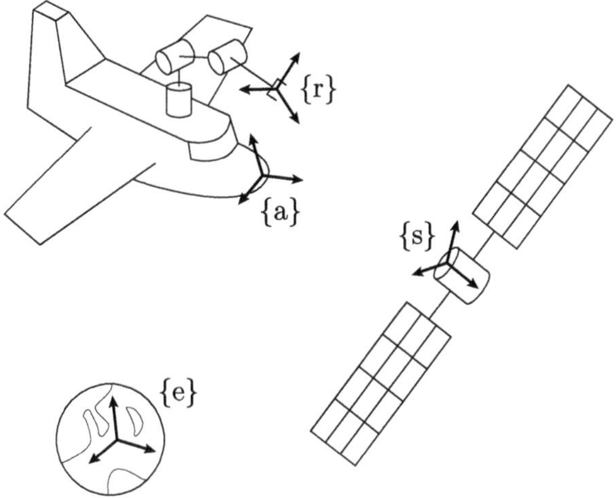

Figure 2.24 Robot arm mounted on spacecraft, the earth, and satellite of Exercise 2.10.

(c) Given a vector $u \in \mathbb{R}^3$, suppose we multiply u by the following rotation matrix:

$$R = \begin{bmatrix} \cos\theta & -\sin\theta & 0 \\ \sin\theta & \cos\theta & 0 \\ 0 & 0 & 1 \end{bmatrix} = \mathrm{Rot}(\hat{z}, \theta).$$

Show that Ru is u rotated about the $\hat{z}$-axis by angle θ. Derive corresponding rotation matrices for $\mathrm{Rot}(\hat{y}, \phi)$ and $\mathrm{Rot}(\hat{x}, \psi)$.

(d) Given the fixed frame with its unit axes denoted by $\{\hat{x}, \hat{y}, \hat{z}\}$, let p be a point whose coordinates are $p = (\frac{1}{\sqrt{3}}, -\frac{1}{\sqrt{6}}, \frac{1}{\sqrt{2}})^\top$. Suppose p is rotated about the $\hat{x}$-axis by $30°$, then about $\hat{y}$-axis by $135°$, and finally about $\hat{z}$-axis by $-120°$. What are the coordinates of point p after these three rotations?

(e) The rotation matrix given by the ZYX Euler angles (θ, ϕ, ψ) is defined as follows:

$$R(\theta, \phi, \psi) = \mathrm{Rot}(\hat{z}, \theta)\mathrm{Rot}(\hat{y}, \phi)\mathrm{Rot}(\hat{x}, \psi).$$

Writing out the entries explicitly, we get

$$R(\theta, \phi, \psi) = \begin{bmatrix} c_\theta c_\phi & c_\theta s_\phi s_\psi - s_\theta c_\psi & c_\theta s_\phi c_\psi + s_\theta s_\psi \\ s_\theta c_\phi & s_\theta s_\phi s_\psi + c_\theta c_\psi & s_\theta s_\phi c_\psi - c_\theta s_\psi \\ -s_\phi & c_\phi s_\psi & c_\phi c_\psi \end{bmatrix},$$

where s_θ is shorthand for $\sin\theta$, c_θ for $\cos\theta$, and so on. Using the above result, find the ZYX Euler angles (θ, ϕ, ψ) for the following rotation matrix:

$$\begin{bmatrix} \frac{1}{2} & \frac{-2+\sqrt{2}}{4} & \frac{2+\sqrt{2}}{4} \\ \frac{1}{2} & \frac{2+\sqrt{2}}{4} & \frac{-2+\sqrt{2}}{4} \\ -\frac{1}{\sqrt{2}} & \frac{1}{2} & \frac{1}{2} \end{bmatrix}.$$

A useful function for this problem is the two-argument arctangent function $\text{atan2}(y, x)$, which calculates the angle between the positive $\hat{x}$-axis and the vector from the origin to point (x, y). The $\text{atan2}(y, x)$ function is the arctangent function $\tan^{-1}\frac{y}{x}$ that also takes into account the signs of x and y, and therefore returns angles in the range $(-\pi, \pi]$ while $\tan^{-1}\frac{y}{x}$ only returns angles in the range $(-\frac{\pi}{2}, \frac{\pi}{2})$. For example, $\text{atan2}(1, 1) = \frac{\pi}{4}$, while $\text{atan2}(-1, -1) = -\frac{3\pi}{4}$.

Exercise 2.12   Euler angles can be defined with respect to different sequences of rotation axis. For example, the XYZ Euler angles (θ, ϕ, ψ) can be defined as

$$R(\theta, \phi, \psi) = \text{Rot}(\hat{x}, \theta)\text{Rot}(\hat{y}, \phi)\text{Rot}(\hat{z}, \psi).$$

(a) Determine the corresponding XYZ Euler angles (θ, ϕ, ψ) for the same rotation matrix from Exercise 2.11:

$$\begin{bmatrix} \frac{1}{2} & \frac{-2+\sqrt{2}}{4} & \frac{2+\sqrt{2}}{4} \\ \frac{1}{2} & \frac{2+\sqrt{2}}{4} & \frac{-2+\sqrt{2}}{4} \\ -\frac{1}{\sqrt{2}} & \frac{1}{2} & \frac{1}{2} \end{bmatrix},$$

and compare these with the ZYX Euler angles you obtained in Exercise 2.11.

(b) Another valid set of Euler angles is the ZYZ Euler angles, defined by

$$R(\theta, \phi, \psi) = \text{Rot}(\hat{z}, \theta)\text{Rot}(\hat{y}, \phi)\text{Rot}(\hat{z}, \psi).$$

What is the relationship between ZYX and ZYZ Euler angles? Provide an explicit formula for the ZYZ Euler angles from a set of given ZYX Euler angles.

(c) A general set of Euler angles can be defined with respect to axes $\{\text{Axis1}, \text{Axis2}, \text{Axis3}\}$ as

$$R(\theta, \phi, \psi) = \text{Rot}(\text{Axis1}, \theta)\text{Rot}(\text{Axis2}, \phi)\text{Rot}(\text{Axis3}, \psi). \tag{2.141}$$

What are the requirements on the three axes in order for (2.141) to be a valid parameterization for rotation matrices?

Exercise 2.13   (a) For two vectors $u, v \in \mathbb{R}^3$, prove that $[u]v = [v]^\top u$, where $[u]$, $[v]$ are the 3×3 skew-symmetric matrix representations of u, v.

(b) For two vectors $u, v \in \mathbb{R}^3$, prove that $[u \times v] = [u][v] - [v][u]$.

(c) For a square matrix $M \in \mathbb{R}^{n \times n}$, prove that M can be uniquely expressed as the sum of a symmetric matrix and a skew-symmetric matrix. (*Note*: A symmetric matrix $A \in \mathbb{R}^{n \times n}$ satisfies the condition $A^\top = A$ and a skew-symmetric matrix $B \in \mathbb{R}^{n \times n}$ satisfies the condition $B^\top = -B$.)

Exercise 2.14   A rigid body is known to rotate about its center of mass with some constant but unknown angular velocity vector $\omega = \omega\hat{\omega}$, where ω is the rate of rotation and unit vector $\hat{\omega}$ is the axis of rotation. Suppose a point p on the rigid body is observed at k time instances $t_1 < t_2 < \cdots < t_k$. What is the minimum number of observations $\{p(t_1), p(t_2), \ldots, p(t_k)\} \in \mathbb{R}^3$ needed to uniquely determine ω?

Exercise 2.15   Consider a rigid body moving in space. The fixed frame $\{s\}$ is attached to the ground and a moving frame $\{b\}$ is attached to the rigid body. Given the rotation matrix R_{sb} describing the orientation of frame {b} as seen from frame {s}, recall that the angular velocity of frame {b} can be obtained as

$$R_{sb}^{\top}\dot{R}_{sb} = [\omega_b],$$
$$\dot{R}_{sb}R_{sb}^{\top} = [\omega_s],$$

where $\omega_b \in \mathbb{R}^3$ is the angular velocity expressed in frame {b} coordinates, and $\omega_s \in \mathbb{R}^3$ is the angular velocity expressed in frame {s} coordinates. Show that the angular acceleration can be obtained as

$$[\dot{\omega}_b] = R_{sb}^{\top}\ddot{R}_{sb} - [\omega_b]^2,$$
$$[\dot{\omega}_s] = \ddot{R}_{sb}R_{sb}^{\top} - [\omega_s]^2,$$

and from this conclude that $\dot{\omega}_s = R_{sb}\dot{\omega}_b$. (*Hint*: Take the time derivative of both sides of $R^{\top}R = I$ to obtain an expression for the time derivative of $R^{\top}$.)

Exercise 2.16 Referring to Figure 2.25, two gears are meshed together and rotate in opposite directions. As shown in the figure, the fixed frame {0} is attached to point O, which is the intersection of the two axes of the gears, a moving frame {1} is attached to gear 1, and a moving frame {2} is attached to gear 2. Gear 1 rotates about the $\hat{z}_1$-axis of frame {1} at a constant rate ω_1, and gear 2 rotates about the $\hat{z}_2$-axis of frame {2} at a constant rate $-\omega_2$ (ω_1 and ω_2 are positive constants). The radius of gear 1 is R and the radius of gear 2 is r. Neglecting the thicknesses of the gears, the distance from frame {0} origin to frame {1} origin is r, and the distance from frame {0} origin to frame {2} origin is R. The figure shows the gears at $t = 0$.
(a) Find the rotation matrix R_{12} describing the orientation of frame {2} as seen from frame {1} for arbitrary time t.
(b) Suppose points P_1 and P_2 are marked on gear 1 and gear 2, respectively. Then, the distance from frame {1} origin to point P_1 is R, and the distance from frame {2} origin to point P_2 is r. For arbitrary time t, express the relative position of point P_2 observed from point P_1 in frame {1} coordinates.
(c) For arbitrary time t, express the relative linear velocity of point P_2 observed from point P_1 in frame {2} coordinates.
(d) For point P_1 and point P_2 to coincide at some time $t = t^*$, what should be the relation between ω_1 and ω_2? Assume that $\frac{\omega_1}{\omega_2} = \frac{r}{R}$.

Exercise 2.17 Consider the hinged door of Figure 2.26. As shown in the figure, the fixed frame {0} is attached to the ground, a moving frame {1} is attached to the door, and a moving frame {2} is attached to the door handle. The distance from frame {0} origin to frame {1} origin is L, and the vector from frame {1} origin to frame {2} origin expressed in frame {1} coordinates is $(0, 0, H)^{\top}$, where L and H are constants. Angles θ_1 and θ_2 are defined as shown in the figure (the door is closed when $\theta_1 = \theta_2 = 0$).
(a) Find the rotation matrices R_{01}, R_{12}, and R_{02} for arbitrary θ_1 and θ_2.
(b) Express the angular velocity of frame {2} for arbitrary θ_1, θ_2, $\dot{\theta}_1$, and $\dot{\theta}_2$ in fixed frame coordinates.
(c) Let point P be the endpoint of the door handle. The distance from frame {2} origin

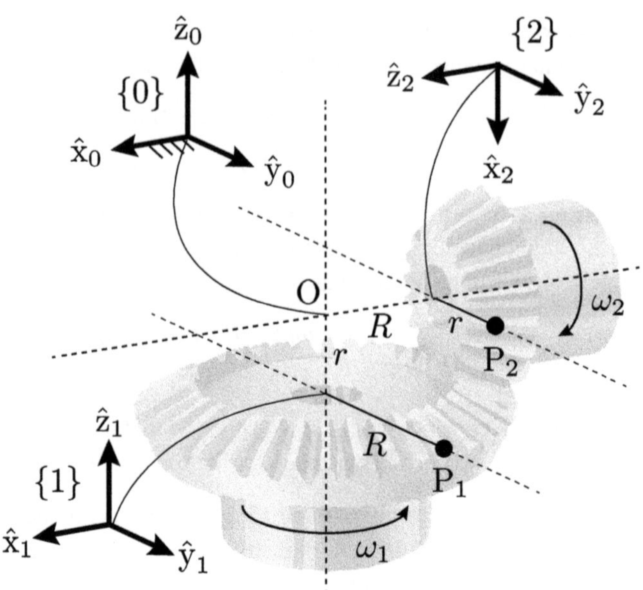

Figure 2.25 Two gears meshed together of Exercise 2.16.

to point P is l, where l is a constant. Express the velocity of P for arbitrary θ_1, θ_2, $\dot{\theta}_1$, and $\dot{\theta}_2$ in fixed frame coordinates.

Exercise 2.18 Referring to Figure 2.27, an asteroid flying by the earth is tracked by a ground radar station on the earth. Attach the fixed frame {s} to the sun, a moving frame {e} to the earth, and a moving frame {g} to the ground station as shown in the figure (both frame {e} and frame {g} rotate with the earth). The earth's rotation axis ($\hat{z}_e$-axis) is tilted $45°$ from the $\hat{z}_s$-axis, and always points in the direction $\hat{x}_s + \hat{z}_s$. At $t = 0$, the orientation of frame {g} is the same as that of the fixed frame {s}, and $\hat{y}_e$ and $\hat{y}_s$ point in the same direction. The earth rotates about its rotation axis at a constant rate $\omega_2 = 2\pi$ rad/day. At the same time, the earth orbits the sun in a circle at a constant rate $\omega_1 = \frac{2\pi}{360}$ rad/day. The earth's radius is R, and the distance between the earth and the sun is L.
(a) Find the rotation matrices R_{se}, R_{eg}, and R_{sg} as functions of t (t is in days).
(b) Find, as a function of t, the angular velocity ω of frame {g} as seen from the fixed frame {s}. Express ω in both frame {s} and frame {g} coordinates.
(c) At the instant shown in the figure, the asteroid path as tracked from the ground station is

$$r_{ga}(t) = -ct\hat{x}_g + H\hat{z}_g,$$

where r_{ga} is the vector from frame {g} origin to the asteroid, and c and H are constants. Find the velocity $v_{sa}(t)$ and acceleration $a_{sa}(t)$ of the asteroid as seen from the fixed frame {s} at $t = 0$.

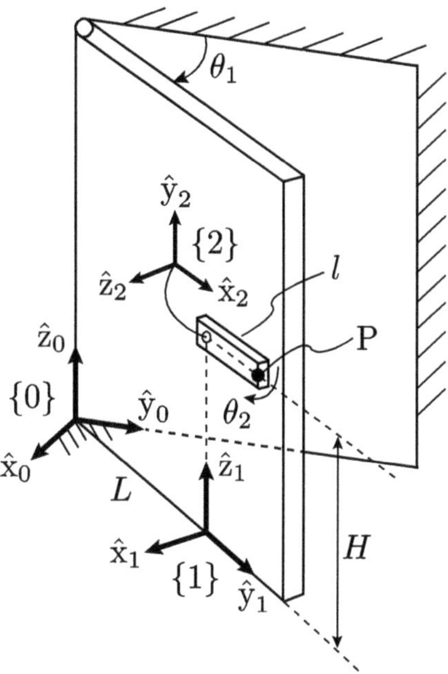

Figure 2.26 Door of Exercise 2.17.

Exercise 2.19   Referring to Figure 2.28, the asteroid that was flying by the earth in Exercise 2.18 now rotates around the earth in a circular orbit at a constant rate ω_2. The earth also rotates about its axis at a constant rate ω_1. The angle between the two rotation axes is $45°$, and the distance between the asteroid and the earth is ρ. Frames $\{e\}$ and $\{a\}$ are moving frames attached to the earth and the asteroid, respectively, as shown in the figure. Assume that $\omega_1 = \omega_2 = 1$.

(a) Find the rotation matrix R_{ea} as a function of t.

(b) At $t = \frac{\pi}{4}$, determine the relative angular velocity of the asteroid as observed from the earth frame. Express your answer in earth frame coordinates.

(c) At $t = \frac{\pi}{4}$, determine the relative linear velocity of the asteroid as observed from the earth frame. Express your answer in earth frame coordinates.

Exercise 2.20   A flying quadrotor is being tracked by a robot moving on the ground as shown in Figure 2.29. The fixed frame $\{o\}$ is attached to ground, a moving frame $\{q\}$ is attached to the quadrotor, and a moving frame $\{r\}$ is attached to the robot. We further assume the following:

- **Quadrotor motion**:
    - Frame $\{q\}$ is attached to the quadrotor such that it moves with constant linear velocity $v = (1, 1, 0)^\top$ expressed in frame $\{o\}$ coordinates, and also rotates about the $\hat{x}_q$-axis of frame $\{q\}$ at $\omega = 1$ rad/s.

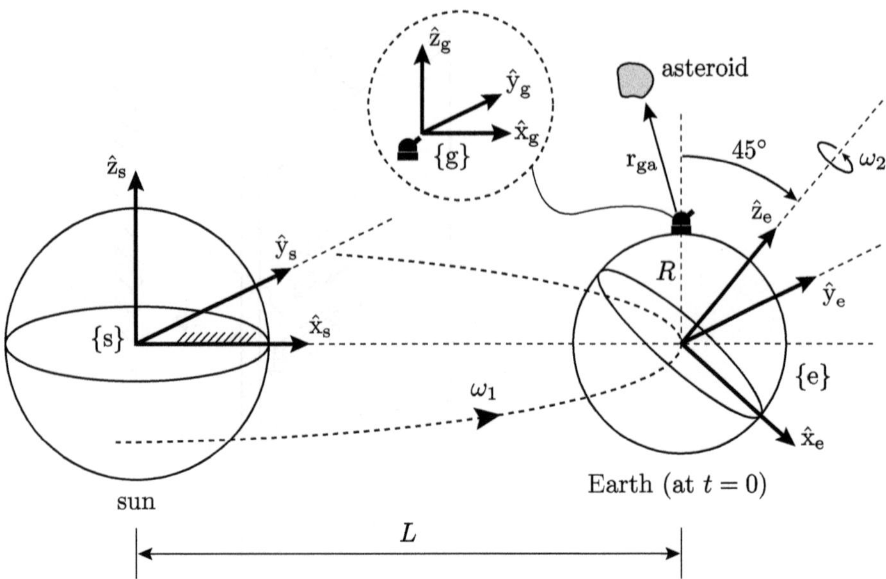

Figure 2.27 A ground station on the earth tracking an asteroid of Exercise 2.18, shown at the instant $t = 0$ (t is in days).

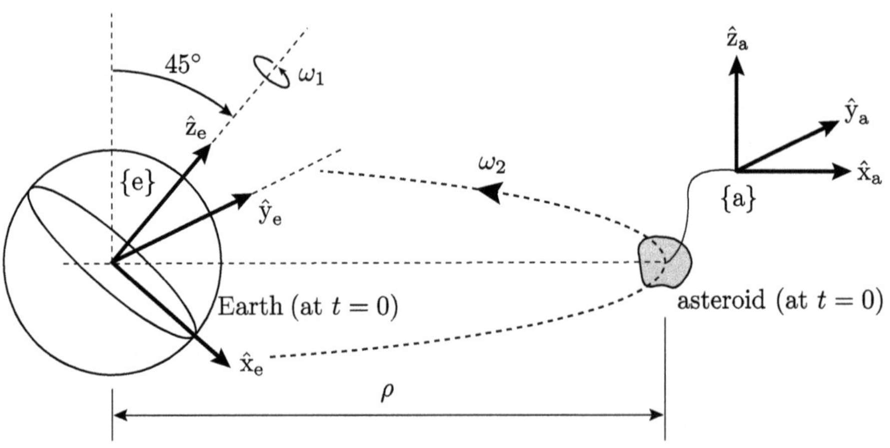

Figure 2.28 Figure for Exercise 2.19, shown at the instant $t = 0$.

- Define $p_{oq} \in \mathbb{R}^3$ to be the vector from frame {o} origin to frame {q} origin, expressed in frame {o} coordinates. Assume that at $t = 0$, the initial value for p_{oq} is given by $p_{oq}(0) = (0, 0, H)^\top$, where H is some given constant.

- Define R_{oq} to be the rotation matrix describing the orientation of frame {q} as seen from frame {o}. Assume that at $t = 0$, the initial value for R_{oq} is given by $R_{oq}(0) = I$, where I denotes the 3×3 identity matrix.

- **Robot motion**:
  - Frame {r} is attached to the robot such that the robot's linear velocity is always in the $\hat{x}_r$-axis direction of frame {r}.
  - Define $p_{or} \in \mathbb{R}^3$ to be the vector from frame {o} origin to frame {r} origin, expressed in frame {o} coordinates. Assume $p_{or}(t)$ is given by $p_{or}(t) = (\sin t, 1 - \cos t, h)^\top$, where h is some given constant.
  - Define R_{or} to be the rotation matrix describing the orientation of frame {r} as seen from frame {o}. Assume that at $t = 0$, the initial value for R_{or} is given by $R_{or}(0) = I$.

(a) Explain the motions of the quadrotor and the robot in terms of their frames' position and orientation.

(b) Find R_{rq} and p_{rq}.

(c) Find the relative linear velocity and the relative angular velocity of the quadrotor frame {q} as seen from the robot frame {r}; express your answers in frame {q} coordinates.

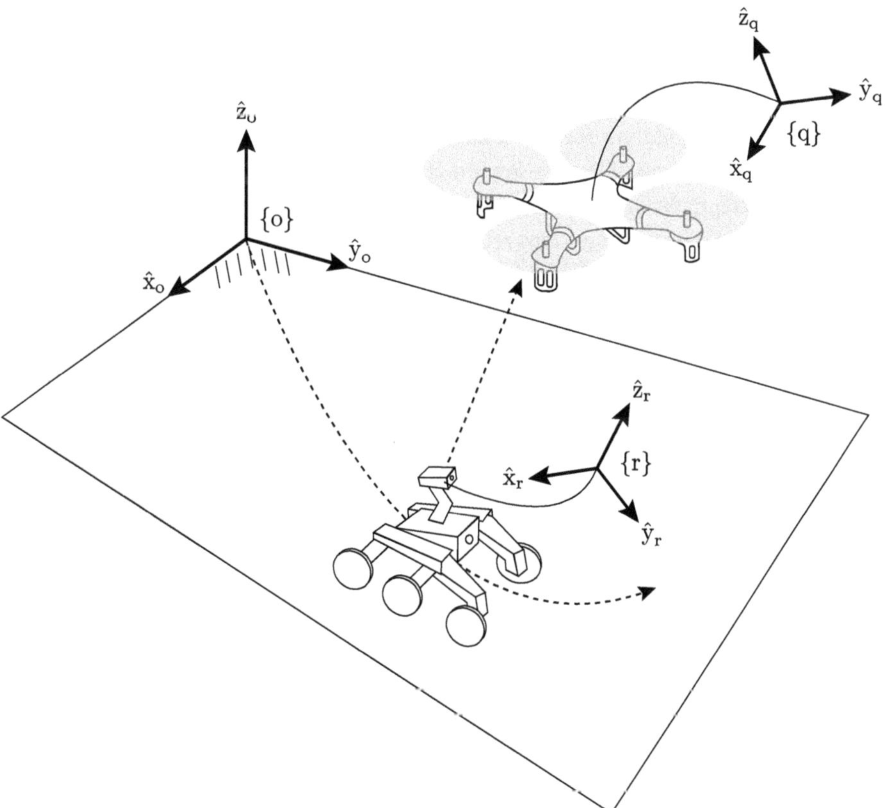

Figure 2.29 A flying quadrotor being tracked by a robot moving on the ground of Exercise 2.20.

3   Particle Dynamics

This chapter begins with the familiar Newton's second law of motion $f = ma$ for a single particle. As mentioned earlier in Chapter 1, a large class of mechanical systems can be modeled as an interconnection of rigid bodies, and for many of these systems the rigid bodies themselves can be reasonably approximated by point masses. The methods described in this chapter are relevant to the analysis of such systems.

The first step in the derivation of the equations of motion for particle systems is to choose an appropriate set of coordinates with which to specify the configuration of each particle – its position, velocity, and acceleration. Sometimes kinematic constraints may exist between particles, in which case a careful kinematic analysis is required to determine an independent set of coordinates for the system. The next step is to draw, for each particle, a **free-body diagram** that captures all the forces applied to the particle. The equations of motion can now be derived for each particle in terms of the chosen coordinates. The two-step procedure of choosing coordinates, then drawing free-body diagrams, also applies to systems of rigid bodies.

Some problems may require solving the equations of motion for the complete motion trajectory of each particle given the applied forces. For such problems the equations of motion, which are a set of differential equations, need to be integrated. In some cases the equations are simple enough to admit closed-form solutions, but for most realistic mechanical systems, complete trajectories can only be obtained through numerical integration.

For other problems the complete motion trajectory may not be necessary. Rather, information about the state of a particle – for example, its position, velocity, or the applied force at a given time instant or location – may be all that is needed. For such problems, methods based on **work** and **energy** can be quite effective: rather than integrating a complicated set of nonlinear differential equations to obtain the complete trajectory, then determining the state of the particle at some given time or configuration, work-energy methods transform the equations of motion into a set of algebraic equations that can be solved directly for the desired information. Work-energy methods are particularly effective when the applied forces are **conservative**; that is, the forces can be expressed as the gradient of some potential function. Familiar examples of conservative forces include gravity, and the force exerted by a spring.

A third class of methods for studying the motion of particles involves **linear and angular momentum** and also **impulse**. Similar to work-energy principles, dynamics principles involving momentum and impulse are derived by transforming the differential equations of motion into a set of vector equations, this time involving the particle velocities. These methods can be useful for problems involving impulsive forces such as impacts from collisions, and other problems where only the state of the particle at a particular time or location is needed.

In Chapter 4, the equations of motion for a rigid body are derived by modeling the body as an infinite collection of rigidly connected particles. In this regard the results derived in this chapter on the angular momentum for a system of particles play a central role in the derivation of the equations of motion for a rigid body.

3.1   Newton's Second Law of Motion

An **inertial reference frame**, or **inertial frame** for short, is a reference frame with respect to which Newton's law $f = ma$ is valid. One obvious example is a fixed frame attached to ground, but so is a frame attached to a bus moving in a straight line with constant speed. If the bus was accelerating, then the frame attached to the bus is no longer an inertial frame (but the fixed frame attached to the ground would still remain an inertial frame).

Whereas in kinematics a distinction is made only between fixed frames and moving frames, for dynamics problems the first order of business is to designate an inertial frame. As illustrated by the bus example, an inertial frame need not necessarily be fixed in the sense of being stationary, but for the purposes of kinematic analysis an inertial frame can always be regarded as a fixed frame. Any frames that are designated as fixed frames in the subsequent derivations and examples are all inertial frames.

With respect to an inertial frame, Newton's second law for a particle of mass m takes the form

$$f = ma, \tag{3.1}$$

where f is the net force vector acting on the particle, and a is the acceleration vector of the particle as measured by an observer at rest in the inertial frame. Choosing the inertial frame to be the fixed frame $\{s\}$, (3.1) can be written in Cartesian coordinates $x \in \mathbb{R}^3$ of the fixed frame as

$$\begin{aligned}
m\ddot{x}_1 &= f_1(x, \dot{x}, t), \\
m\ddot{x}_2 &= f_2(x, \dot{x}, t), \\
m\ddot{x}_3 &= f_3(x, \dot{x}, t).
\end{aligned} \tag{3.2}$$

The components of the force vector $f = (f_1, f_2, f_3)^\top$ can be nonlinear functions of $t, x,$ and $\dot{x}$. Given initial values for x and $\dot{x}$ at t_0, and assuming the f_i are "well behaved" (for example, continuous and differentiable everywhere), the existence of a unique solution trajectory $x(t)$ can be guaranteed for t within some small time interval about t_0. Finding analytic solutions to these differential equations is generally quite difficult, however, and for most realistic mechanical systems, one must resort to numerical integration methods to obtain solutions.

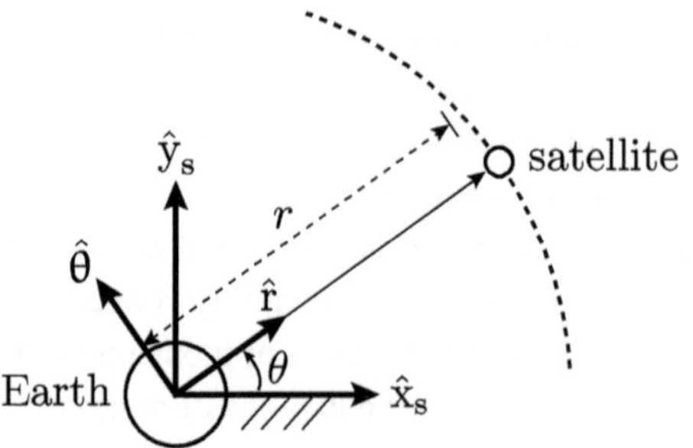

Figure 3.1 A satellite orbiting the earth.

Example 3.1   In this example we derive the equations of motion for a satellite orbiting the earth. Assigning fixed and moving frames as shown in Figure 3.1, use polar coordinates to parameterize the position p of the satellite, and determine the corresponding velocity and acceleration:

$$p = r\hat{r}, \tag{3.3}$$

$$\dot{p} = \dot{r}\hat{r} + r\dot{\theta}\hat{\theta}, \tag{3.4}$$

$$\ddot{p} = (\ddot{r} - r\dot{\theta}^2)\hat{r} + (r\ddot{\theta} + 2\dot{r}\dot{\theta})\hat{\theta}. \tag{3.5}$$

The force of attraction between the satellite and the earth is given by $f = f_r\hat{r}$, where

$$f_r = -\frac{GMm}{r^2}. \tag{3.6}$$

Here G is the gravitational constant, M is the mass of the earth, and m is the mass of the satellite. Newton's second law $f = ma$ can then be written separately in terms of the $\hat{r}$ and $\hat{\theta}$ components as

$$-\frac{GMm}{r^2} = m(\ddot{r} - r\dot{\theta}^2) \rightarrow \ddot{r} = r\dot{\theta}^2 - \frac{GM}{r^2}, \tag{3.7}$$

$$0 = m(r\ddot{\theta} + 2\dot{r}\dot{\theta}) \rightarrow \ddot{\theta} = -\frac{2\dot{r}\dot{\theta}}{r}. \tag{3.8}$$

Note that r = constant and $\theta = \omega t$ (ω constant) is one solution, corresponding to a circular orbit at a constant rate around the earth. Also note that for the constant r solution, the second Equation (3.8), $m(r\ddot{\theta} + 2\dot{r}\dot{\theta}) = 0$, can also be written as $\frac{d}{dt}(mr^2\dot{\theta}) = 0$, or $mr^2\dot{\theta}$ = constant. This latter property is in fact a statement of the principle of conservation of angular momentum, which is discussed later in this chapter.   □

3.1.1   Integrating the Equations of Motion

The equations of motion are a set of second-order differential equations whose solution depends on the given boundary conditions. As seen from the previous satellite example, they can be nonlinear, and may admit closed-form analytic solutions for certain boundary conditions. In most cases, however, the equations will usually not admit general closed-form analytic solutions. One must then resort to numerical integration to obtain solution trajectories.

Mechanical systems whose equations of motion are linear differential equations always have closed-form solutions. In this section we will examine these solutions for the simplest case of a single particle moving on a line, in which $f = ma$ is expressed in Cartesian coordinates $x \in \mathbb{R}$. Solutions to the general multidimensional linear differential equation (both homogeneous and inhomogeneous) will be presented in Chapter 7.

Constant Force

If the applied force $f \in \mathbb{R}$ is constant, the equations of motion become

$$\ddot{x} = \frac{f}{m} = a, \tag{3.9}$$

where a is constant. Assuming $t_0 = 0$ and integrating twice,

$$\dot{x}(t) = \dot{x}(0) + at, \tag{3.10}$$

$$x(t) = x(0) + v(0)t + \frac{1}{2}at^2. \tag{3.11}$$

Writing $\dot{x} = v$, $v(0) = v_0$, $x(0) = x_0$ and solving (3.10) for t and substituting into (3.11), we get

$$v^2 = v_0^2 + 2a(x - x_0), \tag{3.12}$$

which relates the speed v with the displacement x.

Time-Varying Force

If the force depends only on time – that is, $f = f(t)$, the equations of motion become $m\ddot{x} = f(t)$ or $\ddot{x} = a(t)$, $a(t) = f(t)/m$. Setting $v(t) = \dot{x}(t)$ and integrating twice,

$$v(t) = v_0 + \int_0^t a(s)\, ds, \tag{3.13}$$

$$x(t) = x_0 + v_0 t + \int_0^t \int_0^{s_1} a(s_2)\, ds_2\, ds_1. \tag{3.14}$$

Position-Dependent Force

If the force is position-dependent – that is, $f = f(x)$ – the equations of motion become $m\ddot{x} = f(x)$, or $\ddot{x} = a(x)$, $a(x) = f(x)/m$. Since

$$\ddot{x} = \frac{dv}{dt} = \frac{dv}{dx}\frac{dx}{dt} = v\frac{dv}{dx}, \tag{3.15}$$

$\ddot{x} = a(x)$ can also be written

$$v\frac{dv}{dx} = a(x). \tag{3.16}$$

Integrating both sides of $v \, dv = a(x) \, dx$,

$$\frac{1}{2}(v^2 - v_0^2) = \int_{x_0}^{x} a(s) \, ds, \tag{3.17}$$

or

$$v(x) = \left(2 \int_{x_0}^{x} a(s) \, ds + v_0^2\right)^{\frac{1}{2}}. \tag{3.18}$$

Further noting that $v = dx/dt$, or $dt = dx/v$, we can use (3.18) to integrate both sides of $dt = dx/v$:

$$t = t_0 + \int_{x_0}^{x} \left(2 \int_{x_0}^{s_1} a(s_2) \, ds_2 + v_0^2\right)^{-\frac{1}{2}} ds_1. \tag{3.19}$$

Velocity-Dependent Force

If the force is velocity-dependent of the form $f = f(\dot{x})$, the equations of motion become $m\ddot{x} = f(\dot{x})$, which can also be equivalently written as $\ddot{x} = a(\dot{x})$ or $\dot{v} = a(v)$. Writing the latter as $\frac{dv}{a(v)} = dt$ and integrating,

$$\int_{v_0}^{v} \frac{ds}{a(s)} = t - t_0. \tag{3.20}$$

If this integral can be solved for v as a function of v_0 and t, i.e., $v = v(v_0, t)$, then rearranging $v = \frac{dx}{dt}$ as $dx = v(v_0, t) \, dt$ and integrating, x can be obtained as a function of x_0, v_0, t – that is, $x = x(x_0, v_0, t)$.

3.1.2   Solving for Forces and Accelerations

For some problems the entire motion trajectory may not be needed; only the forces and accelerations (and sometimes velocities) at specific instances may be required. In this section we consider a collection of such problems, emphasizing the common steps in formulating and solving such problems: (i) choosing appropriate coordinates for the displacement of each particle, and determining any motion constraints between the particles; (ii) drawing a free-body diagram showing all the forces exerted on each particle; (iii) correctly writing down the equations of motion for each particle, and obtaining the desired solution.

Example 3.2   The pendulum of Figure 3.2 is modeled as a mass m attached to a massless rope of length L. Gravity g acts downward, and at the instant shown in the figure, the tension in the rope is T. To find the velocity and acceleration of the particle at this instant, use tangential-normal coordinates: attach a moving frame with tangential and normal axes $\{\hat{t}, \hat{n}\}$ as shown, and recall that the expression for the velocity $\mathbf{v}$ and acceleration $\mathbf{a}$ in tangential-normal coordinates is

$$\mathbf{v} = v\hat{t}, \tag{3.21}$$

$$\mathbf{a} = \dot{v}\hat{t} + \frac{v^2}{\rho}\hat{n}, \tag{3.22}$$

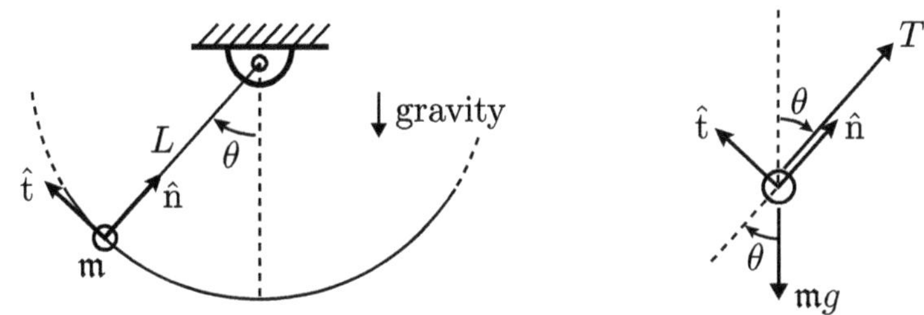

Figure 3.2 Pendulum with free-body diagram.

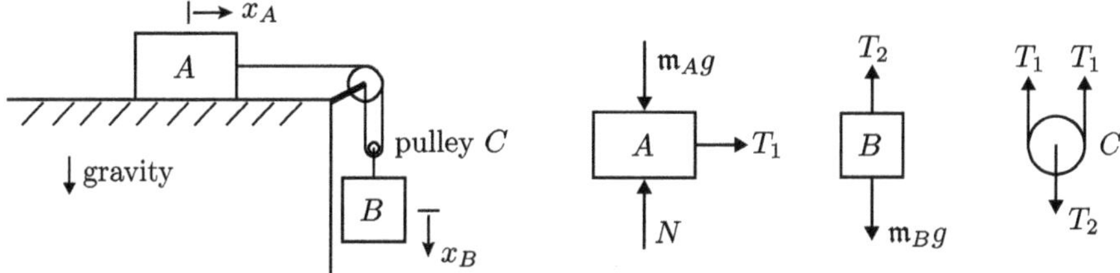

Figure 3.3 Two blocks connected by a pulley.

where ρ is the radius of curvature of the path. Since the pendulum moves along a circle of radius L, it follows that $\rho = L$. Drawing the free body diagram as shown in Figure 3.2, with components of the force and acceleration in the tangential and normal directions, respectively, denoted by subscripts t and n, the equations of motion become

$$f_t = ma_t \rightarrow -mg \sin \theta = ma_t, \tag{3.23}$$

$$f_n = ma_n \rightarrow T - mg \cos \theta = ma_n. \tag{3.24}$$

Solving for a_t and a_n,

$$a_t = -g \sin \theta, \tag{3.25}$$

$$a_n = \frac{T}{m} - g \cos \theta, \tag{3.26}$$

from which the acceleration can be calculated. To find the velocity $\mathbf{v} = v\hat{t}$, match the normal components of the acceleration vector:

$$\frac{T}{m} - g \cos \theta = \frac{v^2}{L} \rightarrow v = \sqrt{L \left(\frac{T}{m} - g \cos \theta \right)}. \tag{3.27}$$

$\square$

Example 3.3   Two blocks A and B are connected to each other by a pulley as shown in Figure 3.3. To find the acceleration of each block and the cord tension, denote by x_A and x_B the linear displacements of the two blocks. Then from the way the pulley is arranged, $x_B = \frac{1}{2}x_A$, so that the accelerations satisfy

$$a_B = \frac{1}{2}a_A. \tag{3.28}$$

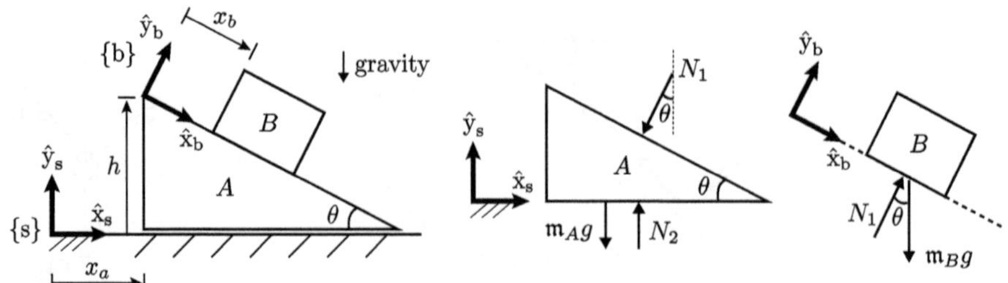

Figure 3.4 A block sliding down an inclined wedge.

To find the cord tension, draw free-body diagrams for blocks A and B and the moving wheel C of the pulley, and write down the equations of motion. For block A, the equations of motion in the horizontal direction become

$$T_1 = m_A a_A, \tag{3.29}$$

while for block B, the equations of motion in the vertical direction are

$$m_B g - T_2 = m_B a_B, \tag{3.30}$$

where m_A and m_B are, respectively, the masses of blocks A and B. For the pulley wheel C, the equations of motion in the vertical direction are

$$T_2 - 2T_1 = m_C a_C = 0 \rightarrow T_2 = 2T_1, \tag{3.31}$$

where the last equality follows from the assumption that m_C is negligibly small and can be set to zero. The system of four linear equations can now be solved for the four unknowns a_A, a_B, T_1, T_2. $\square$

Example 3.4   A block B starts from rest and slides down an inclined wedge A as shown in Figure 3.4 (gravity acts downward). The wedge lies on a flat horizontal surface and is also initially at rest. Assuming all contacts are frictionless, the objective is to find the accelerations of the block and the wedge at the instant shown in the figure. Since both the block and the wedge undergo only rectilinear motion, they can both be modeled as particles. Choose fixed and moving frames {s} and {b} as shown, with {b} attached to the wedge A. The positions of A and B can then be written

$$p_A = x_a \hat{x}_s, \tag{3.32}$$

$$p_B = x_a \hat{x}_s + h \hat{y}_s + x_b \hat{x}_b. \tag{3.33}$$

Referring to the free-body diagram for wedge A, the corresponding equations of motion are

$$-N_1 \sin \theta = m_A \ddot{x}_a \quad (\hat{x}_s\text{-direction}), \tag{3.34}$$

$$N_2 - m_A g - N_1 \cos \theta = 0 \quad (\hat{y}_s\text{-direction}). \tag{3.35}$$

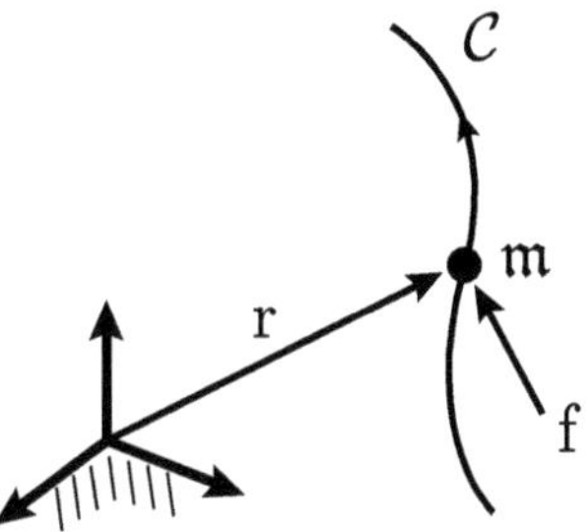

Figure 3.5 Work of a moving particle.

Similarly, referring to the free-body diagram for block B, the corresponding equations of motion are

$$m_B g \sin\theta = m_B(\ddot{x}_b + \ddot{x}_a \cos\theta) \quad (\hat{x}_b\text{-direction}), \tag{3.36}$$

$$N_1 - m_B g \cos\theta = m_B(\ddot{x}_a \sin\theta) \quad (\hat{y}_b\text{-direction}). \tag{3.37}$$

We therefore have four linear equations in the four unknowns $\ddot{x}_a$, $\ddot{x}_b$, N_1, N_2; solving for $\ddot{x}_a$ and $\ddot{x}_b$,

$$\ddot{x}_a = -\frac{m_B \sin\theta \cos\theta}{m_A + m_B \sin^2\theta} \cdot g, \tag{3.38}$$

$$\ddot{x}_b = \frac{(m_A + m_B)\sin\theta}{m_A + m_B \sin^2\theta} \cdot g. \tag{3.39}$$

$\square$

3.2 Work and Energy

Consider a particle of mass m moving in physical space while under the action of a force field f (see Figure 3.5). Denote by r the position of the particle with respect to the fixed frame origin, and let C be the path traced by the particle. The **work** $\mathcal{W}$ done by the force as the particle moves along the path C is defined to be the following path integral:

$$\mathcal{W} = \int_C f \cdot dr. \tag{3.40}$$

To evaluate this integral, parameterize the path C by $r(t)$, where t is a scalar parameter (for example, time or arc length) defined over the range $t_1 \le t \le t_2$. In terms of the fixed frame $\{s\}$,

$$r(t) = x(t)\hat{x}_s + y(t)\hat{y}_s + z(t)\hat{z}_s, \tag{3.41}$$

in which case

$$dr = \dot{r}\, dt = (\dot{x}\hat{x}_s + \dot{y}\hat{y}_s + \dot{z}\hat{z}_s)\, dt. \tag{3.42}$$

Also parameterizing the force by t as $\mathrm{f}(t) = f_x(t)\hat{x}_\mathrm{s} + f_y(t)\hat{y}_\mathrm{s} + f_z(t)\hat{z}_\mathrm{s}$, the work is then evaluated as

$$W = \int_{t_1}^{t_2} f_x \dot{x} + f_y \dot{y} + f_z \dot{z} \, dt. \tag{3.43}$$

It should be apparent from (3.43) that any forces that act normally to the particle's direction of motion contribute no work – the integrand in (3.43) becomes zero – and can be ignored. Also, (3.43) is invariant with respect to different parameterizations of the path. That is, if t is parameterized as $s(t)$, then

$$W = \int_{t_1}^{t_2} \left(f_x \frac{dx}{ds} + f_y \frac{dy}{ds} + f_z \frac{dz}{ds} \right) \frac{ds}{dt} dt$$

$$= \int_{s(t_1)}^{s(t_2)} f_x \frac{dx}{ds} + f_y \frac{dy}{ds} + f_z \frac{dz}{ds} \, ds. \tag{3.44}$$

3.2.1   Work and Kinetic Energy

Now write $\mathrm{f} = m\ddot{\mathrm{r}}$, in which case the work $\mathcal{W}$ can be written

$$W = \int_C \mathrm{f} \cdot d\mathrm{r} = m \int_C \ddot{\mathrm{r}} \cdot d\mathrm{r} = m \int_{t_1}^{t_2} \ddot{\mathrm{r}} \cdot \dot{\mathrm{r}} \, dt. \tag{3.45}$$

Note that $\frac{1}{2}\frac{d}{dt}(\dot{\mathrm{r}} \cdot \dot{\mathrm{r}}) = \ddot{\mathrm{r}} \cdot \dot{\mathrm{r}}$. Also, $\dot{\mathrm{r}} \cdot \dot{\mathrm{r}} = v^2$ (the square of the speed). The work $\mathcal{W}$ can therefore be written

$$W = \frac{m}{2} \int_{t_1}^{t_2} \frac{d}{dt}(\dot{\mathrm{r}} \cdot \dot{\mathrm{r}}) \, dt = \frac{m}{2} \int_{t_1}^{t_2} d(v^2) = \frac{m}{2} \left(v(t_2)^2 - v(t_1)^2 \right). \tag{3.46}$$

Noting that $\frac{1}{2}mv^2$ is the definition of the kinetic energy of the particle, the above can be written

$$W = \mathcal{K}_2 - \mathcal{K}_1, \tag{3.47}$$

where $\mathcal{K}_1$ denotes the kinetic energy of the particle at the initial time t_1, and $\mathcal{K}_2$ is the kinetic energy of the particle at the final time t_2. Equation (3.47) is a general particle dynamics principle relating work with kinetic energy, often referred to as the **work-energy principle**; it is not a new dynamics principle, but rather a consequence of Newton's second law. Because (3.47) is a scalar equation involving the initial and final speeds of the particle, it may be useful for problems that involve finding speeds, as the following example illustrates.

Example 3.5   Consider the particle of mass m attached to a taut massless cord of length L as shown in Figure 3.6(a). The mass is released from some initial state 1 with some speed v_1, and the objective is to find the speed v_2 at a specified final state 2. Gravity acts downward.

Specifying the fixed frame {s} and drawing the free-body diagram as shown in Figure 3.6(b), apply the work-energy principle. Parameterize the path of the particle as follows:

$$\mathrm{r}(s) = -L \cos s \, \hat{x}_\mathrm{s} - L \sin s \, \hat{y}_\mathrm{s}, \tag{3.48}$$

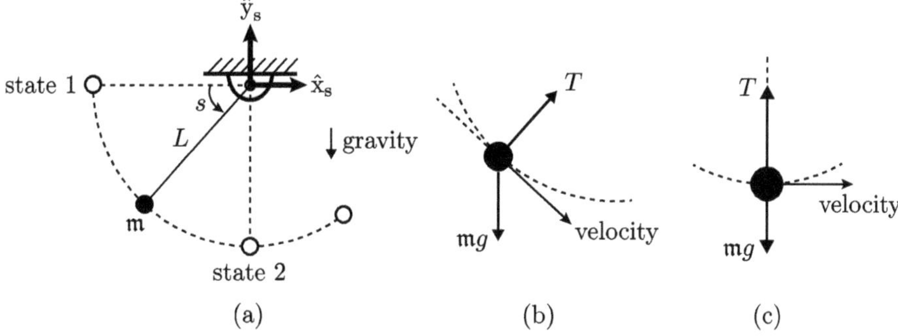

Figure 3.6 Analysis of pendulum using work-energy principle: (a) pendulum. (b) free-body diagram at arbitrary state. (c) free-body diagram at six o'clock configuration.

where the scalar parameter s ranges between $s \in [s_1, s_2]$. The gravitational force is $\mathbf{f} = -mg\hat{y}_s$. The other force exerted on the particle, the cord tension force, does no work since it acts normally to the velocity of the particle. The work $\mathcal{W}$ calculated as a path integral is

$$\mathcal{W} = \int_{s_1}^{s_2} \mathbf{f} \cdot \frac{d\mathbf{r}}{ds}\, ds = \int_{s_1}^{s_2} mgL\cos s\, ds. \tag{3.49}$$

From the work-energy principle, $\mathcal{W} = \mathcal{K}_2 - \mathcal{K}_1 = \frac{1}{2}m(v_2^2 - v_1^2)$. Given values for s_1, s_2, and v_1, one can then determine v_2 from the work-energy principle. For example, if the mass is released from rest at nine o'clock (state 1), then the velocity of the mass as it passes six o'clock (state 2) can be found by setting $s_1 = 0$, $s_2 = \pi/2$, and $v_1 = 0$; solving $\mathcal{W} = \mathcal{K}_2 - \mathcal{K}_1$ for v_2,

$$v_2 = \sqrt{2gL}. \tag{3.50}$$

Because the work-energy principle is a scalar equation, determining vector quantities like the force, acceleration, or velocity will require further analysis of the equations of motion. For example, suppose we seek the cord tension force at state 2. Drawing the free-body diagram at state 2 as shown in Figure 3.6(c), the cord tension force can be determined by recalling that the acceleration in tangential-normal coordinates is

$$\mathbf{a} = \dot{v}\hat{t} + \frac{v^2}{\rho}\hat{n}, \tag{3.51}$$

where the radius of curvature ρ for this example is just the cord length L; the normal component of the acceleration a_n is $\frac{v^2}{L}$. The equations of motion in tangential-normal coordinates is

$$\begin{aligned}
\text{Tangential:} \quad & f_t = 0 = ma_t \quad \rightarrow \quad a_t = 0, \\
\text{Normal:} \quad & f_n = ma_n \quad \rightarrow \quad T - mg = ma_n = 2mg,
\end{aligned} \tag{3.52}$$

from which it follows that $T = 3mg$. $\qquad\qquad\qquad\qquad\qquad\qquad\qquad\square$

3.2.2   Conservative Forces and Potential Energy

Given a force field $\mathrm{f} = f_x(x, y, z)\hat{\mathrm{x}}_\mathrm{s} + f_y(x, y, z)\hat{\mathrm{y}}_\mathrm{s} + f_z(x, y, z)\hat{\mathrm{z}}_\mathrm{s}$ in $\mathbb{R}^3$, if there exists some function $\mathcal{P} : \mathbb{R}^3 \to \mathbb{R}$, $\mathcal{P}(x, y, z)$ differentiable with respect to x, y, z, such that

$$f_x(x, y, z) = -\frac{\partial \mathcal{P}}{\partial x}(x, y, z),$$

$$f_y(x, y, z) = -\frac{\partial \mathcal{P}}{\partial y}(x, y, z), \tag{3.53}$$

$$f_z(x, y, z) = -\frac{\partial \mathcal{P}}{\partial z}(x, y, z),$$

then f is said to be a **conservative force field**, and $\mathcal{P}(x, y, z)$ is its associated **potential energy function**. The notation

$$\nabla \mathcal{P} = \left(\frac{\partial \mathcal{P}}{\partial x}, \ \frac{\partial \mathcal{P}}{\partial y}, \ \frac{\partial \mathcal{P}}{\partial z} \right)^\top \in \mathbb{R}^3, \tag{3.54}$$

with $\nabla \mathcal{P}$ referred to as the **gradient** of $\mathcal{P}$, will be used. Denoting $f = (f_x, f_y, f_z)^\top \in \mathbb{R}^3$, then $f = -\nabla \mathcal{P}$ for a conservative force field f.

Some well-known examples of conservative force fields include gravity, in which $\mathcal{P}(x, y, z) = mgz$ and $-\nabla \mathcal{P} = (0, 0, -mg)^\top$, and the linear spring, in which $\mathcal{P}(x, y, z) = \frac{1}{2}kx^2$ (k is the spring stiffness) and $-\nabla \mathcal{P} = (-kx, 0, 0)^\top$ is the spring force.

While under the action of some force field, a particle of mass m can move from point A to point B in different ways, and the work will generally depend on the path taken. However, if the force field is conservative, then the work does not depend on the path, but only on the initial and final points A and B. To show this, let $\mathcal{P}(x, y, z)$ be the potential function with force $f = -\nabla \mathcal{P}$. Parameterizing the path C from A to B by the curve $(x(t), y(t), z(t))$, $t_1 \le t \le t_2$, then

$$\mathcal{W} = \int_C \mathrm{f} \cdot d\mathrm{r} = \int_{t_1}^{t_2} \left(-\frac{\partial \mathcal{P}}{\partial x}\dot{x} - \frac{\partial \mathcal{P}}{\partial y}\dot{y} - \frac{\partial \mathcal{P}}{\partial z}\dot{z} \right) dt = \int_{t_1}^{t_2} -\frac{d\mathcal{P}}{dt} dt$$

$$= \mathcal{P}|_{t=t_1} - \mathcal{P}|_{t=t_2}. \tag{3.55}$$

The converse of the above statement – if the path integral $\int_C \mathrm{f} \cdot d\mathrm{r}$ is path-independent (that is, the value of the integral is the same for all paths C with the same endpoints), then f is a conservative force – is also true.[1]

An immediate corollary of the above result is that for conservative force fields f, if C is a closed curve (i.e., points A and B are the same, so that the path begins and ends at the same point), then the work $\oint_C \mathrm{f} \cdot d\mathrm{r}$ is zero (the $\oint$ notation is widely used notation for denoting a closed path integral).

Another consequence of the above is that the work-energy principle for conservative forces can also be stated entirely in terms of potential and kinetic energies. Let $\mathcal{K}_i$ and $\mathcal{P}_i$, respectively, be the potential and kinetic energies of the particle at state i, $i = 1, 2$. Then the work-energy principle $\mathcal{W} = \mathcal{K}_2 - \mathcal{K}_1$ can be restated as $\mathcal{P}_1 - \mathcal{P}_2 = \mathcal{K}_2 - \mathcal{K}_1$, or

$$\mathcal{P}_1 + \mathcal{K}_1 = \mathcal{P}_2 + \mathcal{K}_2. \tag{3.56}$$

---

[1] Proof of the converse involves verification of several technical conditions and is not given here.

That is, total energy is conserved for particles moving under the action of a conservative force field.

Example 3.6   Returning to the previous pendulum example of Figure 3.6, at state 1 (the mass is at the nine o'clock position) we have $\mathcal{K}_1 = 0$, $\mathcal{P}_1 = 0$, while at state 2 (the mass is at the six o'clock position) we have $\mathcal{K}_2 = \frac{1}{2}mv_2^2 = mgL$ and $\mathcal{P}_2 = -mgL$. Therefore $\mathcal{P}_1 + \mathcal{K}_1 = \mathcal{P}_2 + \mathcal{K}_2$.   □

Given a force field $f = (f_x, f_y, f_z)^\top$ defined over some three-dimensional domain $\mathcal{D}$, determining whether or not f is conservative may not always be apparent or straightforward. For this purpose, **Stokes Theorem** can be helpful. Stokes Theorem can be stated as follows: Given a bounded two-dimensional surface S with a closed curve C as its boundary, and a vector field f defined over a bounded region of $\mathbb{R}^3$ containing S,

$$\oint_C \mathbf{f} \cdot dr = \int\int_S \langle \nabla \times \mathbf{f}, \hat{n} \rangle \, dA, \tag{3.57}$$

where the right-hand side is a surface integral over S: $\langle \cdot, \cdot \rangle$ denotes the inner product, $\nabla \times \mathbf{f}$ is the curl of the vector field f, $\hat{n}$ in the integrand denotes the unit normal to the surface. Recall that the curl of f can be calculated as

$$\nabla \times \mathbf{f} = \begin{vmatrix} \hat{x}_s & \hat{y}_s & \hat{z}_s \\ \frac{\partial}{\partial x} & \frac{\partial}{\partial y} & \frac{\partial}{\partial z} \\ f_x & f_y & f_z \end{vmatrix} = \left(\frac{\partial f_z}{\partial y} - \frac{\partial f_y}{\partial z}, \frac{\partial f_x}{\partial z} - \frac{\partial f_z}{\partial x}, \frac{\partial f_y}{\partial x} - \frac{\partial f_x}{\partial y} \right)^\top, \tag{3.58}$$

or using the skew-symmetric matrix notation as

$$\nabla \times \mathbf{f} = \begin{bmatrix} 0 & -\frac{\partial}{\partial z} & \frac{\partial}{\partial y} \\ \frac{\partial}{\partial z} & 0 & -\frac{\partial}{\partial x} \\ -\frac{\partial}{\partial y} & \frac{\partial}{\partial x} & 0 \end{bmatrix} \begin{bmatrix} f_x \\ f_y \\ f_z \end{bmatrix}. \tag{3.59}$$

If f is a conservative force vector field, then the path integral on the left-hand side must be zero for all closed paths C. It can also be verified by calculation that $\nabla \times \mathbf{f} = 0$ if f is conservative. For the converse, we state without proof the following result: If $\nabla \times \mathbf{f} = 0$ everywhere over some simply connected domain[2] $\mathcal{D}$ of $\mathbb{R}^3$, then f is a conservative force vector field.

Example 3.7   (a) The force vector field $(y^2z^3, 2xyz^3, 3xy^2z^2)^\top$ is conservative, and can be generated as the gradient of the potential function $\mathcal{P}(x, y, z) = -xy^2z^3$.
(b) The force vector field $(3x^2z, z^2, x^3 + 2yz)^\top$ is conservative and can be generated from the potential function $\mathcal{P}(x, y, z) = -(x^3z + yz^2)$.
(c) The force vector field $(x, y, 5x)^\top$ is not conservative, since its curl is $(0, -5, 0)$.
(d) The force vector field $\frac{1}{x^2+y^2+z^2}(x, y, z)^\top$ has curl zero, and is well defined everywhere except at the origin $(0, 0, 0)^\top$. The region $\mathcal{D}$ corresponding to $\mathbb{R}^3$ with the origin removed is simply connected – any closed curve in $\mathcal{D}$ can be shrunk to a point – and therefore

---

[2] $\mathcal{D}$ is a *simply connected domain* if any closed curve in $\mathcal{D}$ can always be shrunk to a point.

this force vector field is conservative. The potential function generating this force vector field is

$$P(x, y, z) = -\frac{1}{2}\ln(x^2 + y^2 + z^2). \tag{3.60}$$

$\square$

Example 3.8   A well-known example of a vector field whose curl is zero but is not conservative is given by

$$f(x, y, z) = \left(-\frac{y}{x^2 + y^2}, \frac{x}{x^2 + y^2}, 0\right)^\top. \tag{3.61}$$

The curl $\nabla \times f$ is zero everywhere, but note that f is not well defined for points (x, y, z) lying on the z-axis (the first two components of f are of the form $0/0$). The region $\mathcal{D}$ corresponding to $\mathbb{R}^3$ with the z-axis removed is not simply connected. A calculation reveals that any path integral taken about a curve wrapping around the origin will be 2π, and not zero.

$\square$

3.3   Momentum and Impulse

We now consider a third class of methods for studying the motion of particles that is particularly useful for problems involving impulsive forces and impacts. Like the work-energy principles derived earlier, the basic dynamic principles involving impulse and momentum are directly derived from the equations of motion.

3.3.1   Linear Momentum and Impulse

The **linear momentum** of a particle is just the product of its mass and velocity, $\mathrm{m}v$. Writing Newton's second law as $f = \mathrm{m}\dot{v}$, then integrating both sides over some time interval $[t_1, t_2]$,

$$\int_{t_1}^{t_2} f\, dt = \mathrm{m} \int_{t_1}^{t_2} \dot{v}\, dt = \mathrm{m}v_2 - \mathrm{m}v_1, \tag{3.62}$$

where $v_1 = v(t_1)$, $v_2 = v(t_2)$. The integral on the left-hand side of (3.62), denoted $\mathcal{I}$, is the **linear impulse** of the force f:

$$\mathcal{I} = \int_{t_1}^{t_2} f(t)\, dt. \tag{3.63}$$

Equation (3.62) is a statement of the **principle of linear momentum and impulse**: The change in linear momentum of a particle over some time interval $[t_1, t_2]$ is equal to the impulse applied to the particle over this same interval:

$$\mathcal{I} = \mathrm{m}v_2 - \mathrm{m}v_1. \tag{3.64}$$

Note that (3.64) is a vector equation involving a single particle; for systems involving multiple particles, (3.64) should be written for each particle.

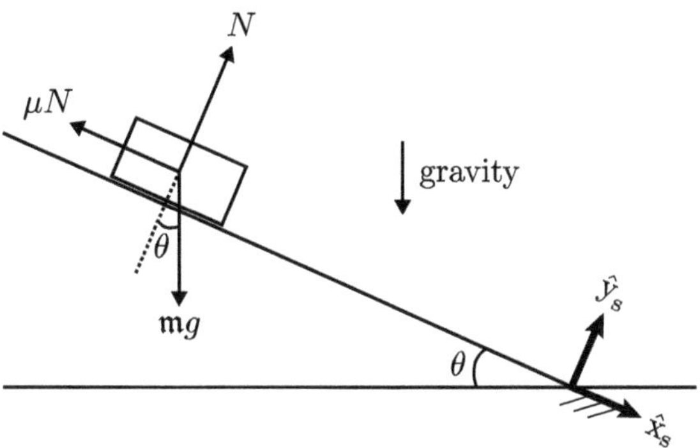

Figure 3.7 A block sliding down an inclined plane.

Example 3.9   A wheeled vehicle going downhill suddenly brakes, causing the wheels to lock up and the vehicle to eventually slide to a complete stop. This problem can be modeled as a block sliding down an inclined plane, but this time with friction force causing the block to eventually come to a complete stop. We will determine the time required for the vehicle (block) to come to a complete stop, and also the total distance that the vehicle (block) slides. The solution will be obtained using all three methods: (i) direct analysis of the equations of motion; (ii) the work-energy method; (iii) the impulse-momentum method.

First, assign a fixed reference frame and draw a free-body diagram as shown in Figure 3.7, and derive the equations of motion for the vehicle in the chosen coordinates: letting μ be the coefficient of friction, and N be the normal force exerted by the vehicle on the hill, the friction force is then of magnitude μN, directed opposite to the direction of motion. Then the equations of motion are given by

$$\sum f_y = m\ddot{y} \rightarrow N - mg\cos\theta = 0 \rightarrow N = mg\cos\theta, \tag{3.65}$$

$$\sum f_x = m\ddot{x} \rightarrow mg\sin\theta - \mu N = m\ddot{x} \rightarrow g(\sin\theta - \mu\cos\theta) = \ddot{x}. \tag{3.66}$$

Integrating the latter,

$$\dot{x}(t) = \int g(\sin\theta - \mu\cos\theta)\, dt = -g(\mu\cos\theta - \sin\theta)t + v_0. \tag{3.67}$$

To determine the time for the vehicle to come to a complete stop after the brakes are applied, set $\dot{x}(t_f) = 0$ in (3.67) and solve for t_f,

$$t_f = \frac{v_0}{g(\mu\cos\theta - \sin\theta)}. \tag{3.68}$$

To find the stopping distance, integrate (3.67) again to obtain

$$x(t) = -\frac{g}{2}(\mu\cos\theta - \sin\theta)t^2 + v_0 t. \tag{3.69}$$

The distance traveled is then

$$d = x(t_f) = \frac{v_0^2}{2g(\mu \cos\theta - \sin\theta)}. \tag{3.70}$$

We now use the work-energy principle. First, the stopping distance can be determined via $\mathcal{W} = \mathcal{K}_2 - \mathcal{K}_1$, with $\mathcal{K}_2 = 0$, $\mathcal{K}_1 = \frac{1}{2}mv_0^2$, and

$$\mathcal{W} = -mg(\mu \cos\theta - \sin\theta) \cdot d, \tag{3.71}$$

where d denotes the stopping distance. Solving for d,

$$d = \frac{v_0^2}{2g(\mu \cos\theta - \sin\theta)} \tag{3.72}$$

as before. Substituting this value for $d = x(t_f)$ into (3.69) and solving for t_f leads to the same result as (3.68).

We now use the principle of impulse-momentum. In the y-direction,

$$\int_0^{t_f} f_y \, dt = mv_y(t_f) - mv_y(0) = 0, \tag{3.73}$$

since $v_y(t) = 0$ for all t. Setting $f_y = N - mg\cos\theta$ in (3.73) and integrating leads to

$$(N - mg\cos\theta) \cdot t_f = 0 \rightarrow N = mg\cos\theta. \tag{3.74}$$

Similarly, in the x-direction we have

$$\int_0^{t_f} f_x \, dt = mv_x(t_f) - mv_x(0), \tag{3.75}$$

$$\int_0^{t_f} (mg\sin\theta - \mu mg\cos\theta) \, dt = 0 - mv_0. \tag{3.76}$$

Solving the above for t_f, we get the same result as in (3.68). $x(t_f)$ can then be obtained similarly from (3.69). $\qquad\square$

As a general rule of thumb, when the force f is given as a function of time, $f(t)$, then either direct solution of the equations of motion, or impulse-momentum methods, tend to be effective. If f is specified as a function of the displacement x, $f(x)$, then work-energy methods tend to be effective.

We conclude this section with a discussion of the **Dirac delta** function. An impulsive force f is a very large force applied over a very short time interval. A **unit impulse force** by definition is an impulsive force with unit magnitude; the magnitude of a unit impulse force applied at time $t = \tau$, denoted $f_\epsilon(t - \tau)$, is then required to satisfy

$$f_\epsilon(t - \tau) = \begin{cases} \frac{1}{\epsilon} & t \in [\tau - \frac{\epsilon}{2}, \tau + \frac{\epsilon}{2}], \\ 0 & \text{otherwise}, \end{cases} \tag{3.77}$$

where ϵ is the width of a rectangle of height $1/\epsilon$. Observe that the area of this rectangle is always one. The Dirac delta function $\delta(t - \tau)$ is defined as the limit

$$\delta(t - \tau) = \lim_{\epsilon \to 0} f_\epsilon(t - \tau). \tag{3.78}$$

The Dirac delta function can alternatively be defined as follows:

$$\delta(t - \tau) = \begin{cases} 0, & t \neq \tau, \\ \infty, & t = \tau, \end{cases} \tag{3.79}$$

with the additional requirement that $\int_{-\infty}^{\infty} \delta(t)\, dt = 1$. Observe that $\delta(t-\tau)$ is not, strictly speaking, a function in the usual mathematical sense. One way in which the Dirac delta function is useful is that impulsive forces of different magnitudes can be represented as follows:

$$\mathcal{I}(t - \tau) = I \cdot \delta(t - \tau), \tag{3.80}$$

where $I > 0$ represents the magnitude of the impulse $\mathcal{I}$ applied at time $t = \tau$.

3.3.2 Angular Momentum and Impulse

Consider a particle of mass m moving with velocity v. Let the position of the particle with respect to the fixed frame origin O be r. The **angular momentum** of the particle is defined to be

$$h_o = r \times mv. \tag{3.81}$$

Taking derivatives of both sides, we get

$$\dot{h}_o = \dot{r} \times mv + r \times m\dot{v} = r \times m\ddot{r}, \tag{3.82}$$

where we use the fact that $\dot{r} = v$ and $v \times v = 0$. Also from Newton's second law $f = m\ddot{r}$, we can write

$$r \times f = r \times m\ddot{r} = \dot{h}_o. \tag{3.83}$$

Noting that $r \times f$ is the moment of the force f about the point O, we use the notation $m_o = r \times f$. Therefore

$$m_o = \dot{h}_o. \tag{3.84}$$

The above relation will be central to deriving the dynamics of a rotating rigid body, but for now, the case where the applied force $f = 0$ (and therefore $m_o = 0$) offers some physical intuition: when the applied force is zero, the particle moves in such a way that its angular momentum is conserved:

$$\dot{h}_o = 0. \tag{3.85}$$

That is, the angular momentum h_o is constant in both magnitude and direction. Equation (3.85) is one version of the **conservation of angular momentum**, in this case for a single particle whose angular momentum is defined with respect to a fixed point O. Other versions of the angular momentum conservation principle, for example, involving multiple particles, appear later. We emphasize that this "law" is not a new dynamic principle, but merely a consequence of Newton's laws, in this case an instance of (3.84) with $m_o = 0$.

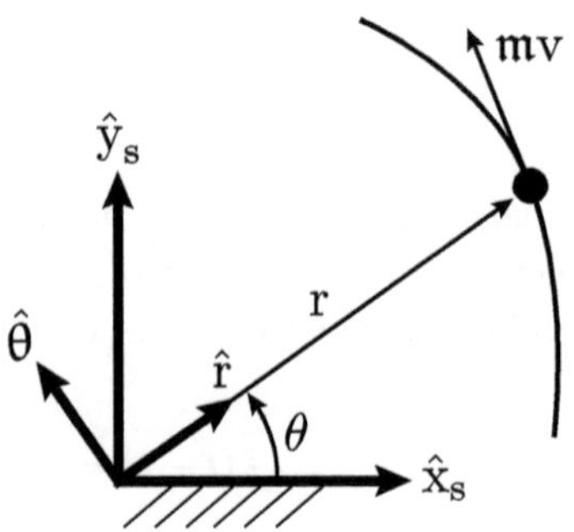

Figure 3.8 Angular momentum in polar coordinates.

Example 3.10 We now work out the formula for the angular momentum in polar coordinates. Referring to Figure 3.8, suppose a particle moves in the $\{\hat{x}_s, \hat{y}_s\}$ plane, and its position and velocity are described in polar coordinates:

$$r = r\hat{r}, \tag{3.86}$$

$$v = \dot{r}\hat{r} + r\dot{\theta}\hat{\theta}. \tag{3.87}$$

The angular momentum about the fixed frame origin O is then

$$h_o = r \times mv = mr^2\dot{\theta}\hat{z}_s. \tag{3.88}$$

Observe that for planar motions, both h_o and m_o are always normal to the $\{\hat{x}_s, \hat{y}_s\}$ plane. Assume all forces acting on the particle are in the radial direction ($\pm\hat{r}$). Then $r \times f = 0$, so h_o is conserved and $mr^2\dot{\theta}$ is constant for all t. Recall that this was precisely the condition derived in our earlier example on the motion of a satellite orbiting the earth at a constant rate. In the case of general $f = f_r\hat{r} + f_\theta\hat{\theta}$,

$$r \times f = r f_\theta \hat{z}_s = \frac{d}{dt}(mr^2\dot{\theta})\hat{z}_s = \left(2mr\dot{r}\dot{\theta} + mr^2\ddot{\theta}\right)\hat{z}_s. \tag{3.89}$$

$\square$

3.4   Collisions and Impact

We now analyze the dynamics of collision between two bodies. To simplify the analysis, the two bodies will be modeled as two smooth spheres, so that for the purposes of analysis we may treat the colliding bodies as particles. This example not only serves as a useful approximation to collision between more complex objects, but also demonstrates how impulse-momentum techniques can be applied in the analysis of collision and impact problems. For a mass m experiencing a very large force $f(t)$ over short time duration $[t_1, t_2]$, its associated impulse $\mathcal{I} = \int_{t_1}^{t_2} f(t)\,dt$ is related to the particle velocity v_1 and v_2 at t_1 and t_2, respectively, by $\mathcal{I} = mv(t_2) - mv(t_1)$.

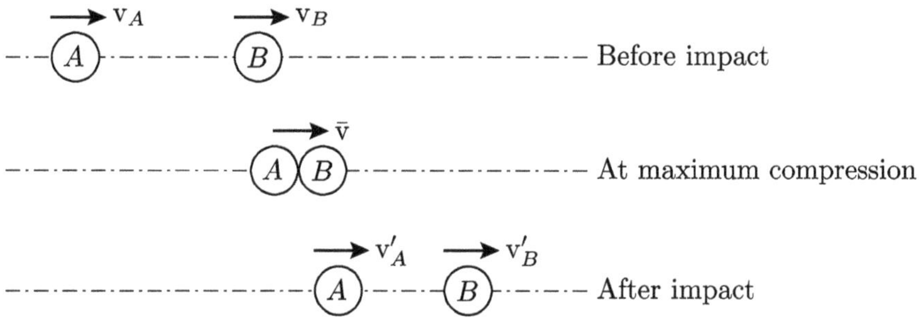

Figure 3.9 Direct central impact between two spheres.

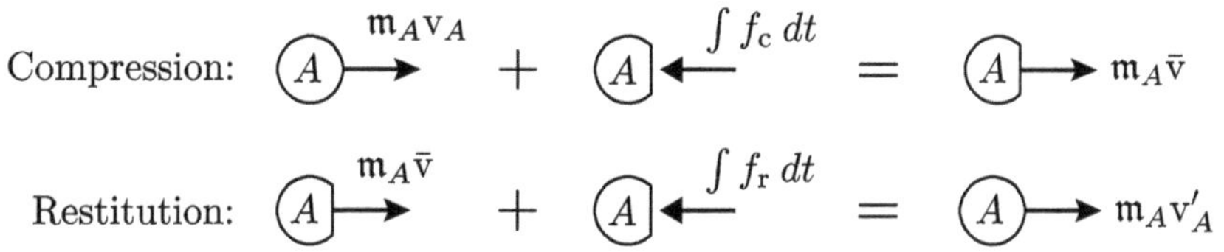

Figure 3.10 Impulse and momentum during deformation and restitution.

3.4.1 Direct Central Impact

First consider the simple case of a direct central impact, "head-on" collision between the two spheres. Referring to Figure 3.9, the two spheres of mass m_A and m_B initially move at velocities v_A and v_B. Let I be the impulse of the collision experienced by sphere A, in which case the impulse experienced by sphere B is $-I$ (from Newton's third law of motion). Denote the velocities of A and B immediately after impact by v'_A and v'_B, respectively. From the principle of linear momentum and impulse,

$$I = m_A v'_A - m_A v_A \quad \text{(mass } A\text{)}, \tag{3.90}$$
$$-I = m_B v'_B - m_B v_B \quad \text{(mass } B\text{)}. \tag{3.91}$$

It then follows that

$$m_A v_A + m_B v_B = m_A v'_A + m_B v'_B. \tag{3.92}$$

Equation (3.92) is a statement of the principle of conservation of linear momentum – if the net external force on a system of particles is zero, then total linear momentum is conserved. This principle is examined in more detail in the next section on the dynamics of a system of particles. Since for the direct central impact case all velocities are along the same line, the above vector quantities can be replaced by their scalar magnitudes:

$$m_A v_A + m_B v_B = m_A v'_A + m_B v'_B. \tag{3.93}$$

Given v_A and v_B, another relation is needed to determine the velocities v'_A and v'_B. We now examine more closely how particle A behaves during the period of **compression** immediately upon impact, and during the period of **restitution** immediately afterwards as the shape of A returns to that of a sphere. Referring to Figure 3.10, let f_c denote the force applied by B on A during the period of compression, and $\bar{v}$ the velocity of A after reaching maximum compression. From the principle of momentum and impulse,

$$\int f_c \, dt = m_A \bar{v} - m_A v_A. \tag{3.94}$$

After maximum compression, A now enters the period of restitution and its shape begins to return to that of a sphere. During the period of restitution, let f_r denote the force applied by B on A. Then again from the principle of momentum and impulse,

$$\int f_r \, dt = m_A v'_A - m_A \bar{v}, \tag{3.95}$$

where the integral is over the duration of the period of restitution, from maximum compression until A separates from B.

In general, $\int f_c \, dt \geq \int f_r \, dt$. The **coefficient of restitution** e is defined as

$$e = \frac{\left| \int f_r \, dt \right|}{\left| \int f_c \, dt \right|}. \tag{3.96}$$

The value of e always lies between 0 and 1, with 1 indicating a perfectly elastic collision (the approach and separation velocities are the same) and 0 indicating a perfectly inelastic collision (the two bodies stick to each other, and kinetic energy is lost in the form of internal friction). The coefficient of restitution depends on several factors including material properties of A and B, as well as their shapes, sizes, and impact velocities. In most of the examples and problems in this book, the coefficient of restitution will be assumed given as a constant.

From (3.94)–(3.96) we obtain

$$e = \frac{\bar{v} - v'_A}{v_A - \bar{v}}. \tag{3.97}$$

A similar analysis for B leads to

$$e = \frac{v'_B - \bar{v}}{\bar{v} - v_B}. \tag{3.98}$$

Now, for any real scalars a, b, c, d, if $\frac{a}{b} = \frac{c}{d}$, then it follows that $\frac{a}{b} = \frac{c}{d} = \frac{a+c}{b+d}$. Using this general identity we have

$$e = \frac{v'_B - v'_A}{v_A - v_B}, \tag{3.99}$$

or equivalently,

$$v'_B - v'_A = e(v_A - v_B). \tag{3.100}$$

Equation (3.100) offers an experimental method for estimating the coefficient of restitution between two bodies. Equation (3.100) also provides the extra condition needed to determine the post-impact velocities v'_A and v'_B.

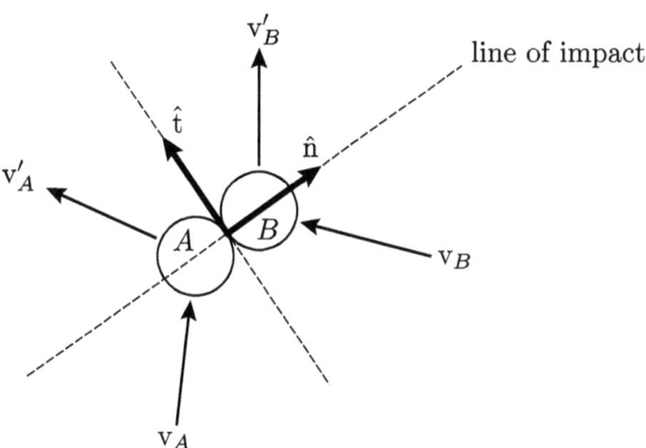

Figure 3.11 Oblique impact.

3.4.2 Oblique Impact

Now consider the case when A and B are moving in different directions when collision occurs. Referring to Figure 3.11, the velocity vectors v_A, v_B, v'_A, v'_B are no longer colinear, but the identity

$$m_A v_A + m_B v_B = m_A v'_A + m_B v'_B \tag{3.101}$$

still holds. Choose a fixed frame with axes labeled $\hat{t}$ (tangential to the two bodies at the point of contact between A and B) and $\hat{n}$ (normal to the two bodies at the point of contact – $\hat{n}$ is also said to be along the **line of impact**) as shown in the figure, and assume the coefficient of restitution e, and the initial velocity vectors v_A, v_B are given. We now determine v'_A and v'_B. First, assuming no friction at the point of contact, all forces due to the collision are normal to the two bodies at the point of contact; that is, the impact and restitutive forces are along the line of impact. Therefore in the $\hat{t}$-direction we have

$$v_{tA} = v'_{tA}, \tag{3.102}$$
$$v_{tB} = v'_{tB}. \tag{3.103}$$

In the $\hat{n}$-direction we have

$$m_A v_{nA} + m_B v_{nB} = m_A v'_{nA} + m_B v'_{nB}, \tag{3.104}$$
$$v'_{nB} - v'_{nA} = e(v_{nA} - v_{nB}). \tag{3.105}$$

These last two linear equations can now be solved for v'_{nA} and v'_{nB}:

$$v'_{nA} = \frac{m_A - e m_B}{m_A + m_B} v_{nA} + \frac{(1+e) m_B}{m_A + m_B} v_{nB}, \tag{3.106}$$
$$v'_{nB} = \frac{(1+e) m_A}{m_A + m_B} v_{nA} + \frac{m_B - e m_A}{m_A + m_B} v_{nB}. \tag{3.107}$$

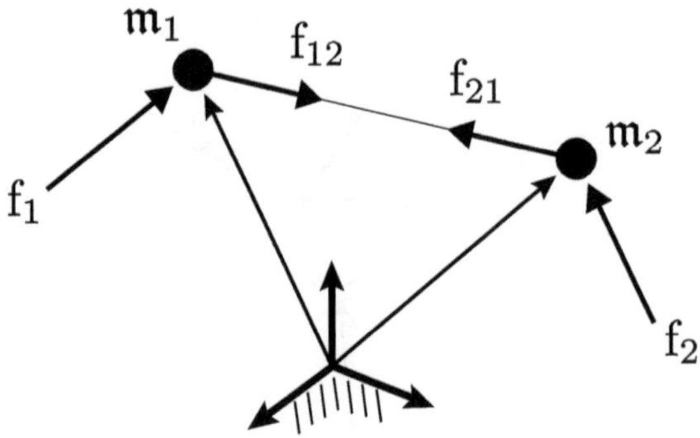

Figure 3.12 Two particles exerting equal and opposite forces on each other.

3.5   Dynamics of a System of Particles

As a first step toward deriving the equations of the motion for a single rigid body, in this section we will derive the equations of motion for a system of particles. We do this by applying Newton's second ($f = ma$) and third (for every action there is an equal and opposite reaction) laws of motion to a system of particles, resulting in equations that describe the motion of the system's center of mass, and also the total angular momentum of the system. In Chapter 4 we will model a rigid body as an infinite collection of particles, and extend the previous equations for the center of mass and angular momentum of a system of particles to the rigid body. The equations of motion for systems of particles, as well as corresponding versions of work-energy and impulse-momentum methods, are also useful in their own right, especially when analyzing systems that can be modeled as a finite number of particles.

3.5.1   Equations of Motion for the Mass Center

Our primary motivation for considering systems of particles is to eventually model a rigid body as a collection of an infinite number of particles, connected to each other by massless rods of fixed length so as to preserve rigidity. As a first step, consider two particles of mass m_1 and m_2 as shown in Figure 3.12. The two particles are assumed to exert equal and opposite forces on each other. This can be satisfied, for example, by connecting the two particles by a rigid massless rod, or a linear spring, or simply by assuming the usual gravitational forces of attraction between two masses. Writing down the equations of motion for each mass,

$$\text{mass 1: } f_1 + f_{21} = m_1 a_1, \tag{3.108}$$

$$\text{mass 2: } f_2 + f_{12} = m_2 a_2, \tag{3.109}$$

where f_{ij} is the force exerted on particle j by particle i, and f_j is the net external force exerted on particle j while excluding f_{ij}. From Newton's third law – for every action there is an equal and opposite reaction – it follows that $f_{12} = -f_{21}$. Summing (3.108) and (3.109) together, we get

$$f_1 + f_2 + (f_{21} + f_{12}) = f_1 + f_2 = m_1 a_1 + m_2 a_2, \tag{3.110}$$

where we use the fact that $f_{12} + f_{21} = 0$.

Now generalize the above result to a system of N interconnected particles. For each particle the equations of motion are of the form

$$f_i + \sum_{j=1}^{N} f_{ji} = m_i \ddot{r}_i, \quad i = 1, \ldots, N, \tag{3.111}$$

where r_i is the position of particle i with respect to the fixed (inertial) frame origin. Add the N equations to get

$$\sum_{i=1}^{N} m_i \ddot{r}_i = \sum_{i=1}^{N} f_i + \sum_{i=1}^{N} \sum_{j=1}^{N} f_{ji}. \tag{3.112}$$

Since $f_{ij} + f_{ji} = 0$, the last term on the right-hand side is zero. Further denote by f the sum of all external forces $\sum_{i=1}^{N} f_i$:

$$f = \sum_{i=1}^{N} f_i. \tag{3.113}$$

Equation (3.112) then reduces to $\sum_{i=1}^{N} m_i \ddot{r}_i = f$. Defining

$$m = m_1 + m_2 + \cdots + m_N, \tag{3.114}$$

$$r_c = \frac{1}{m} (m_1 r_1 + m_2 r_2 + \cdots + m_N r_N), \tag{3.115}$$

the equations of motion then become

$$f = m \ddot{r}_c. \tag{3.116}$$

The vector r_c is the **center of mass**; the above are the equations of motion for the center of mass given the net external force f on the system of particles.

Again, we remind the reader that Equation (3.116) holds not only for systems in which the particles are connected by rigid massless rods, but for any system in which the relation $f_{ij} = -f_{ji}$ is valid, for example, if f_{ij} is the gravitational force of attraction applied by m_i on m_j.

3.5.2   Work and Energy

Express the position r_i of particle i as the vector sum $r_i = r_c + r_i'$, where r_c is the position of the center of the mass of the system of particles and r_i' the vector from the center of mass to particle i. Recall from the work-energy principle that the work $\mathcal{W}_i$ done by particle i in moving from init_i to final_i is equal to $\mathcal{K}_{\text{final}_i} - \mathcal{K}_{\text{init}_i}$, the difference in kinetic energy between the final and initial positions:

$$\mathcal{W}_i = \frac{1}{2}m_i\langle \dot{\mathbf{r}}_i, \dot{\mathbf{r}}_i\rangle\Big|_{\text{init}_i}^{\text{final}_i} = \frac{1}{2}m_i\left(\langle \dot{\mathbf{r}}_c, \dot{\mathbf{r}}_c\rangle + 2\langle \dot{\mathbf{r}}_c, \dot{\mathbf{r}}_i'\rangle + \langle \dot{\mathbf{r}}_i', \dot{\mathbf{r}}_i'\rangle\right)\Big|_{\text{init}_i}^{\text{final}_i}, \tag{3.117}$$

where we use the $\langle \cdot, \cdot \rangle$ notation to denote the inner product. In the above denote by v_c^2 the square of the speed of the center of mass: $v_c^2 = \langle \dot{\mathbf{r}}_c, \dot{\mathbf{r}}_c\rangle$. Since the total work $\mathcal{W}$ done by the system of particles is just the sum of the work done by each of the particles, we have

$$\begin{aligned}
\mathcal{W} &= \sum_{i=1}^{N} \mathcal{W}_i \\
&= \frac{1}{2}mv_c^2\Big|_{\text{init}_c}^{\text{final}_c} + \left\langle \dot{\mathbf{r}}_c, \sum_{i=1}^{N} m_i\dot{\mathbf{r}}_i'\right\rangle\Big|_{\text{init}_i}^{\text{final}_i} + \frac{1}{2}\sum_{i=1}^{N} m_i\langle \dot{\mathbf{r}}_i', \dot{\mathbf{r}}_i'\rangle\Big|_{\text{init}_i}^{\text{final}_i} \\
&= \frac{1}{2}mv_c^2\Big|_{\text{init}_c}^{\text{final}_c} + \frac{1}{2}\sum_{i=1}^{N} m_i v_i'^2\Big|_{\text{init}_i}^{\text{final}_i}, \tag{3.118}
\end{aligned}$$

where we use the fact that $\sum_{i=1}^{N} m_i\dot{\mathbf{r}}_i' = 0$ (from the definition of center of mass), and denote $\langle \dot{\mathbf{r}}_i', \dot{\mathbf{r}}_i'\rangle = v_i'^2$. Denoting by $\mathcal{K}$ the total kinetic energy of the system of particles, i.e.,

$$\mathcal{K} = \frac{1}{2}mv_c^2 + \frac{1}{2}\sum_{i=1}^{N} m_i v_i'^2, \tag{3.119}$$

we can still write the work-energy relation for a system of particles in the same way as before:

$$\mathcal{W} = \mathcal{K}_{\text{final}} - \mathcal{K}_{\text{init}}. \tag{3.120}$$

It is important to note that the total kinetic energy $\mathcal{K}$ involves not only the velocity v_c of the mass center, but also the relative velocities v_i' of all the particles with respect to the center of mass. If the particles happen to be points on a rigid body, and the rigid body undergoes a pure translation (i.e., the rigid body maintains the same orientation throughout the entire motion), then the kinetic energy is simply $\frac{1}{2}mv_c^2$. If, on the other hand, the rigid body undergoes a change in orientation during the motion, then the $\frac{1}{2}\sum_{i=1}^{N} m_i v_i'^2$ term will be nonzero (for example, consider the case in which the rigid body rotates about its center of mass, with the center of mass stationary).

Finally, as with the single particle case, note that if all forces applied to the system of particles are conservative, as before, the principle of conservation of energy applies:

$$\mathcal{K}_{\text{init}} + \mathcal{P}_{\text{init}} = \mathcal{K}_{\text{final}} + \mathcal{P}_{\text{final}}, \tag{3.121}$$

where $\mathcal{P}_{\text{init}}$ and $\mathcal{P}_{\text{final}}$ are the total potential energies at the initial and final configurations, respectively.

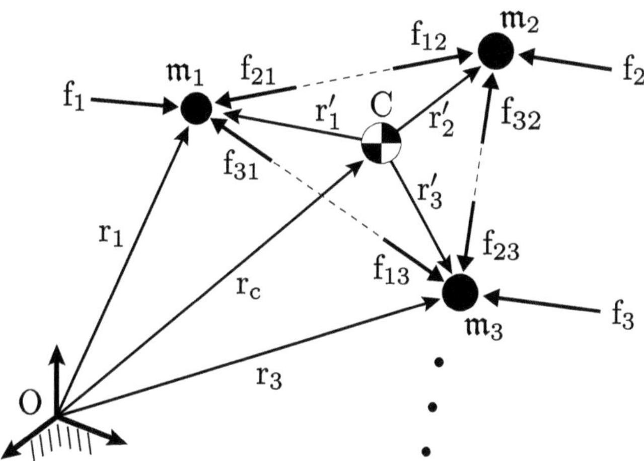

Figure 3.13 Angular momentum for a system of particles.

3.5.3 Linear and Angular Momentum

Recall that for a single particle of mass m moving with velocity v, its linear momentum is simply mv. The linear momentum for a system of particles is defined to be the sum of the linear momenta for each of the particles. More precisely, letting r_i denote the position of particle i and m_i its mass, the linear momentum of the system is defined as

$$l_{\text{sys}} = \sum_{i=1}^{N} m_i \dot{r}_i = m\dot{r}_{\text{c}}, \tag{3.122}$$

where we make use of the relation $mr_{\text{c}} = m_1 r_1 + \cdots + m_N r_N$. We use the notation

$$l_{\text{c}} = m\dot{r}_{\text{c}}. \tag{3.123}$$

Taking derivatives of both sides,

$$\dot{l}_{\text{c}} = m\ddot{r}_{\text{c}} = f, \tag{3.124}$$

i.e., the time derivative of the linear momentum is equal to the net external force $f = \sum_{i=1}^{N} f_i$ applied to the system of particles.

We now consider the angular momentum of a system of particles. Referring to Figure 3.13, first calculate the angular momentum about some fixed point O (in the figure O is taken to be the origin of the fixed frame, but it can be any arbitrary fixed point in space). The total angular momentum about O is

$$h_{\text{o}} = \sum_{i=1}^{N} r_i \times m_i \dot{r}_i. \tag{3.125}$$

Taking derivatives of both sides,

$$\dot{h}_o = \sum_{i=1}^{N} \dot{r}_i \times m_i \dot{r}_i + r_i \times m_i \ddot{r}_i. \tag{3.126}$$

Note that the first term, $\dot{r}_i \times m_i \dot{r}_i$, is always zero, while from (3.111)–(3.112) the second term can be replaced by $r_i \times f_i$, which is the moment about O contributed by particle i. Labeling by m_o the sum of all the moments about O contributed by the particles, i.e., $m_o = \sum_{i=1}^{N} r_i \times f_i$, we can write

$$m_o = \dot{h}_o. \tag{3.127}$$

Since the linear force-momentum equation $f = \dot{l}_c$ is expressed in terms of the mass center, let us also calculate the angular momentum about the mass center. Referring again to Figure 3.13, the angular momentum h_c about the system mass center is

$$\begin{aligned}
h_c &= \sum_{i=1}^{N} r_i' \times m_i \dot{r}_i \\
&= \sum_{i=1}^{N} r_i' \times m_i (\dot{r}_c + \dot{r}_i') \\
&= \left(\sum_{i=1}^{N} m_i r_i' \right) \times \dot{r}_c + \sum_{i=1}^{N} r_i' \times m_i \dot{r}_i' \\
&= \sum_{i=1}^{N} r_i' \times m_i \dot{r}_i',
\end{aligned} \tag{3.128}$$

where we make use of the fact that $\sum_{i=1}^{N} m_i r_i' = 0$ (the mass center relation). Taking derivatives of both sides with respect to time,

$$\begin{aligned}
\dot{h}_c &= \sum_{i=1}^{N} \dot{r}_i' \times m_i \dot{r}_i' + \sum_{i=1}^{N} r_i' \times m_i \ddot{r}_i' \\
&= \sum_{i=1}^{N} r_i' \times m_i (\ddot{r}_i - \ddot{r}_c) \\
&= \sum_{i=1}^{N} r_i' \times m_i \ddot{r}_i - \left(\sum_{i=1}^{N} m_i r_i' \right) \times \ddot{r}_c \\
&= \sum_{i=1}^{N} r_i' \times f_i.
\end{aligned} \tag{3.129}$$

That is, $\dot{h}_c$ is the sum of the moments about the mass center contributed by each particle; denoting the total moment about the mass center by m_c, we have

$$m_c = \dot{h}_c, \tag{3.130}$$

together with $f = \dot{l}_c$.

Having derived the equations $f = \dot{l}_c$ and $m_c = \dot{h}_c$ (and also $m_o = \dot{h}_o$ for fixed point O), for a system of particles, we now restate the principles of **conservation of linear momentum** and also **angular momentum**. In the former case, if the total force f acting on a system of particles is zero, then $\dot{l}_c = 0$ and the linear momentum l_c is constant. For the latter, if the net angular momentum m_c (m_o) is zero, then the angular momentum h_c (h_o) of the system is constant. Note that for a single particle, l_c being constant implies that the particle moves with constant velocity, but that for a system of particles, the individual particles can have different velocities; only the center of mass needs to move with constant linear velocity. Similar reasoning also applies for constant h_c (h_o).

Example 3.11   An asteroid moving with constant velocity is observed heading toward earth. To prevent any potential collisions, explosives are planted at its center and detonated, generating radial forces that cause the asteroid to break up into three pieces. The asteroid can be modeled as a system of three particles of mass m_1, m_2, m_3, initially clustered together and moving with the same constant velocity $\bar{v}$, but forced apart by the radial forces of the explosion.

Choose an inertial frame whose origin O coincides with the asteroid when it breaks apart, and denote the positions of the three particles by r_1, r_2, r_3. The center of mass r_c is then characterized by

$$m r_c = m_1 r_1 + m_2 r_2 + m_3 r_3, \tag{3.131}$$

where $m = m_1 + m_2 + m_3$. Because the vector sum of the radial forces generated by the explosion is zero, the net external force and net external moment (with respect to any fixed point, and also the center of mass) applied to the particles are zero; both the linear momentum and angular momentum of the three-particle system are therefore constant.

The linear momentum of the center of mass before and after the explosion are, respectively, $m\bar{v}$ and $m_1 v_1 + m_2 v_2 + m_3 v_3$. The angular momentum with respect to the inertial frame origin O before and after the explosion are, respectively, zero and $r_1 \times m_1 v_1 + r_2 \times m_2 v_2 + r_3 \times m_3 v_3$. Conservation of linear and angular momentum then implies

$$m\bar{v} = m_1 v_1 + m_2 v_2 + m_3 v_3, \tag{3.132}$$
$$0 = r_1 \times m_1 v_1 + r_2 \times m_2 v_2 + r_3 \times m_3 v_3. \tag{3.133}$$

(a) Assume the explosion occurs at $t = 0$. At some later time $t = T$, the positions of m_1 and m_2 are measured to be $r_1(T)$ and $r_2(T)$, respectively, and it is desired to determine the location $r_3(T)$ of m_3. The position of the center of mass at $t = T$ is

$$r_c(T) = r_c(0) + \bar{v}T = \bar{v}T. \tag{3.134}$$

From (3.131) it follows that

$$r_3(T) = (m\bar{v}T - m_1 r_1(T) - m_2 r_2(T)) / m_3. \tag{3.135}$$

(b) Collecting the relevant equations governing linear and angular momentum of the three-particle system, for all $t \geq 0$ we have

$$m_1 r_1 + m_2 r_2 + m_3 r_3 = m \bar{v} t, \tag{3.136}$$

$$m_1 v_1 + m_2 v_2 + m_3 v_3 = m \bar{v}, \tag{3.137}$$

$$m_1 (r_1 \times v_1) + m_2 (r_2 \times v_2) + m_3 (r_3 \times v_3) = 0. \tag{3.138}$$

Note that (3.138) is (3.133) written in slightly modified form. Equations (3.136)–(3.138) represent a system of nine equations in the six vectors r_i, v_i, $i = 1, 2, 3$ (a total of 18 variables). $\qquad\square$

3.6   Summary

- An **inertial frame** is one that is not undergoing any acceleration. Newton's laws of motion are only valid with respect to an inertial frame.
- **Newton's second law** for a particle of mass m takes the form $f = ma$, where f is the net force acting on the particle, and a is the acceleration vector of the particle as measured by an observer at rest in the inertial frame.
- In the case of linear motion of the particle, Newton's second law can be written $f = m\ddot{x}$, where $x \in \mathbb{R}$ is the linear displacement of the particle, $\dot{x}$ its velocity, and $\ddot{x}$ its acceleration. Given a **time-varying force** $f(t)$, the equations of motion become $\ddot{x} = f(t)/m = a(t)$, which can be integrated as follows:

$$\dot{x}(t) = v(t) = v_0 + \int_0^t a(s)\, ds, \tag{3.139}$$

$$x(t) = x_0 + v_0 t + \int_0^t \int_0^{s_1} a(s_2)\, ds_2\, ds_1. \tag{3.140}$$

If the force is **position-dependent** of the form $f = f(x)$, the equations of motion become $\ddot{x} = f(x)/m = a(x)$, which upon integration becomes

$$v(x) = \left(2 \int_{x_0}^x a(s)\, ds + v_0^2 \right)^{\frac{1}{2}}. \tag{3.141}$$

Rewriting $dx = v(x)\, dt$ as $\frac{dx}{v(x)} = dt$ and integrating further leads to

$$t = t_0 + \int_{x_0}^x \left(2 \int_{x_0}^{s_1} a(s_2)\, ds_2 + v_0^2 \right)^{-\frac{1}{2}} ds_1. \tag{3.142}$$

Finally, if the force is **velocity-dependent** of the form $f = f(\dot{x})$, the equations of motion become $\ddot{x} = a(\dot{x})$, or $\dot{v} = a(v)$. Integrating,

$$\int_{v_0}^v \frac{ds}{a(s)} = t - t_0. \tag{3.143}$$

- For a particle of mass m moving along a path C in physical space while under the action of a force field f, its **work** $\mathcal{W}$ is defined to be the following path integral:

$$\mathcal{W} = \int_C f \cdot dr. \tag{3.144}$$

Parameterizing the path C by $r(t)$, $t_1 \leq t \leq t_2$, as

$$r(t) = x(t)\hat{x}_s + y(t)\hat{y}_s + z(t)\hat{z}_s, \tag{3.145}$$

and the force as $f(t) = f_x(t)\hat{x}_s + f_y(t)\hat{y}_s + f_z(t)\hat{z}_s$, the work integral becomes

$$\mathcal{W} = \int_{t_1}^{t_2} f_x\dot{x} + f_y\dot{y} + f_z\dot{z}\, dt. \tag{3.146}$$

Any force that acts normally to the particle's direction of motion contributes no work. Work is invariant with respect to different parameterizations of the path; i.e., work depends only on the shape of the path, and not how the particle moves along the path with respect to time.
- The **kinetic energy** of a particle of mass m moving with speed v is given by $\mathcal{K} = \frac{1}{2}mv^2$.
- Given a particle whose velocity at time t_i is v_i, $i = 1, 2$, the work done by the particle is related to its kinetic energy via the scalar equation

$$\mathcal{W} = \mathcal{K}_2 - \mathcal{K}_1, \tag{3.147}$$

where $\mathcal{K}_i = \frac{1}{2}mv_i^2$.
- Given a position-dependent force field $f = f_x(x, y, z)\hat{x}_s + f_y(x, y, z)\hat{y}_s + f_z(x, y, z)\hat{z}_s$ in $\mathbb{R}^3$, if there exists some function $\mathcal{P}\colon \mathbb{R}^3 \to \mathbb{R}$, $\mathcal{P}(x, y, z)$ differentiable with respect to x, y, z, such that

$$f_x(x, y, z) = -\frac{\partial \mathcal{P}}{\partial x}(x, y, z),$$
$$f_y(x, y, z) = -\frac{\partial \mathcal{P}}{\partial y}(x, y, z), \tag{3.148}$$
$$f_z(x, y, z) = -\frac{\partial \mathcal{P}}{\partial z}(x, y, z),$$

then f is said to be a **conservative force field**, and $\mathcal{P}(x, y, z)$ its associated **potential function**. The notation

$$\nabla\mathcal{P} = \left(\frac{\partial \mathcal{P}}{\partial x}, \frac{\partial \mathcal{P}}{\partial y}, \frac{\partial \mathcal{P}}{\partial z}\right)^\top \in \mathbb{R}^3, \tag{3.149}$$

with $\nabla\mathcal{P}$ referred to as the **gradient** of $\mathcal{P}$, is also used. Denoting $f = (f_x, f_y, f_z)^\top \in \mathbb{R}^3$, then $f = -\nabla\mathcal{P}$ for a conservative force field f. Examples of conservative force fields include gravity, i.e., $\mathcal{P}(x, y, z) = mgz$ and $-\nabla\mathcal{P} = (0, 0, -mg)^\top$, and the linear spring, i.e., $\mathcal{P}(x, y, z) = \frac{1}{2}kx^2$ for spring stiffness k, in which case $-\nabla\mathcal{P} = (-kx, 0, 0)^\top$ is the spring force.

- If a particle moves under the action of a conservative force field, its work does not depend on the path, but only on the initial and final positions. Conversely, if $\int_C f \cdot dr$ is path-independent, i.e., the same for all paths C with the same endpoints, then f is a conservative force. As a consequence, for any conservative force field f, if C is a closed curve – i.e., the path begins and ends at the same point – then the work $\oint_C f \cdot dr$ is zero (the $\oint$ notation is used to denote a closed path integral).

- The **work-energy principle** for conservative forces can also be stated entirely in terms of potential and kinetic energies as

$$\mathcal{P}_1 + \mathcal{K}_1 = \mathcal{P}_2 + \mathcal{K}_2, \tag{3.150}$$

where $\mathcal{P}_i$ and $\mathcal{K}_i$, respectively, denote the potential and kinetic energies at state $i = 1, 2$.

- Given a force field $f = (f_x, f_y, f_z)^\top$ defined over some three-dimensional domain $\mathcal{D}$, **Stokes Theorem** can be used to determine whether or not f is conservative: Given a bounded two-dimensional surface $\mathcal{S}$ with a closed curve C as its boundary, and a vector field f defined over a bounded region of $\mathbb{R}^3$ containing $\mathcal{S}$,

$$\oint_C f \cdot dr = \int\int_{\mathcal{S}} \langle \nabla \times f, \hat{n} \rangle \, dA, \tag{3.151}$$

where the right-hand side is a surface integral over $\mathcal{S}$: $\langle \cdot, \cdot \rangle$ denotes the inner product, $\nabla \times f$ is the curl of the vector field f, and $\hat{n}$ denotes the unit normal to the surface. The curl of f can be calculated as

$$\nabla \times f = \left(\frac{\partial f_z}{\partial y} - \frac{\partial f_y}{\partial z}, \frac{\partial f_x}{\partial z} - \frac{\partial f_z}{\partial x}, \frac{\partial f_y}{\partial x} - \frac{\partial f_x}{\partial y} \right). \tag{3.152}$$

A necessary (but not sufficient) condition for f to be a conservative force vector field is that its curl $\nabla \times f = 0$. Conversely, if $\nabla \times f = 0$ everywhere over some simply connected domain[3] $\mathcal{D}$ of $\mathbb{R}^3$, then f is a conservative force vector field.

- The **linear momentum** of a particle of mass m moving with velocity vector v is mv. If the particle moves under the action of a force f, then the force and momentum are related via

$$\mathcal{I} = \int_{t_1}^{t_2} f(t) \, dt = mv_2 - mv_1, \tag{3.153}$$

where $\mathcal{I}$ is the **linear impulse** of the force f, and $v_i = v(t_i)$, $i = 1, 2$.

- The **angular momentum** of a particle of mass m with respect to some fixed point O is

$$h_o = r \times mv, \tag{3.154}$$

---

[3] $\mathcal{D}$ is a simply connected domain if any closed curve in $\mathcal{D}$ can be shrunk to a point.

where r is the vector from point O to the particle and v is the velocity of the particle. If the particle is subject to a force f, the moment generated by f with respect to point O is $m_o = r \times f$. Then

$$m_o = \dot{h}_o. \tag{3.155}$$

If the applied force f is zero, the particle moves in such a way that its angular momentum is conserved, i.e., $\dot{h}_o = 0$.

- Suppose two particles A and B of mass m_A and m_B collide. Let v_A, v_B be the velocities immediately before impact, and v'_A, v'_B be the velocities immediately after impact. Then from the conservation of momentum,

$$m_A v_A + m_B v_B = m_A v'_A + m_B v'_B. \tag{3.156}$$

- Suppose two particles A and B experience a direct (head-on) collision along the same line: let v_A, v_B be the (scalar) velocities immediately before impact, and v'_A, v'_B be the (scalar) velocities immediately after impact. The **coefficient of restitution** e between A and B is defined as

$$e = \frac{v'_B - v'_A}{v_A - v_B}. \tag{3.157}$$

The coefficient of restitution depends on factors such as the shape, sizes, and material properties of A and B.

- Given N particles of mass m_i, $i = 1, \ldots, N$, let r_i be the position of particle i with respect to the inertial frame origin, f_{ji} be the force exerted by particle j on particle i (with the underlying assumption that $f_{ji} + f_{ij} = 0$ for all i, j), and f_i be the net external force on particle i (excluding all internal forces f_{ji}). Then for each particle i, the equations of motion are of the form

$$f_i + \sum_{j=1}^{N} f_{ji} = \ddot{r}_i, \quad i = 1, \ldots, N. \tag{3.158}$$

Summing the N equations, we get

$$f = m\ddot{r}_c, \tag{3.159}$$

where $m = m_1 + \cdots + m_N$ is the total mass, and $r_c = (m_1 r_1 + \cdots + m_N r_N)/m$ is the center of mass. The above equation holds for any system of particles in which $f_{ij} = -f_{ji}$.

- The total kinetic energy for the previous system of N particles is given by

$$\mathcal{K} = \frac{1}{2} m v_c^2 + \frac{1}{2} \sum_{i=1}^{N} m_i v_i'^2, \tag{3.160}$$

where $v_c^2 = \langle \dot{r}_c, \dot{r}_c \rangle$, and $v_i'^2 = \langle \dot{r}_i', \dot{r}_i' \rangle$ with $r_i = r_c + r_i'$. The work-energy relation for the system of particles can then be written in the same way as

$$\mathcal{K} = \mathcal{K}_{\text{final}} - \mathcal{K}_{\text{init}}, \tag{3.161}$$

where the kinetic energy term involves not only the velocity v_c of the mass center, but also the relative velocities v_i' of all the particles. If all the external forces applied to the system of particles are conservative, then from the conservation of energy,

$$\mathcal{K}_{\text{init}} + \mathcal{P}_{\text{init}} = \mathcal{K}_{\text{final}} + \mathcal{P}_{\text{final}}, \tag{3.162}$$

where $\mathcal{P}_{\text{init}}$ and $\mathcal{P}_{\text{final}}$ are the total potential energies at the initial and final configurations, respectively.

- For the previous system of N particles, the linear momentum of the system center of mass can be written

$$l_c = m\dot{r}_c. \tag{3.163}$$

Taking derivatives of the above leads to

$$\dot{l}_c = m\ddot{r}_c = f, \tag{3.164}$$

where f is the net force applied to the system of particles.

- The angular momentum for the previous system of N particles about some fixed point O is given by

$$h_o = \sum_{i=1}^{N} r_i \times m_i \dot{r}_i. \tag{3.165}$$

Taking derivatives of both sides leads to

$$\dot{h}_o = \sum_{i=1}^{N} r_i \times m_i \ddot{r}_i = \sum_{i=1}^{N} r_i \times f_i = m_o. \tag{3.166}$$

- For the previous system of N particles, define the angular momentum about the mass center as

$$h_c = \sum_{i=1}^{N} r_i' \times m_i \dot{r}_i. \tag{3.167}$$

Taking derivatives of both sides leads to

$$\dot{h}_c = \sum_{i=1}^{N} r_i' \times f_i = m_c. \tag{3.168}$$

This equation, together with the linear momentum equation $f = \dot{l}_c$, describes the motion of the center of mass of the system of N particles.

- The principle of **conservation of linear (angular) momentum** for a system of particles states that when the force $f = 0$ (moment $m_c = 0$), then the linear momentum l_c (angular momentum h_c) is constant.

3.7   Exercises

Exercise 3.1   As shown in Figure 3.14, block A of mass m_A and block B of mass m_B are connected to each other by a cable and pulley fixed to an inclined wedge. The mass and the angle of the inclined wedge are m_W and θ, respectively. The coefficient of friction between blocks A and B is μ_1, and the coefficient of friction between the wedge and block B is μ_2. Gravity g acts downward and the wedge is fixed to the ground.
(a) Draw free-body diagrams and derive the equations of motion for blocks A and B.
(b) Assume $m_A = 10$ kg, $m_B = 20$ kg, $g = 10$ m/s^2, $\mu_1 = 0.1$, $\mu_2 = 0.2$, and $\theta = 45°$. Find the cable tension and the acceleration of the blocks.
(c) Now assume that the wedge can move freely on the frictionless ground. Draw free-body diagrams and derive the equations of motion for the blocks and the wedge. Is it possible to derive the acceleration of the wedge and the cable tension? Explain why, based on the number of equations and number of unknowns.

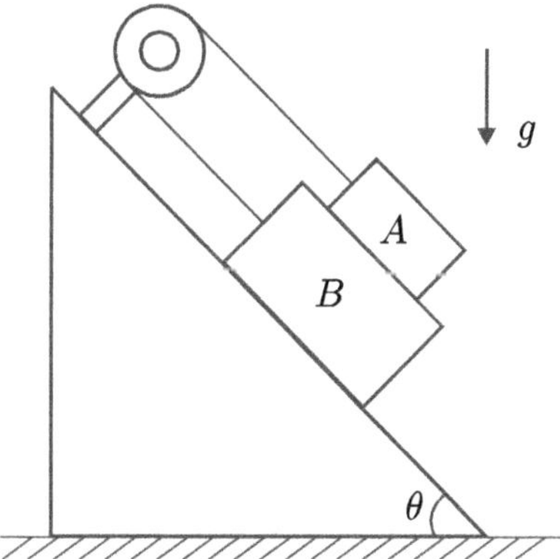

Figure 3.14 Two blocks connected by a cable and pulley of Exercise 3.1.

Exercise 3.2   A ball of mass m is launched from the origin of the fixed frame $\{\hat{x}, \hat{y}\}$ with initial velocity $v_0 = v_{x,0}\,\hat{x} + v_{y,0}\,\hat{y}$, where $v_{x,0}$ and $v_{y,0}$ are positive constants. Due to air resistance, a drag force of $f_d = -\gamma v$ acts on the ball, where v is the velocity of the ball and γ is a positive constant. Gravity g acts in the $-\hat{y}$-direction.
(a) Find the position and velocity of the ball as functions of time t ($t = 0$ when the ball is launched from the origin).
(b) Explain the position and velocity of the ball as t goes to infinity.

Exercise 3.3   A particle of mass $m = 1$ kg moves in the $\{x, y\}$ plane, where $\{\hat{x}, \hat{y}\}$ is the fixed frame. The position of the particle is given as $p = x\,\hat{x} + y\,\hat{y}$. Gravity g acts in the $-\hat{y}$-direction. Suppose an external force of

$$f = -av - by\,\hat{y}$$

is applied to the particle, where a and b are positive constants and v is the velocity of the particle.

(a) Derive the equations of motion for the particle.

(b) Assume that the initial position of the particle is $x(0) = y(0) = 0$ and the initial velocity of the particle is $\dot{x}(0) = \dot{y}(0) = 1$. Find the position p of the particle as a function of time t.

(c) Explain the position of the particle as t goes to infinity.

Exercise 3.4   Figure 3.15 shows a satellite of mass m orbiting the earth of mass M with radius r. The gravitational force acting on the satellite is given by $f = -\frac{GMm}{r^2}\hat{r}$, where G is the gravitational constant.

(a) Derive the equations of motion for the satellite in polar coordinates.

(b) Show that $r = r_0$ and $\theta = w_0 t$, where r_0 and w_0 are constants, can be a solution.

(c) For initial radius r_0 and initial angular velocity w_0, derive the differential equation for $r = r(\theta)$ rather than for the function of t, i.e., $r = r(t)$. Replace r with $u = \frac{1}{r}$ and derive the differential equation for $u(\theta)$. (*Hint*: Use $\frac{d}{dt}(r^2\dot{\theta}) = 2r\dot{r}\dot{\theta} + r^2\ddot{\theta}$.)

(d) Show that $r(t) = \left(\left(\frac{1}{r_0} - \frac{GM}{r_0^4 w_0^2}\right)\cos\theta(t) + \frac{GM}{r_0^4 w_0^2}\right)^{-1}$ with $\theta(t)$ satisfying $\dot{\theta} = r_0^2 w_0 \left(\left(\frac{1}{r_0} - \frac{GM}{r_0^4 w_0^2}\right)\cos\theta + \frac{GM}{r_0^4 w_0^2}\right)^2$ and $\theta(0) = 0$ can be a solution $\left(r_0 < \left(\frac{2GM}{w_0^2}\right)^{\frac{1}{3}}\right)$.

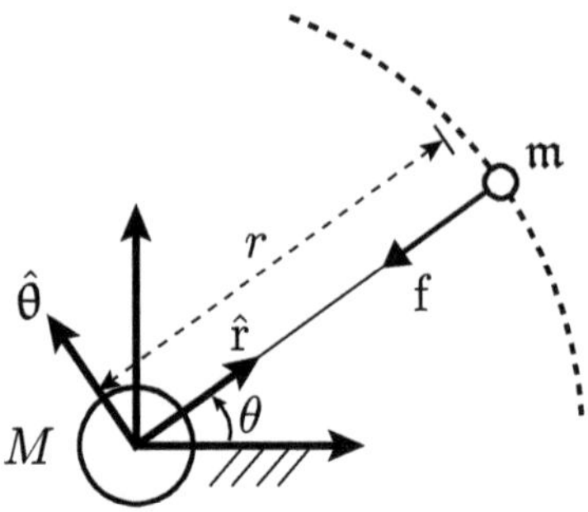

Figure 3.15 A satellite orbiting the earth of Exercise 3.4.

Exercise 3.5   A particle of mass m is connected to the end of a long massless cord that is wrapped around a cylinder of radius R as shown in Figure 3.16. Assume that the cord does not slip, and there is no gravity. The cylinder rotates counterclockwise at a constant rate w rad/s, with the particle in contact with the cylinder at $t = 0$. Let $l(t)$ be the length of the part of the cord that is not in contact with the cylinder, with $l(0) = 0$ and $\dot{l}(0) = Rw$.

(a) Find $l(t)$ as a function of time t. (*Hint*: Parameterize the position of the particle and derive the equations of motion. Also, $\frac{d}{dt}(x\dot{x}) = x\ddot{x} + \dot{x}^2$ may be helpful.)

(b) As shown in the figure, point U is the point on the cylinder where the part of the cord that is not in contact with the cylinder begins. What is the rate of rotation of point U?

(c) Find the cord tension $T(t)$ as a function of time.

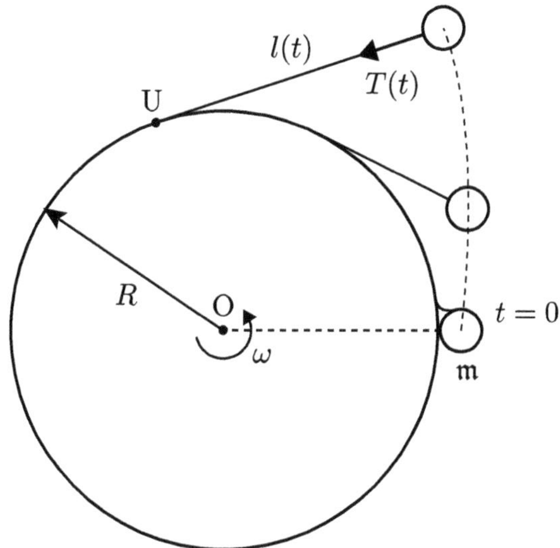

Figure 3.16 Particle attached to a rotating cylinder by a cord of Exercise 3.5.

Exercise 3.6   Figure 3.17 shows a model of a jumping insect. Block A of mass m and block B of mass M are connected by a linear spring of stiffness k and undeformed length L. The spring is initially completely compressed and then suddenly released. The spring length after the release is x. Gravity g acts downward.

(a) Before block B jumps off the floor (loses contact, i.e., the normal force acting on the block vanishes), draw free-body diagrams and derive the equations of motion for block A and block B.

(b) Assume that the spring stiffness k is large enough to make block B jump off the floor. What is the length x_0 of the spring right before block B loses contact with the floor?

(c) Now assume $M = 10$ kg, $k = 200$ N/m, $L = 1$ m, and $g = 10$ m/s^2. Determine the range of m that block B is able to jump off.

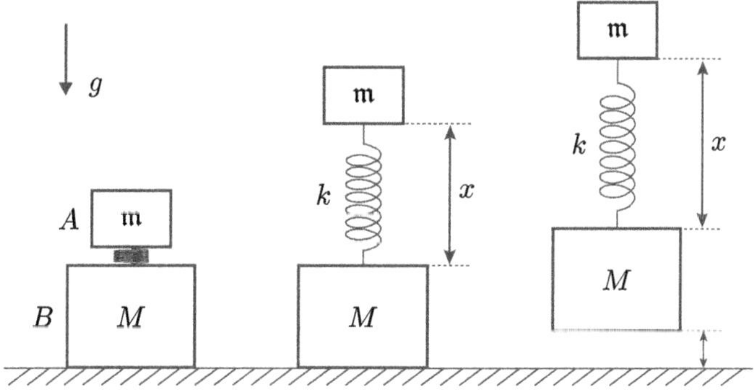

Figure 3.17 Jumping insect robot model of Exercise 3.6.

Exercise 3.7   As shown in Figure 3.18, a block of mass m sits on top of a half-sphere of radius R. The block begins to slide down the sphere with a negligibly small initial speed (assume that the surface of the sphere is frictionless). Angle θ is defined as shown in the figure. Gravity g acts downward.

 (a) Draw a free-body diagram for the block, and derive the equations of motion as a second-order differential equation in θ.

(b) At what angle θ does the block leave the surface of the sphere? Also, find the speed of the box when it leaves the surface.

(c) After the box leaves the surface, express the position of the box in time t ($t = 0$ when contact is lost). Express your answer in the fixed frame $\{\hat{x}, \hat{y}\}$ coordinates. Assume $R = 10$ m and $g = 10$ m/s^2.

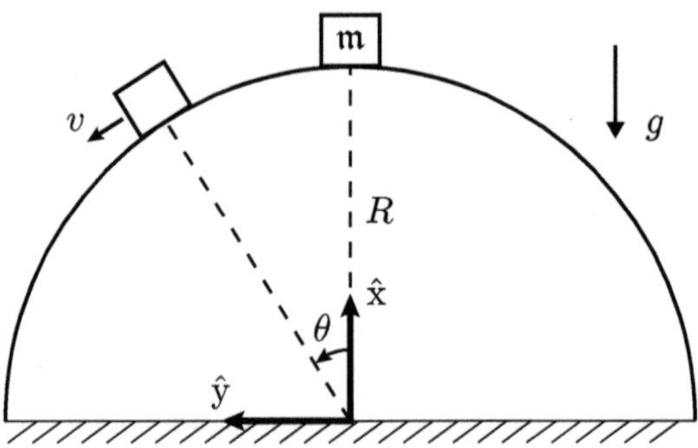

Figure 3.18 Block sliding down on a half-sphere of Exercise 3.7.

Exercise 3.8   As shown in Figure 3.19, a ball of mass m rolls down a curve in the $\{x, y\}$ plane, where the shape of the curve is given as $y = x^2$. The ball is initially at rest at (x_0, x_0^2) and gravity g acts in the $-y$-direction.

(a) When $x = x_1$ ($|x_1| < |x_0|$), draw a free-body diagram for the ball and derive the equations of motion. Also, what is the velocity of the ball?

(b) Assume $x_0 = \sqrt{20}$. Find the magnitude of the normal force acting on the ball when $x = \sqrt{2}$. (*Hint*: For a curve given as $y = f(x)$, the radius of curvature of the curve is given as

$$\rho = \left| \frac{(1 + (\frac{dy}{dx})^2)^{\frac{3}{2}}}{\frac{d^2 y}{dx^2}} \right|.$$

Using tangential-normal coordinates may be helpful.)

Exercise 3.9   As shown in Figure 3.20, a particle of mass m slides inside a frictionless cone of angle β. The fixed frame $\{\hat{x}_s, \hat{y}_s, \hat{z}_s\}$ is attached to the bottom of the cone as shown in the figure, and gravity g acts in the $-\hat{z}_s$-direction. The position p of the particle is expressed in coordinates (r, θ) as

$$p = r \cos\theta \, \hat{x}_s + r \sin\theta \, \hat{y}_s + r \cot\beta \, \hat{z}_s.$$

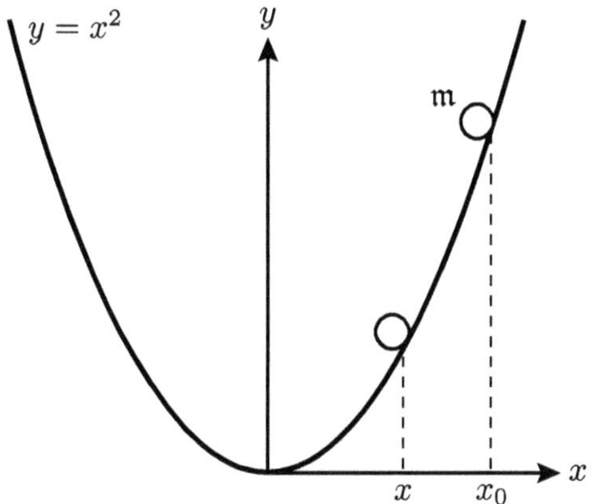

Figure 3.19 Ball rolling down a curve of Exercise 3.8.

(a) Define a moving frame $\{\hat{z}, \hat{r}, \hat{\theta}\}$ by

$$\hat{z} = \hat{z}_s,$$
$$\hat{r} = \cos\theta\, \hat{x}_s + \sin\theta\, \hat{y}_s,$$
$$\hat{\theta} = \hat{z} \times \hat{r}.$$

Express the position, velocity, and acceleration of the particle using coordinates (r, θ) in moving frame coordinates.

(b) Derive the equations of motion for the particle in the form of

$$\ddot{r} = f(r, \theta, \dot{r}, \dot{\theta}),$$
$$\ddot{\theta} = g(r, \theta, \dot{r}, \dot{\theta}),$$

where f and g are functions in terms of r, θ, $\dot{r}$, and $\dot{\theta}$.

(c) Suppose the particle is moving with velocity $\mathbf{v} = v_H\,\hat{\theta}$ at height H from the ground. In order for the height H of the particle not to change, at what speed v_H should the particle move? Also, express the force exerted on the cone by the particle in moving frame coordinates.

Exercise 3.10 Now the particle of Exercise 3.9 starts at the top edge of the cone with initial speed v_1 in the direction tangential to the cone edge as shown in Figure 3.21. Because of gravity, the particle spirals down to the bottom of the cone, where the height change from the top of the cone is h. The initial radius of the cone is r_1.

(a) Suppose the speed of the particle is increased from v_1 to v_2. Find the height change h.

(b) The angle of descent φ is defined as shown in the figure. Find the angle of descent φ when the height of the particle is decreased by h.

Exercise 3.11 Block A of mass m_A is connected to block B of mass m_B by a cable and pulley as shown in Figure 3.22. Assume that $m_A = m_B = m$, and the cable and

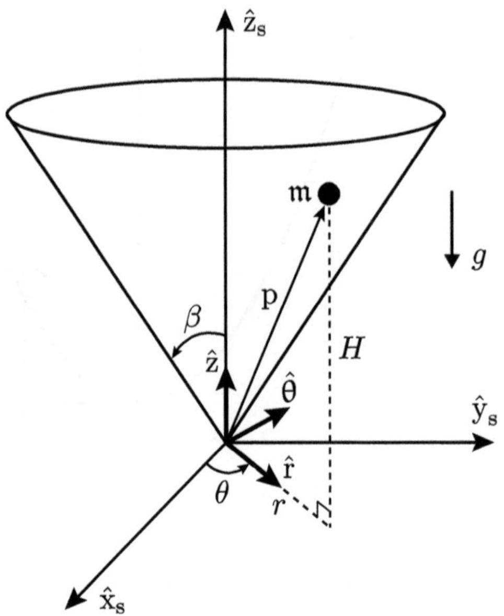

Figure 3.20 Particle sliding inside a cone of Exercise 3.9.

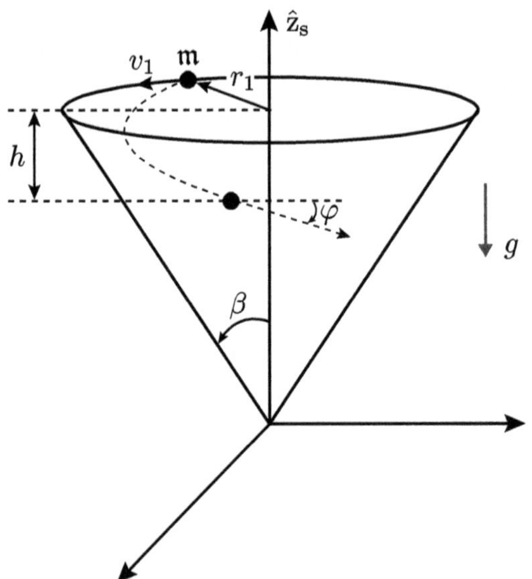

Figure 3.21 Particle spiraling down of Exercise 3.10.

pulley are both massless. The surface is frictionless, and gravity g acts downward. Cable length r from the pulley to block A, cable angle θ, and displacements x_A and x_B of the blocks from their initial positions are defined as shown in the figure. Both blocks are initially at rest when $\theta = 30°$.

(a) Draw free-body diagrams and derive the equations of motion for blocks A and B.
(b) Show that the following equalities hold: (i) $r\dot{\theta} = v_A \sin\theta$, (ii) $v_A \cos\theta = v_B$, and
(iii) $a_A \cos\theta = a_B + r\dot{\theta}^2$, where $v_A = \dot{x}_A$, $v_B = \dot{x}_B$, $a_A = \ddot{x}_A$, and $a_B = \ddot{x}_B$. (*Hint:* Use polar coordinates.)
(c) Find the cable tension T and the accelerations a_A and a_B of the blocks at the initial state.
(d) At what angle θ does block A lose contact with the surface? Derive an equation of the form $f(\sin\theta) = 0$, where $f(\sin\theta)$ is a function in terms of $\sin\theta$.

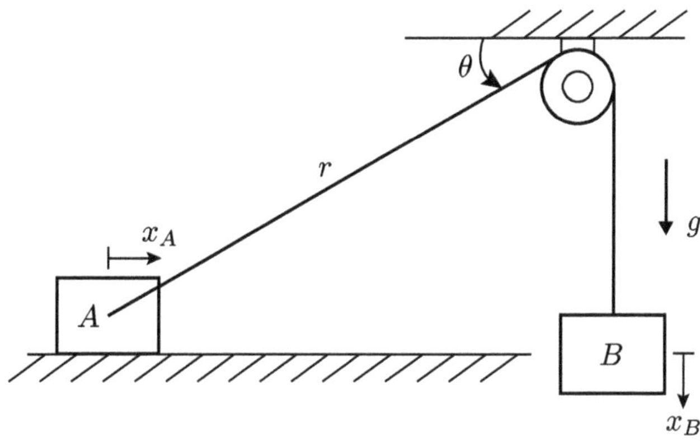

Figure 3.22 Two blocks connected by a cable and pulley of Exercise 3.11.

Exercise 3.12   As shown in Figure 3.23, a particle of mass m is connected to a rotary joint with a massless rod of length $2L$. Angle θ of the rotary joint is defined as shown in the figure. One end of a massless linear spring is attached $2L$ above the rotary joint, and the other end of the spring is attached to the center of the rod. When $\theta = 60°$, the system is in equilibrium, and the spring and the rod are perpendicular to each other. Gravity g acts downward, and the spring stiffness is $k = \frac{2mg}{L}$. The particle is launched from the equilibrium position of the system with initial speed v in the direction that θ increases.
(a) Find the undeformed length of the spring.
(b) Determine the minimum initial speed v of the particle that allows the particle to reach the position of $\theta = 180°$.

Exercise 3.13   Determine if the following force vector fields are conservative and explain why. If possible, find the potential function $\mathcal{P}$ such that $f = -\nabla\mathcal{P}$.
(a) Force vector field f given by

$$f(x, y) = (2xe^{xy} + x^2 y e^{xy})\, \hat{x} + (x^3 e^{xy} + 2y)\, \hat{y}.$$

(b) Force vector field f given by

$$f(x, y) = -x\frac{\sqrt{x^2 + y^2} - 1}{\sqrt{x^2 + y^2}}\, \hat{x} - y\frac{\sqrt{x^2 + y^2} - 1}{\sqrt{x^2 + y^2}}\, \hat{y}.$$

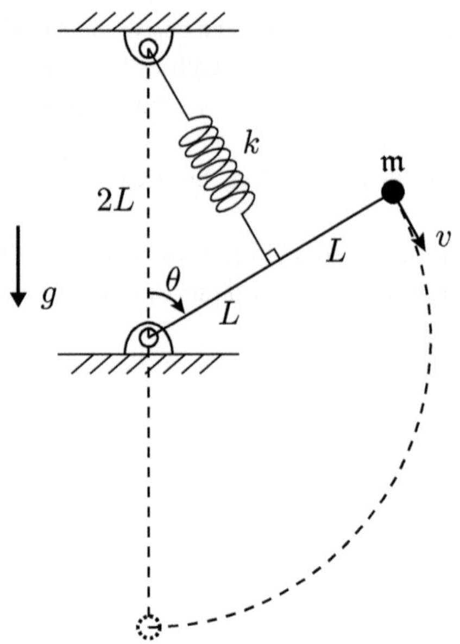

Figure 3.23 System in equilibrium of Exercise 3.12.

(c) Force vector field f given by

$$\mathbf{f}(x, y, z) = \frac{2y}{x^2 + y^2}\,\hat{\mathbf{x}} - \frac{2x}{x^2 + y^2}\,\hat{\mathbf{y}} + 1\,\hat{\mathbf{z}}.$$

Exercise 3.14   Figure 3.24 shows a snowboard halfpipe constructed from a frictionless circular tube of radius R and mass M, attached to the wall at point A. A snowboarder of mass m starts at point A with initial speed v_0 and slides down the halfpipe. The fixed frame $\{\hat{\mathbf{x}}, \hat{\mathbf{y}}\}$ is attached to point A, and gravity g acts downward.
(a) Express the force exerted on the halfpipe by the snowboarder as a function of θ, where θ is defined as shown in the figure.
(b) Find the moment at point A exerted by the snowboarder as a function of θ. What is the maximum moment that point A must be able to support?
(c) A strong wind suddenly blows through the halfpipe and exerts a force of

$$\mathbf{f} = \frac{\mathsf{m}g}{2R^2}\left(2xy\,\hat{\mathbf{x}} + (x^2 + 3y^2)\,\hat{\mathbf{y}}\right)$$

to the snowboarder at $x\,\hat{\mathbf{x}} + y\,\hat{\mathbf{y}}$. What is the minimum initial speed v_0 in order for the snowboarder to be able to reach the other side of the halfpipe? (*Hint*: Use $\sin 2\theta = 2\sin\theta\cos\theta$.)

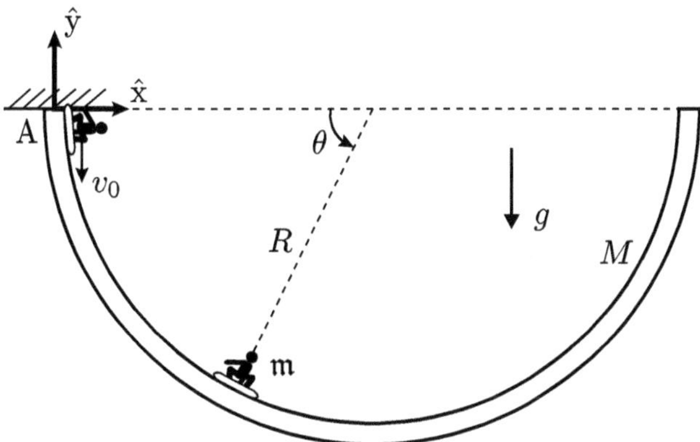

Figure 3.24 Snowboarder on a circular halfpipe of Exercise 3.14.

Exercise 3.15 A spring-powered toy cannon is shown in Figure 3.25(a). The toy cannon can be modeled as a mass-spring system on an inclined plane as shown in Figure 3.25(b). A block of mass $m = 2$ kg is loaded at the system's equilibrium position, at which the distance from the block to the end of the cannon is $L = 1$ m. Assume there is no friction between the block and the inclined plane. The angle of the cannon from the ground is $\theta = 30°$, the linear spring is massless with spring stiffness $k = 50$ N/m, and gravity acts downward with acceleration $g = 10$ m/s^2.
(a) At the system's equilibrium position as shown in Figure 3.25(b), determine how much the spring is compressed from its undeformed length.
(b) The block is released after the spring is compressed until the distance from the block to the end of the cannon is $2L$ as shown in Figure 3.25(c). Assuming that the initial speed of the block is zero, find the speed of the block when it reaches the end of the cannon.
(c) Now suppose there is friction between the block and the inclined plane with friction coefficient $\mu = \frac{7\sqrt{3}}{60}$. Find the speed of the block when it reaches the end of the cannon.
(d) Let the coefficient of friction μ be arbitrary. What is the range of μ such that the block is able to reach the end of the cannon?

Exercise 3.16 As shown in Figure 3.26, a block of mass $m = 1$ kg attached to a linear spring is launched from point A with initial speed $v_0 = 30$ m/s. The linear spring is massless with spring stiffness $k = 20$ N/m, and the undeformed length of the spring is $l_0 = 5$ m (the spring is initially in its undeformed state). Gravity acts downward with acceleration $g = 10$ m/s^2.
(a) Assume the surface is frictionless. Find the speed of the block when it passes through points B and C, respectively.
(b) Now assume there is friction between the block and the surface with friction coefficient $\mu = 1$. Find the speed of the block when it passes through points B and C, respectively. (*Hint*: The following formula may be useful:

$$\int \sec\theta \, d\theta = \ln(\sec\theta + \tan\theta).$$

You may also use the approximation $\ln 3 \approx 1.1$.)

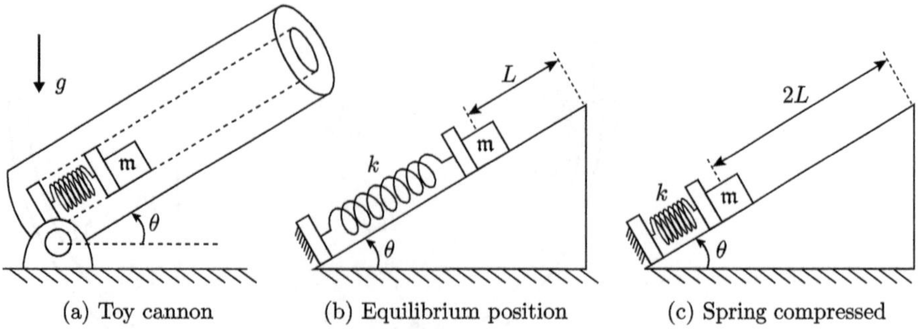

Figure 3.25 Toy cannon modeled as a mass-spring system of Exercise 3.15.

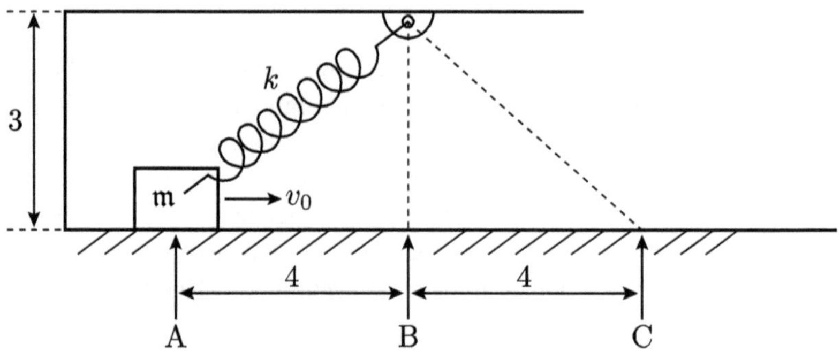

Figure 3.26 Mass-spring system of Exercise 3.16.

Exercise 3.17 As shown in Figure 3.27, a slider of mass $m = 1$ kg slides along a frictionless quarter-circle guide OAB of radius $R = 2$ m. One end of a massless linear spring is attached to the slider and the other end is attached to point C, where the distance from point O to point C is R. The spring stiffness is $k = 50$ N/m and the undeformed length of the spring is $\sqrt{2}R$. The slider is launched with initial velocity $v_0 = v_0\,\hat{x}$, where $v_0 = 10$ m/s. The position of the slider is parameterized by angle θ as shown in the figure.

(a) Find the speed of the slider as a function of θ. Note that with the given constants, the slider does not stop until it gets to the ground.

(b) Find θ at which the speed of the slider is the smallest. Derive an equation of the form $f(\theta) = 0$, where $f(\theta)$ is a function of θ.

(c) Now assume that the guide exerts an electromagnetic resistance force on the slider in the opposite direction to the velocity of the slider. The magnitude of the resistance force is $f_r(\theta) = \gamma \cos\theta$, where $\gamma = 10$ N. Find the speed $v(\theta)$ of the slider as a function of θ. Again, the slider does not stop until it gets to the ground.

(d) Express the normal force exerted on the slider by the guide in terms of θ and $v(\theta)$. Is there a moment when the normal force becomes zero during the slide? Explain why. You may use the approximation $\sqrt{2} \approx 1.4$.

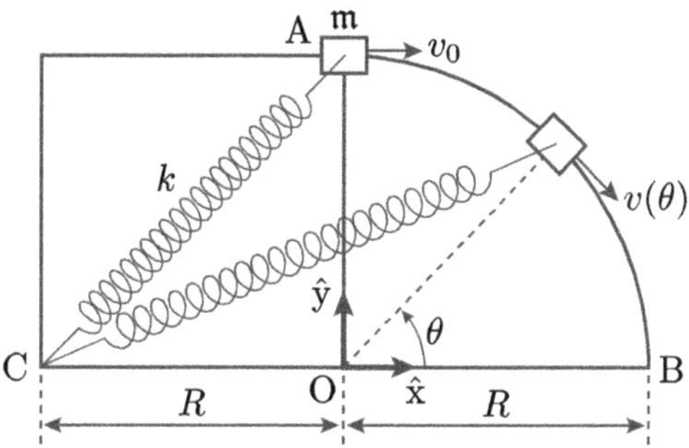

Figure 3.27 Slider sliding along a quarter-circle guide of Exercise 3.17.

Exercise 3.18   As shown in Figure 3.28, a ball of mass $m = 1$ kg is connected to the end of a massless cord that passes through a vertical tube. Gravity acts downward with acceleration $g = 10$ m/s^2. At the initial state shown in the figure, the ball rotates at a constant speed along a circular path of radius $r_1 = 1$ m, and the length of the cord from the ball to the entrance of the tube is $l_1 = 2$ m. By pulling the cord through the tube, the cord length is suddenly shortened from l_1 to $l_2 = 1$ m. The state after the cord has been shortened is considered the final state.
(a) Find the cord tension T_1 at the initial state.
(b) Determine the radius r_2 of the circular path at the final state. Derive an equation of the form $f(r_2) = 0$, where $f(r_2)$ is a function of r_2.
(c) Express the cord tension T_2 at the final state in terms of r_2.

Exercise 3.19   When walking, a human leg can be modeled as an inverted spring pendulum as shown in Figure 3.29. The contact between the foot and the ground can be modeled as a rotary joint with angle θ as defined in the figure, connected to a mass m by a linear spring of stiffness k and undeformed length l_0. Let l be the deformed length of the spring. The mass follows a trajectory with initial speed v_0 and angle ϕ_0 as shown in the figure, and the initial contact angle between the foot and the ground is θ_0. A polar coordinate $\{\hat{r}, \hat{\theta}\}$ is attached to the rotary joint as shown in the figure.
(a) Assuming no gravity, find a general expression for $\dot{\theta}$ as a function of l.
(b) Again assuming no gravity, derive the equation for obtaining the minimum length l_m of the spring when it is maximally compressed. You do not need to solve the equation for l_m.

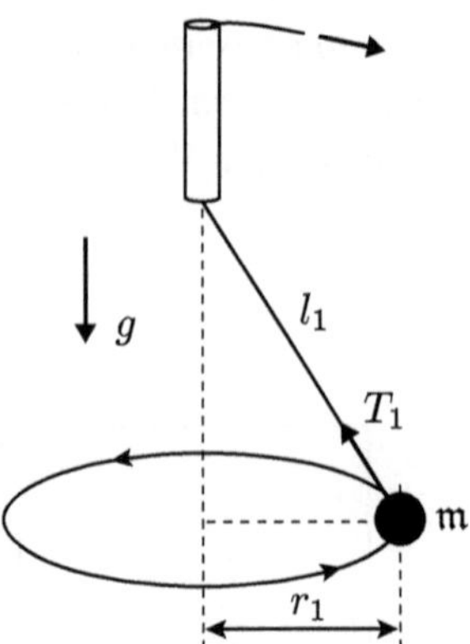

Figure 3.28 Ball rotating along a circular path of Exercise 3.18.

(c) Now gravity acts downward with acceleration $g = 10$ m/s^2. Set $k = 100$ N/m, $m = 1$ kg, $l_0 = 1$ m, $v_0 = 4$ m/s, $\theta_0 = 30°$, and $\phi_0 = 45°$. Derive the equations of motion for the mass in terms of l and θ.

(d) Under the same conditions as (c), when the spring is maximally compressed, the speed of the mass is 2 m/s, and the $\hat{\theta}$-direction acceleration of the mass is -5 m/s^2. When the spring is maximally compressed, find the spring length l_m, contact angle θ_m, and the angle's time derivatives $\dot{\theta}_m$ and $\ddot{\theta}_m$.

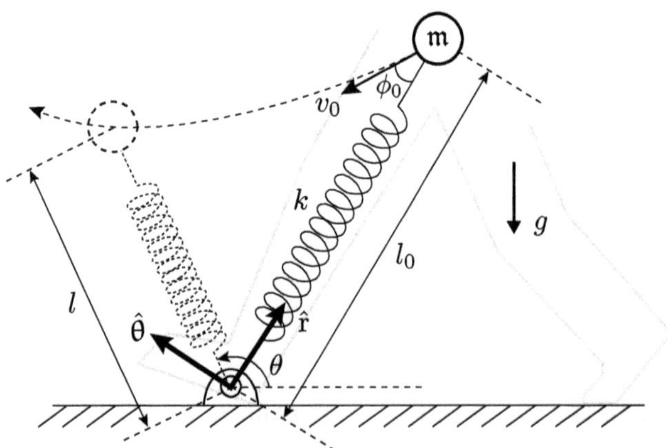

Figure 3.29 Inverted spring pendulum model of Exercise 3.19.

Exercise 3.20   As shown in Figure 3.30, the moon of mass M is orbiting the earth of mass M_E at constant speed v_m with radius R. Assuming $M_E \gg M$ so that the earth can be considered as a fixed point, the fixed frame $\{\hat{x}, \hat{y}\}$ is attached to the earth. A comet of mass m flies along the $\hat{x}$-direction with constant speed v_c and hits the moon when the orientation of the moon is θ, where θ is defined as shown in the figure. After the impact, the comet becomes one with the moon.

(a) For arbitrary θ, find the linear impulse $\mathcal{I}$ exerted on the moon by the comet. Also, find the velocity of the moon right after the impact.

(b) Draw the trace of $\mathcal{I}$ in fixed frame coordinates for all $\theta \in (0, 2\pi)$, and find the maximum possible magnitude of $\mathcal{I}$ and the corresponding θ. (*Hint*: Use $\cos(\theta - \frac{\pi}{2}) = \sin\theta$, $\sin(\theta - \frac{\pi}{2}) = -\cos\theta$.)

(c) Right after the impact, find the magnitude of the angular momentum of the moon (which is now one with the comet) about the origin.

(d) Assume that the impact occurred when $\theta = 90°$. Let the distance between the moon and the earth after the impact be $r(t)$, where $r(0) = R$. The gravitational constant of the force of attraction between the moon and the earth is G. Express $\ddot{r}(t)$ as a function of r.

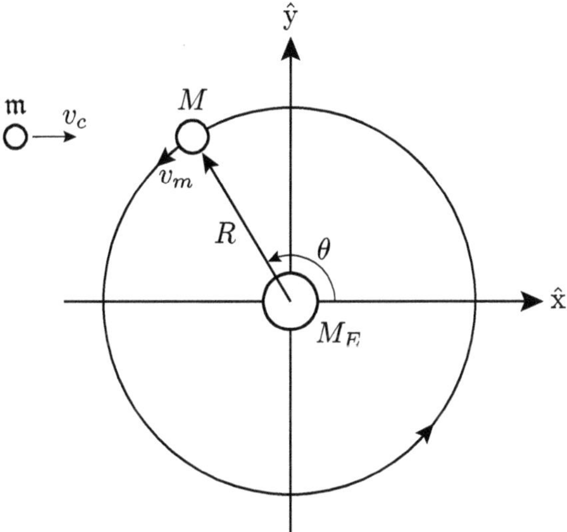

Figure 3.30 A comet hitting the moon of Exercise 3.20.

Exercise 3.21 As shown in Figure 3.31, ball A of mass m drops and strikes the triangular block B of mass $3m$. The angle θ of the triangular block is defined as shown in the figure, and the speed of the ball at the moment of impact is v. Assume there is no friction, and the coefficient of restitution between ball A and block B is e.

(a) Find the velocity of block B right after the collision.

(b) Ball A flies horizontally right after the collision. Assuming $e = \frac{4}{9}$, what is the block angle θ?

Exercise 3.22 As shown in Figure 3.32, frictionless wheeled cart B of mass m_B is initially at rest, and ball A of mass m_A flies horizontally with speed v_0 and strikes the cart. After the impact, the ball flies straight up with speed v_A, while the cart moves backward with speed v_B. The nose angle of the cart θ is defined as shown in the figure.

(a) Express v_A and v_B in terms of m_A, m_B, v_0, and θ.

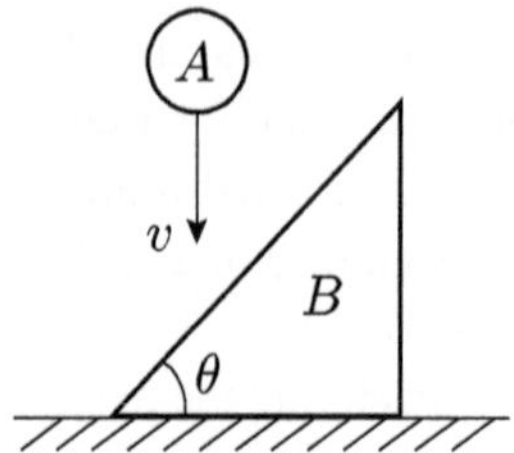

Figure 3.31 Collision between a ball and a triangular block of Exercise 3.21.

(b) Show that the coefficient of restitution e_1 between the ball and the cart is

$$e_1 = \frac{m_A}{m_B} + \cot^2 \theta.$$

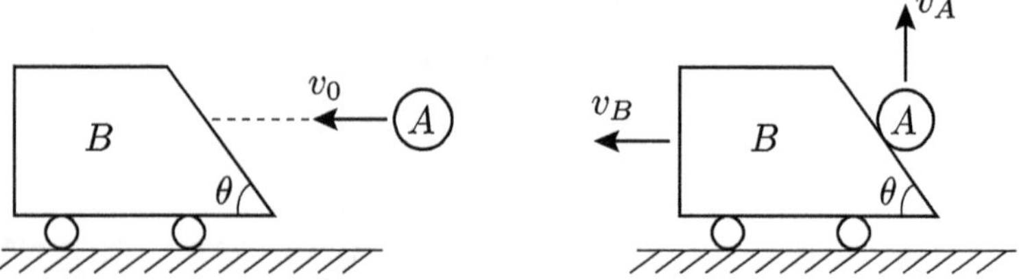

Figure 3.32 Collision between a ball and a wheeled cart of Exercise 3.22.

Exercise 3.23   Now the collision of Exercise 3.22 is modeled as a sequence of collisions between three spheres A, B, and C, each representing the ball, cart, and ground, as shown in Figure 3.33. After the first impact between spheres A and B, sphere B moves with speed v'_B and sphere A moves with speed v'_A, where the coefficient of restitution between spheres A and B is e_1 derived in Exercise 3.22(b). After the second impact between spheres B and C, sphere B moves backward horizontally with speed v''_B.
(a) Express v'_B in terms of m_A, m_B, v_0, e_1, and θ. Does sphere A fly straight up? Explain why.
(b) Express v''_B in terms of m_A, m_B, v_0, and θ.
(c) Derive the coefficient of restitution e_2 between spheres B and C in terms of m_B, m_C, and θ.
(d) Comparing v_B obtained in Exercise 3.22(a) with v''_B obtained in (b), which is bigger, and why?

Exercise 3.24   As shown in Figure 3.34, a spaceship of mass M (not including fuel mass) starts to lose its fuel of mass m at the rate of $\frac{dm}{dt} = -km^2$, where k is a positive constant. The initial mass of the fuel is m_0 and the initial speed of the spaceship is v_0. The fuel is always emitted with speed v_e and the emission is always perpendicular to

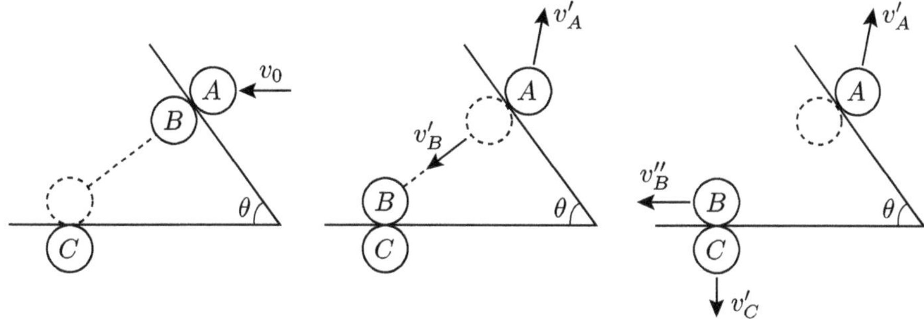

Figure 3.33 Collision between three spheres of Exercise 3.23.

the velocity of the spaceship. Assume there is no spinning due to the leakage and no thrust.

(a) Express the mass m of the fuel left in the spaceship in time t ($t = 0$ when the leakage starts).
(b) Derive the speed of the spaceship in t.
(c) Derive the radius of curvature ρ of the spaceship's trajectory at t. Assume $k = 1\ \mathrm{kg}^{-1}\mathrm{s}^{-1}$, $\mathrm{m}_0 = 1$ kg, $M = 10$ kg, $v_0 = 2$ m/s, and $v_e = 4$ m/s.
(d) Find ρ after the fuel is completely emitted, and explain the motion of the spaceship.

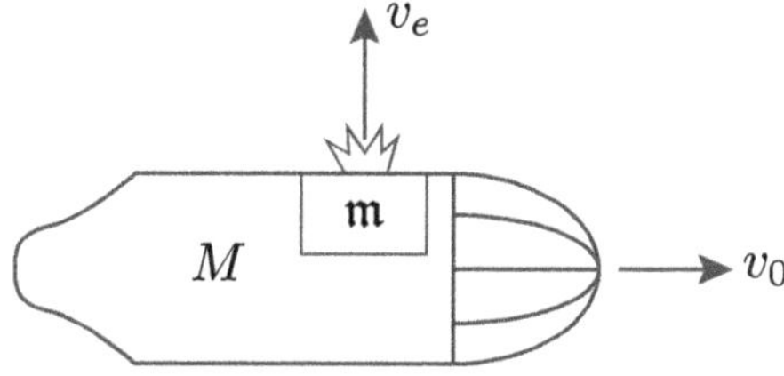

Figure 3.34 Spaceship of Exercise 3.24.

Exercise 3.25 As shown in Figure 3.35, a pendulum modeled as a mass attached to the end of a massless rod is swinging back and forth. The fixed frame $\{\hat{x}, \hat{y}, \hat{z}\}$ is attached to the other end of the rod as shown in the figure. Suddenly due to an internal explosion, the mass splits into eight pieces, where all the pieces have an identical mass of m. Right before the explosion, the velocity of the mass is $v_0 = v_{0,x}\hat{x} + v_{0,y}\hat{y} + v_{0,z}\hat{z}$ and the angle of the rod is θ as defined in the figure. The volume and rotation of the particles can be neglected. The velocity $(v_{i,x}, v_{i,y}, v_{i,z})$, $i = 1, \ldots, 8$ of each particle right after the explosion expressed in fixed frame coordinates is as follows (m/s):

i	$(v_{i,x}, v_{i,y}, v_{i,z})$
1	$(1, 1, 0)$
2	$(-1, 1, 0)$
3	$(-1, 3, 0)$
4	$(1, 3, 0)$
5	$(1, 1, -2)$
6	$(-1, 1, -2)$
7	$(-1, 3, -2)$
8	$(1, 3, -2)$

(a) Find v_0 and θ of the pendulum right before the explosion.

(b) Assuming $m = 1$ kg, derive the energy generated by the explosion. The loss of energy in the form of heat or sound during the explosion can be neglected.

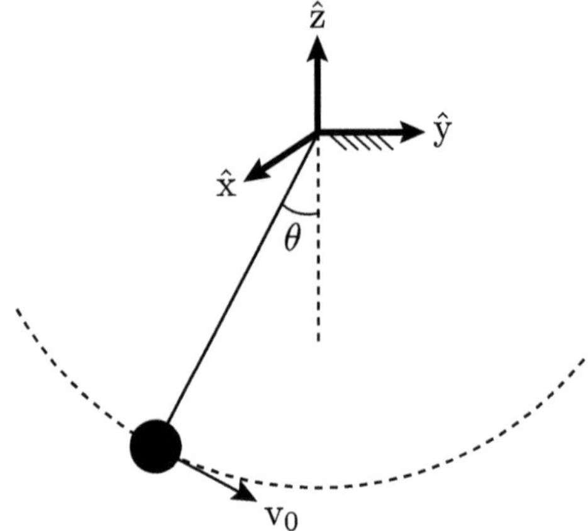

Figure 3.35 Swinging pendulum before the explosion of Exercise 3.25.

Exercise 3.26   Figure 3.36 shows two disks A and B of mass $m_A = m$ and $m_B = 2m$, respectively, connected by a massless rod and sliding on a frictionless table. The fixed frame $\{\hat{x}, \hat{y}\}$ is attached to the table origin O as shown in the figure. At $t = 0$, the two disks are rotating counterclockwise at a rate of ω_0 about their center of mass G, with G located at $(0, 5)$ and moving with velocity v_0. The rod suddenly breaks at $t = 0$; it is observed that disk A moves linearly with velocity $v_A = v_A \hat{y}$ at a distance $\sqrt{3}$ from the $\hat{y}$-axis, while disk B moves with velocity $v_B = v_B \cos 30° \, \hat{x} - v_B \sin 30° \, \hat{y}$ ($v_A > 0$, $v_B > 0$). Disk B later bounces off the lower wall at $(4\sqrt{3}, 0)$.

(a) Determine the initial velocity v_0 of G in terms of v_A, v_B, and m.

(b) Find the rotation rate ω_0 of the rod right before the rod breaks.

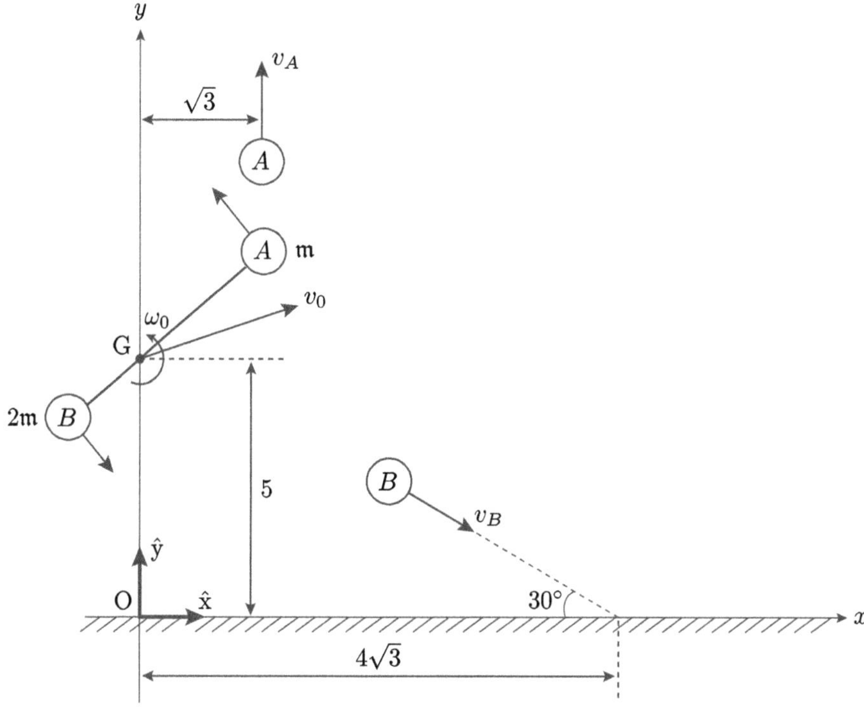

Figure 3.36 Trajectories of disks A and B after the rod breaks of Exercise 3.26.

Exercise 3.27   As shown in Figure 3.37, after disk B of Exercise 3.26 bounces off the lower wall, it is struck by disk C of mass $m_C = 2m$ moving horizontally on the line $y = 4$ with velocity $v_C = -v_C\,\hat{x}$ ($v_C > 0$). The coefficient of restitution between disk B and the lower wall is e_1, and that between disks B and C is $e_2 = 0.5$. Assume that the line of impact between disks B and C is parallel to the $\hat{x}$-axis. Disk B then bounces off the upper wall at $y = 18$ and reaches the origin O as shown in the figure, where collision with the upper wall is perfectly elastic.

 (a) Find the velocity of disk B after it bounces off the lower wall. Also, determine the location at which the collision between disks B and C occurs.

(b) Determine the speed v_C of disk C before it collides with disk B. What is the velocity of disk C after the impact?

Exercise 3.28   Figure 3.38 shows a sliding pendulum. A slider of mass $m_1 = 2$ kg moves horizontally along a frictionless guide and a ball of mass $m_2 = 2$ kg is connected to the slider by a massless cord of length $l = 1$ m. The fixed frame $\{\hat{x}, \hat{y}\}$ is attached to the guide, and the position x of the slider and the angle θ of the cord are defined as shown in the figure. The ball is released from rest at $\theta = 90°$. Gravity acts downward with acceleration $g = 10$ m/s^2.

(a) Draw free-body diagrams and derive the equations of motion for the slider and the ball.

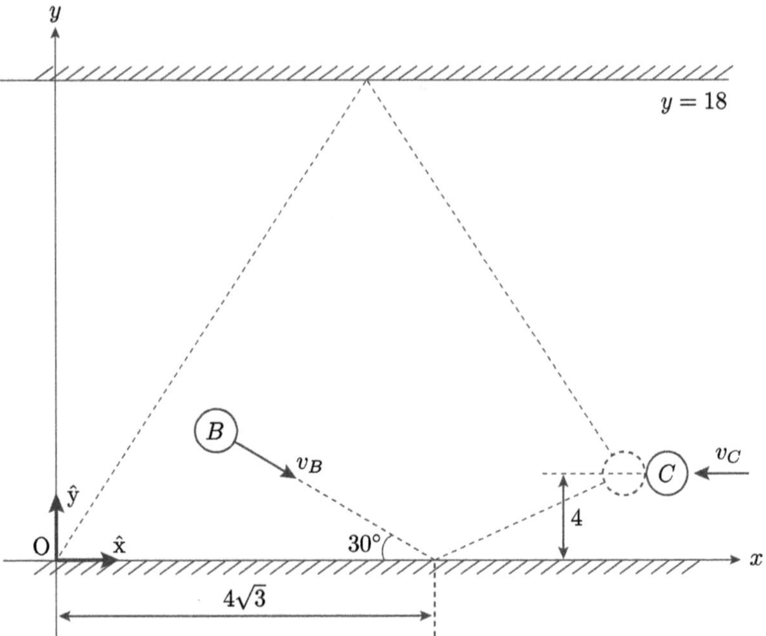

Figure 3.37 Trajectory of disk B after it bounces off the lower wall of Exercise 3.27.

(b) Find the speed of the slider as a function of θ.
(c) Find the cord tension when the ball reaches $\theta = 0°$.

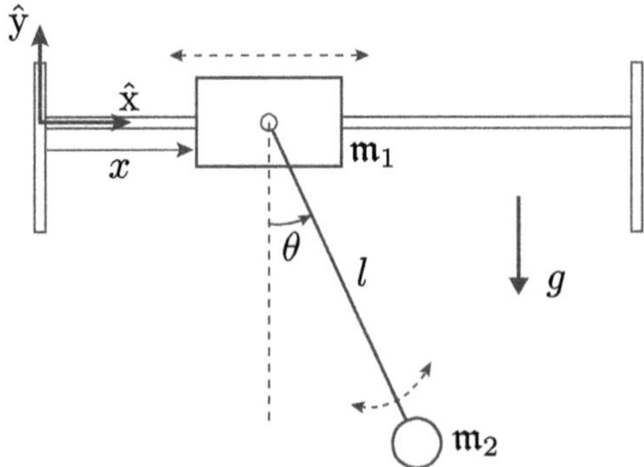

Figure 3.38 Sliding pendulum of Exercise 3.28.

Exercise 3.29   As shown in Figure 3.39, two disks B and C connected by a massless rod of length l (considered the dumbbell system) are at rest on a frictionless plane. The fixed frame $\{\hat{x}, \hat{y}\}$ is attached to the initial position of disk B, and the initial angle θ of the rod is defined as shown in the figure. Disk A slides with velocity $v_A = v\,\hat{x}$ and strikes disk B, where the line of impact is parallel to the $\hat{x}$-axis, and the coefficient of restitution is e. All the disks have an identical mass of m.

(a) Find the velocity of disk A right after the collision.

(b) Assume that $e = \frac{5}{8}$ and that disk A stops after the collision. Find the velocity of the dumbbell system's center of mass after the collision. Also, at what rate does the dumbbell system rotate about its center of mass?

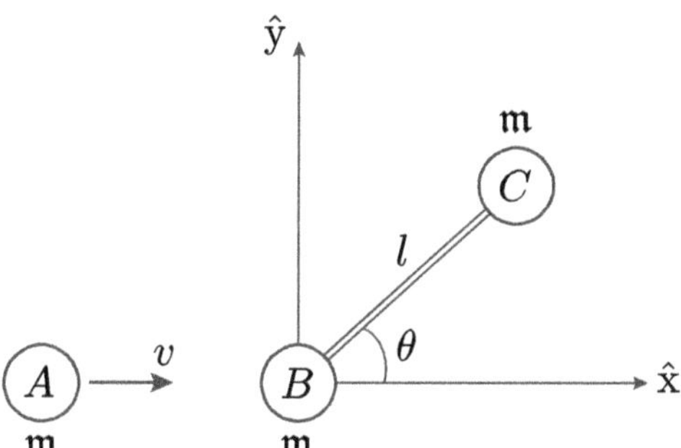

Figure 3.39 Disk striking a dumbbell system of Exercise 3.29.

Exercise 3.30 Suppose in Example 3.11 that arbitrary values for r_1, r_2, v_3 in Equations (3.136)–(3.138) are given at some fixed time t. Is it possible to solve (3.136)–(3.138) for v_1, v_2, r_3 uniquely?

4   Dynamics of a Rigid Body

The focus of this chapter is on deriving the equations of motion for a single rigid body. Building on the equations of motion derived in Chapter 3 for a system of particles, a rigid body is modeled as a system composed of an infinite number of particles: the particles are assumed to be interconnected by massless rods, with forces between particles acting in equal and opposite directions in accordance with Newton's third law.

After deriving the **inertia matrix** of a rigid body, the corresponding equations of motion are then formulated as a pair of equations: one relates the motion of the center of mass with external applied forces, the other relates the rotational motion of the rigid body with external applied moments. These equations are most easily derived by attaching a body frame at the center of mass, and expressing the dynamics in terms of the center of mass frame coordinates. However, for the commonly occurring case of a rigid body rotating about some fixed point – for example, the rigid body may be attached to ground by a rotating joint – quite often the moment equations are more conveniently expressed in terms of a body frame whose origin overlaps the fixed point rather than the center of mass. We examine both versions of the moment equations.

Deriving the equations of motion for a single rigid body is the first step in the analysis of **multibody systems**, which is the subject of Chapter 5. The standard procedure is to draw a free-body diagram for each rigid body comprising the multibody system, and to derive the corresponding equations of motion accordingly using the methods of this chapter. For each rigid body, when taking into account the forces and moments exerted by contacting bodies, the equations of motion of one body will need to be expressed in terms of reference frames attached to adjacent contacting bodies.

Toward this end, in Chapter 5 we will present an alternative formulation of rigid body dynamics based on **twists** and **wrenches**. The key underlying idea involves merging the force and moment equations of motion into a single system of equations. The twist–wrench formalism offers a simple and concise set of rules for how physical quantities like forces, moments, velocities and accelerations transform under a change of reference frame, and is particularly effective for multibody system dynamics. The contents of this chapter are an essential prerequisite to developing the tools for twist–wrench dynamic analysis.

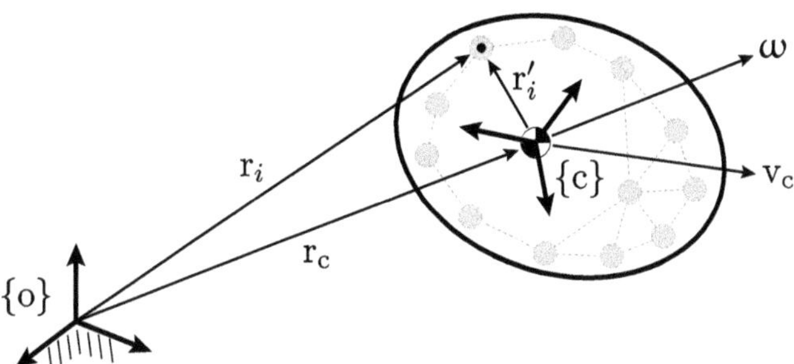

Figure 4.1 A rigid body regarded as a collection of particles. Frame $\{c\}$ attached to the center of mass has angular velocity ω and linear velocity v_c.

4.1 The Force Equation for the Center of Mass

Referring to Figure 4.1, assign the fixed frame $\{o\}$ with origin at O, and regard the rigid body as a collection of an infinite number of rigidly connected particles. Particle i, located at r_i, has mass m_i; the total mass of the rigid body is $m = \sum_i m_i$. Attach a body-fixed frame $\{c\}$ to the rigid body's center of mass (recall that the body's center of mass is the point r_c such that $mr_c = \sum_i m_i r_i$). Let r_i' be the vector from the center of mass to particle i, so that

$$r_i = r_c + r_i'. \tag{4.1}$$

Substituting (4.1) into $mr_c = \sum_i m_i r_i$, it follows that

$$\sum_i m_i r_i' = 0. \tag{4.2}$$

Writing Newton's second law for each particle i,

$$f_i + \sum_j f_{ji} = m_i \ddot{r}_i, \quad i = 1, 2, \ldots, \tag{4.3}$$

where f_i is the external force applied to particle i, and f_{ji} is the force exerted by particle j on particle i ($j \neq i$). Summing these equations over i,

$$\sum_i \left(f_i + \sum_j f_{ji} \right) = \sum_i f_i = \sum_i m_i \ddot{r}_i, \tag{4.4}$$

where we use the fact that $f_{ji} + f_{ij} = 0$ for all i, j. Furthermore, since $m = \sum_i m_i$ and $r_c = (\sum_i m_i r_i)/m$, we have

$$f = \sum_i f_i = m\ddot{r}_c. \tag{4.5}$$

We now write down (4.5) explicitly in terms of coordinates of frame $\{c\}$. Denote the unit axes of frame $\{c\}$ by $\{\hat{x}_c, \hat{y}_c, \hat{z}_c\}$, and express the applied force f as

$$f = f_x \hat{x}_c + f_y \hat{y}_c + f_z \hat{z}_c. \tag{4.6}$$

Also express ω, the angular velocity of frame $\{c\}$, as

$$\omega = \omega_x \hat{x}_c + \omega_y \hat{y}_c + \omega_z \hat{z}_c, \tag{4.7}$$

and the velocity $v_c = \dot{r}_c$ of the center of mass as

$$v_c = v_x \hat{x}_c + v_y \hat{y}_c + v_z \hat{z}_c. \tag{4.8}$$

The acceleration a_c is then

$$\begin{aligned}
a_c &= \dot{v}_x \hat{x}_c + \dot{v}_y \hat{y}_c + \dot{v}_z \hat{z}_c + v_x \dot{\hat{x}}_c + v_y \dot{\hat{y}}_c + v_z \dot{\hat{z}}_c \\
&= \dot{v}_x \hat{x}_c + \dot{v}_y \hat{y}_c + \dot{v}_z \hat{z}_c + \omega \times v_c.
\end{aligned} \tag{4.9}$$

Defining the four column vectors $f_c = (f_x, f_y, f_z)^\top$, $v_c = (v_x, v_y, v_z)^\top$, $\dot{v}_c = (\dot{v}_x, \dot{v}_y, \dot{v}_z)^\top$ and $\omega_c = (\omega_x, \omega_y, \omega_z)^\top$, $f = ma_c$ can now be explicitly written as

$$f_c = m(\dot{v}_c + \omega_c \times v_c). \tag{4.10}$$

4.2   The Moment Equation with Respect to the Center of Mass

Recall that for a system of N particles, the moment equation with respect to the center of mass is given by

$$m_c = \sum_{i=1}^{N} r_i' \times f_i = \dot{h}_c, \tag{4.11}$$

where r_i' is the vector from the center of mass to particle i, f_i is the external force applied to particle i, h_c is the angular momentum with respect to the center of mass and $\dot{h}_c$ its time derivative. These equations also hold for a rigid body modeled as a system of particles. We now examine how these equations behave in the limit as the number of particles increases to infinity.

Referring again to Figure 4.1, suppose frame $\{c\}$ attached to the rigid body's center of mass moves with linear velocity v_c and angular velocity ω. The angular momentum of the rigid body with respect to its center of mass is given by

$$h_c = \sum_i r_i' \times m_i \dot{r}_i = \sum_i r_i' \times m_i (\dot{r}_c + \dot{r}_i') = \left(\sum_i m_i r_i' \right) \times \dot{r}_c + \sum_i r_i' \times m_i \dot{r}_i'. \tag{4.12}$$

Noting that $\sum_i m_i r_i' = 0$ from the center of mass relation (4.2), h_c simplifies to

$$h_c = \sum_i r_i' \times m_i \dot{r}_i'. \tag{4.13}$$

Writing r_i' in frame $\{c\}$ coordinates as $r_i' = x_i' \hat{x}_c + y_i' \hat{y}_c + z_i' \hat{z}_c$, note because of the rigid body assumption that the x_i', y_i', z_i' are all constant. It therefore follows that $\dot{r}_i' = \omega \times r_i'$, and the angular momentum h_c simplifies to

$$h_c = \sum_i m_i (r_i' \times (\omega \times r_i')) = \sum_i -m_i (r_i' \times (r_i' \times \omega)). \tag{4.14}$$

We now write down the components of h_c in frame $\{c\}$ coordinates. First recall that given $r = (x, y, z)^\top \in \mathbb{R}^3$,

$$[r] = \begin{bmatrix} 0 & -z & y \\ z & 0 & -x \\ -y & x & 0 \end{bmatrix} \tag{4.15}$$

denotes its 3×3 skew-symmetric matrix representation. Recall that the cross-product $u \times v$ of two vectors $u, v \in \mathbb{R}^3$ can be written $u \times v = [u]v$. Finally, note that $[r]^\top = -[r]$, and

$$[r][r]^\top = -[r]^2 = \begin{bmatrix} y^2 + z^2 & -xy & -xz \\ -xy & x^2 + z^2 & -yz \\ -xz & -yz & x^2 + y^2 \end{bmatrix}. \tag{4.16}$$

With these preliminaries, let $r_i' = x_i'\hat{x}_c + y_i'\hat{y}_c + z_i'\hat{z}_c$, $\omega = \omega_x\hat{x}_c + \omega_y\hat{y}_c + \omega_z\hat{z}_c$, $h_c = h_x\hat{x}_c + h_y\hat{y}_c + h_z\hat{z}_c$, and define

$$r_i' = (x_i', y_i', z_i')^\top, \quad \omega_c = (\omega_x, \omega_y, \omega_z)^\top, \quad h_c = (h_x, h_y, h_z)^\top. \tag{4.17}$$

The angular momentum h_c in frame $\{c\}$ coordinates is then given by

$$h_c = \left(\sum_i -m_i [r_i']^2 \right) \omega_c, \tag{4.18}$$

which when further expanded becomes

$$h_c = \begin{bmatrix} h_x \\ h_y \\ h_z \end{bmatrix} = \begin{bmatrix} I_{xx} & -I_{xy} & -I_{xz} \\ -I_{xy} & I_{yy} & -I_{yz} \\ -I_{xz} & -I_{yz} & I_{zz} \end{bmatrix} \begin{bmatrix} \omega_x \\ \omega_y \\ \omega_z \end{bmatrix} = \mathcal{I}_c \omega_c, \tag{4.19}$$

with

$$I_{xx} = \sum_i m_i(y_i'^2 + z_i'^2), \quad I_{xy} = \sum_i m_i x_i' y_i', \quad I_{xz} = \sum_i m_i x_i' z_i',$$

$$I_{yy} = \sum_i m_i(x_i'^2 + z_i'^2), \quad I_{yz} = \sum_i m_i y_i' z_i', \quad I_{zz} = \sum_i m_i(x_i'^2 + y_i'^2). \tag{4.20}$$

The 3×3 matrix $\mathcal{I}_c$ is the **inertia matrix** of the rigid body with respect to frame $\{c\}$; the I_{xx}, I_{yy}, and I_{zz} terms are called the **moments of inertia**, while the I_{xy}, I_{yz}, and I_{xz} terms are called the **products of inertia**.

Modeling a rigid body as a collection of an infinite number of particles, the summations in (4.20) can be replaced by integrals, and the particle mass m_i is replaced by the infinitesimal mass-volume

$$m_i \mapsto \rho(x_i', y_i', z_i') \, dx_i' \, dy_i' \, dz_i', \tag{4.21}$$

where $\rho(x', y', z')$ is the mass density at point (x', y', z') of the rigid body. The elements of the inertia matrix $\mathcal{I}_c$ then become

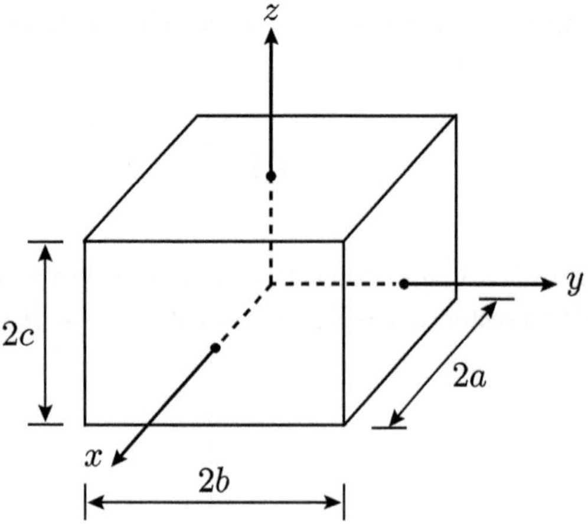

Figure 4.2 The inertia matrix for a rectangular box.

$$I_{xx} = \int \int \int_{\mathcal{B}} \rho(x, y, z)(y^2 + z^2)\, dx\, dy\, dz, \tag{4.22}$$

$$I_{yy} = \int \int \int_{\mathcal{B}} \rho(x, y, z)(x^2 + z^2)\, dx\, dy\, dz, \tag{4.23}$$

$$I_{zz} = \int \int \int_{\mathcal{B}} \rho(x, y, z)(x^2 + y^2)\, dx\, dy\, dz, \tag{4.24}$$

$$I_{xy} = \int \int \int_{\mathcal{B}} \rho(x, y, z)xy\, dx\, dy\, dz, \tag{4.25}$$

$$I_{yz} = \int \int \int_{\mathcal{B}} \rho(x, y, z)yz\, dx\, dy\, dz, \tag{4.26}$$

$$I_{xz} = \int \int \int_{\mathcal{B}} \rho(x, y, z)xz\, dx\, dy\, dz, \tag{4.27}$$

where $\mathcal{B}$ denotes the volume occupied by the rigid body.

Example 4.1   In this example we calculate the elements of the inertia matrix for the rectangular block of Figure 4.2 with respect to the reference frame attached to the center of mass as shown. Assuming the mass density function ρ is constant throughout and total mass $\mathfrak{m} = 8abc\rho$, the moments of inertia are

$$I_{xx} = \sum_i \mathfrak{m}_i(y_i^2 + z_i^2) = \int_{-c}^{c} \int_{-b}^{b} \int_{-a}^{a} (y^2 + z^2)\rho\, dx\, dy\, dz$$

$$= \frac{8}{3}\rho abc(b^2 + c^2) = \frac{\mathfrak{m}}{3}(b^2 + c^2), \tag{4.28}$$

$$I_{yy} = \frac{8}{3}\rho abc(a^2 + c^2) = \frac{\mathfrak{m}}{3}(a^2 + c^2), \tag{4.29}$$

$$I_{zz} = \frac{8}{3}\rho abc(a^2 + b^2) = \frac{\mathfrak{m}}{3}(a^2 + b^2), \tag{4.30}$$

while the products of inertia are $I_{xy} = I_{yz} = I_{xz} = 0$. $\qquad\qquad\square$

Example 4.2   Setting the height $c = 0$ in the rectangular block of Figure 4.2, we can obtain the corresponding inertia for a rectangular plate. Assume as before a constant mass density ρ and total mass $m = 4ab\rho$. The moment of inertia I_{xx} is given by

$$I_{xx} = \sum_i m_i y_i^2 = \int_{-b}^{b} \int_{-a}^{a} \rho y^2 dx\, dy = \left.\left.\frac{\rho x y^3}{3}\right|_{x=-a}^{x=a}\right|_{y=-b}^{y=b} = \frac{4}{3}\rho ab^3 = \frac{mb^2}{3}. \quad (4.31)$$

Following a similar calculation,

$$I_{yy} = \frac{4}{3}\rho ba^3 = \frac{ma^2}{3}, \quad (4.32)$$

while

$$I_{zz} = \frac{4\rho}{3}(a^3 b + ab^3) = \frac{m}{3}(a^2 + b^2). \quad (4.33)$$

Although the rectangular plate itself is two-dimensional, it still has moments of inertia in all three spatial directions. A larger moment of inertia along a given axis indicates greater resistance to rotation about that axis. $\qquad\square$

Continuing with our expansion of the angular momentum equation $m_c = \dot{h}_c$, differentiating $h_c = h_x \hat{x}_c + h_y \hat{y}_c + h_z \hat{z}_c$ with respect to t leads to

$$\dot{h}_c = \dot{h}_x \hat{x}_c + \dot{h}_y \hat{y}_c + \dot{h}_z \hat{z}_c + \omega \times h_c. \quad (4.34)$$

Since $h_c = (h_x, h_y, h_z)^\top = I_c \omega_c$ with I_c constant, it follows that

$$\dot{h}_c = \begin{bmatrix} \dot{h}_x \\ \dot{h}_y \\ \dot{h}_z \end{bmatrix} = I_c \dot{\omega}_c. \quad (4.35)$$

Letting $m_c = m_x \hat{x}_c + m_y \hat{y}_c + m_z \hat{z}_c$ with $m_c = (m_x, m_y, m_z)^\top$, the moment equation $m_c = \dot{h}_c$ expressed in frame $\{c\}$ coordinates is

$$m_c = I_c \dot{\omega}_c + \omega_c \times I_c \omega_c. \quad (4.36)$$

We now make the following observation about $\dot{\omega}_c$. The angular acceleration vector α is defined as the time-derivative of ω:

$$\alpha = \frac{d}{dt}\omega = \dot{\omega}_x \hat{x}_c + \dot{\omega}_y \hat{y}_c + \dot{\omega}_z \hat{z}_c + \omega_x \dot{\hat{x}}_c + \omega_y \dot{\hat{y}}_c + \omega_z \dot{\hat{z}}_c. \quad (4.37)$$

Replacing $\dot{\hat{x}}_c$ by $\omega \times \hat{x}_c$, $\dot{\hat{y}}_c$ by $\omega \times \hat{y}_c$, and $\dot{\hat{z}}_c$ by $\omega \times \hat{z}_c$, the angular acceleration α then simplifies to

$$\alpha = \dot{\omega}_x \hat{x}_c + \dot{\omega}_y \hat{y}_c + \dot{\omega}_z \hat{z}_c + \omega \times \omega = \dot{\omega}_x \hat{x}_c + \dot{\omega}_y \hat{y}_c + \dot{\omega}_z \hat{z}_c. \quad (4.38)$$

We have just shown that $\dot{\omega}_c$ is also the frame $\{c\}$ representation of the angular acceleration vector, and for that reason we will also write (4.36) in the equivalent form

$$m_c = I_c \alpha_c + \omega_c \times I_c \omega_c, \quad (4.39)$$

where $\alpha_c = \dot{\omega}_c$.

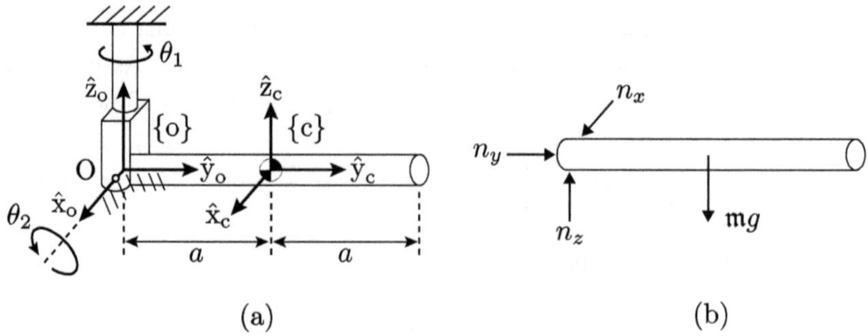

Figure 4.3 (a) A spherical pendulum; (b) Free-body diagram.

Example 4.3   In this example we derive the equations of motion for the spherical pendulum of Figure 4.3(a), consisting of a rod of mass m and length $2a$ rotating about two orthogonal axes. Assign the fixed inertial frame {o} at the intersection of the two axes of rotation, and a body frame {c} at the rod center of mass as shown. The inertia matrix of the rod with respect to {c} is

$$\mathcal{I}_c = \begin{bmatrix} \mu & 0 & 0 \\ 0 & 0 & 0 \\ 0 & 0 & \mu \end{bmatrix}, \quad \mu = \frac{1}{3}ma^2. \tag{4.40}$$

The angular velocity of the moving frame {c} is then

$$\begin{aligned} \omega &= \dot{\theta}_1 \hat{z}_o + \dot{\theta}_2 \hat{x}_c \\ &= \dot{\theta}_2 \hat{x}_c + \dot{\theta}_1 \sin\theta_2 \hat{y}_c + \dot{\theta}_1 \cos\theta_2 \hat{z}_c, \end{aligned} \tag{4.41}$$

where we use the fact that $\hat{z}_o = \sin\theta_2 \hat{y}_c + \cos\theta_2 \hat{z}_c$. Set $\omega_c = (\dot{\theta}_2, \dot{\theta}_1 \sin\theta_2, \dot{\theta}_1 \cos\theta_2)^\top$ and $h_c = h_x \hat{x}_c + h_y \hat{y}_c + h_z \hat{z}_c$. Then

$$\begin{bmatrix} h_x \\ h_y \\ h_z \end{bmatrix} = \mathcal{I}_c \omega_c = \begin{bmatrix} \mu\dot{\theta}_2 \\ 0 \\ \mu\dot{\theta}_1 \cos\theta_2 \end{bmatrix}, \tag{4.42}$$

or $h_c = \mu\dot{\theta}_2 \hat{x}_c + \mu\dot{\theta}_1 \cos\theta_2 \hat{z}_c$. Then

$$\dot{h}_c = \mu\ddot{\theta}_2 \hat{x}_c + (\mu\ddot{\theta}_1 \cos\theta_2 - \mu\dot{\theta}_1\dot{\theta}_2 \sin\theta_2)\hat{z}_c + \mu\dot{\theta}_2 \dot{\hat{x}}_c + \mu\dot{\theta}_1 \cos\theta_2 \dot{\hat{z}}_c. \tag{4.43}$$

Noting that

$$\dot{\hat{x}}_c = \omega \times \hat{x}_c = \dot{\theta}_1 \cos\theta_2 \hat{y}_c - \dot{\theta}_1 \sin\theta_2 \hat{z}_c, \tag{4.44}$$

$$\dot{\hat{z}}_c = \omega \times \hat{z}_c = \dot{\theta}_1 \sin\theta_2 \hat{x}_c - \dot{\theta}_2 \hat{y}_c, \tag{4.45}$$

we have

$$\dot{h}_c = \mu(\ddot{\theta}_2 + \dot{\theta}_1^2 \sin\theta_2 \cos\theta_2)\hat{x}_c + \mu(\ddot{\theta}_1 \cos\theta_2 - 2\dot{\theta}_1\dot{\theta}_2 \sin\theta_2)\hat{z}_c. \tag{4.46}$$

To calculate the moment m_c about the $\{c\}$ frame origin, observe that the force due to gravity applied to the mass center, mg, does not contribute to m_c; only the reaction force $n_x\hat{x}_c + n_y\hat{y}_c + n_z\hat{z}_c$ at the joint center contributes to m_c (see Figure 4.3(b)). Then

$$m_c = -a\hat{y}_c \times (n_x\hat{x}_c + n_y\hat{y}_c + n_z\hat{z}_c) = -n_z a\hat{x}_c + n_x a\hat{z}_c. \tag{4.47}$$

Setting $m_c = \dot{h}_c$ and matching components,

$$n_x = \frac{\mu}{a}(\ddot{\theta}_1 \cos\theta_2 - 2\dot{\theta}_1\dot{\theta}_2 \sin\theta_2), \tag{4.48}$$

$$n_z = -\frac{\mu}{a}(\ddot{\theta}_2 + \dot{\theta}_1^2 \sin\theta_2 \cos\theta_2). \tag{4.49}$$

Note that Equations (4.48) and (4.49) can also be obtained directly from $m_c = \mathcal{I}_c\dot{\omega}_c + \omega_c \times \mathcal{I}_c\omega_c$, where m_c is now set to $m_c = (-n_z a, 0, n_x a)^\top$. These equations of motion cannot be integrated to obtain a solution trajectory, since n_x and n_z are as yet unknown. For this purpose we need the equation $\sum f = f = m\ddot{r}_c$ for the center of mass, where r_c is the vector from the frame $\{o\}$ origin to the center of mass. Observe that

$$\sum f = -mg(\sin\theta_2\hat{y}_c + \cos\theta_2\hat{z}_c) + n_x\hat{x}_c + n_y\hat{y}_c + n_z\hat{z}_c$$
$$= n_x\hat{x}_c + (n_y - mg\sin\theta_2)\hat{y}_c + (n_z - mg\cos\theta_2)\hat{z}_c. \tag{4.50}$$

To calculate $\ddot{r}_c$, differentiate $r_c = a\hat{y}_c$ twice:

$$\dot{r}_c = \omega \times a\hat{y}_c = -a\dot{\theta}_1 \cos\theta_2\hat{x}_c + a\dot{\theta}_2\hat{z}_c, \tag{4.51}$$

$$\ddot{r}_c = a(2\dot{\theta}_1\dot{\theta}_2 \sin\theta_2 - \ddot{\theta}_1 \cos\theta_2)\hat{x}_c + a(-\dot{\theta}_2^2 - \dot{\theta}_1^2 \cos^2\theta_2)\hat{y}_c$$
$$+ a(\ddot{\theta}_2 + \dot{\theta}_1^2 \cos\theta_2 \sin\theta_2)\hat{z}_c. \tag{4.52}$$

Matching components in the equation $f = m\ddot{r}_c$,

$$n_x = ma(2\dot{\theta}_1\dot{\theta}_2 \sin\theta_2 - \ddot{\theta}_1 \cos\theta_2), \tag{4.53}$$

$$n_y = -ma(\dot{\theta}_2^2 + \dot{\theta}_1^2 \cos^2\theta_2) + mg\sin\theta_2, \tag{4.54}$$

$$n_z = ma(\ddot{\theta}_2 + \dot{\theta}_1^2 \sin\theta_2 \cos\theta_2) + mg\cos\theta_2. \tag{4.55}$$

Note that Equations (4.53)–(4.55) can also be obtained directly from $f_c = m(\dot{v}_c + \omega_c \times v_c)$, where v_c is now set to $v_c = (-a\dot{\theta}_1 \cos\theta_2, 0, a\dot{\theta}_2)^\top$. Matching the expressions for n_x and n_z with those obtained earlier, we get the following pair of equations:

$$\ddot{\theta}_1 = 2\dot{\theta}_1\dot{\theta}_2 \tan\theta_2, \tag{4.56}$$

$$\ddot{\theta}_2 = -\dot{\theta}_1^2 \sin\theta_2 \cos\theta_2 - \frac{3g}{4a} \cos\theta_2. \tag{4.57}$$

As seen from above, deriving the equations of motion with respect to the center of mass frame $\{c\}$ for this example requires considerable work. We return to this example in the next section, and show that by choosing a moving frame located at the frame $\{o\}$ origin rather than the mass center, the equations of motion can be derived with slightly less calculation. However, if the joint reaction forces (n_x, n_y, n_z) are needed, then the calculations made here cannot be avoided. $\qquad\square$

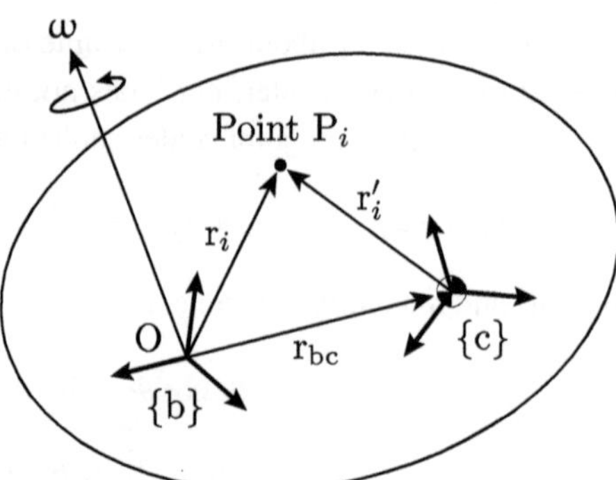

Figure 4.4 A rigid body undergoing rotations about a fixed point O. Frame {b} is attached to the rigid body at O and undergoes angular velocity ω.

4.3   The Moment Equation with Respect to a Fixed Point in Space

Recall that for a system of particles, the moment equations can also be written with respect to a fixed point O instead of its center of mass. Once again modeling a rigid body as a collection of rigidly connected particles, if the rigid body undergoes only rotations about some fixed point in space, then the corresponding moment equations with respect to the fixed point can assume a much simpler form than the moment equations with respect to the center of mass.

Consider a rigid body of mass m rotating about some fixed point O in space with angular velocity ω. Viewing the rigid body as a system of particles, let m_i be the mass of particle i, and r_i be the vector from O to particle i. The velocity $v_i = \dot{r}_i$ of particle i is then given by $v_i = \omega \times r_i$. The angular momentum of the rigid body with respect to O is then

$$h_o = \sum_i r_i \times m_i v_i = \sum_i m_i r_i \times (\omega \times r_i) = \sum_i -m_i r_i \times (r_i \times \omega). \qquad (4.58)$$

We now express these equations in terms of a body frame {b} attached to the body at O, but rotating with the body (i.e., the origin of frame {b} is fixed at O, but the frame rotates with the same angular velocity ω as the rigid body) (see Figure 4.4). Writing

$$h_o = h_x \hat{x}_b + h_y \hat{y}_b + h_z \hat{z}_b, \qquad (4.59)$$

$$\omega = \omega_x \hat{x}_b + \omega_y \hat{y}_b + \omega_z \hat{z}_b, \qquad (4.60)$$

$$r_i = x_i \hat{x}_b + y_i \hat{y}_b + z_i \hat{z}_b, \qquad (4.61)$$

and $h_b = (h_x, h_y, h_z)^\top$, $\omega_b = (\omega_x, \omega_y, \omega_z)^\top$, $r_i = (x_i, y_i, z_i)^\top$, (4.58) can be written

$$h_b = \left(-\sum_i m_i [r_i]^2 \right) \omega_b = I_b \omega_b, \qquad (4.62)$$

where $I_b = -\sum_i m_i [r_i]^2$ is the inertia matrix for the rigid body with respect to frame {b}.

In principle one could evaluate $\mathcal{I}_b$ in the same way as we did for $\mathcal{I}_c$ in (4.19)–(4.20), by taking appropriate integrals for the moments and products of inertia. This is not necessary; it is in fact possible to obtain $\mathcal{I}_b$ directly from $\mathcal{I}_c$ via a simple formula that we now derive.

Referring to Figure 4.4, given a point P_i on the body, let $r_i' \in \mathbb{R}^3$ be its coordinates in the center of mass frame $\{c\}$ (i.e., r_i' is the vector r_i' expressed in frame $\{c\}$ coordinates). Similarly, let $r_i \in \mathbb{R}^3$ be the coordinates for P_i in frame $\{b\}$ (i.e., r_i is the vector r_i expressed in frame $\{b\}$ coordinates). Since r_i and r_i' are coordinates for the same point P_i in physical space, from (2.64) we can write

$$r_i = R_{bc} r_i' + r_{bc}, \tag{4.63}$$

where R_{bc} is the rotation matrix describing the orientation of frame $\{c\}$ relative to frame $\{b\}$, and $r_{bc} \in \mathbb{R}^3$ is the vector from frame $\{b\}$ to frame $\{c\}$, expressed in frame $\{b\}$ coordinates (i.e., r_{bc} is the vector r_{bc} expressed in frame $\{b\}$ coordinates). Let $r_{cb}' \in \mathbb{R}^3$ be the vector $-\mathrm{r}_{bc}$ expressed in frame $\{c\}$ coordinates (or equivalently, $-r_{cb}'$ is r_{bc} expressed in frame $\{c\}$). Then from (2.62), r_{cb}' and r_{bc} are related by

$$r_{bc} = -R_{bc} r_{cb}'. \tag{4.64}$$

Equation (4.63) therefore becomes

$$r_i = R_{bc}(r_i' - r_{cb}'). \tag{4.65}$$

The inertia matrix $\mathcal{I}_b$ can now be defined as follows:

$$\mathcal{I}_b = -\sum_i \mathrm{m}_i \, [r_i]^2 = -\sum_i \mathrm{m}_i \left[R_{bc}(r_i' - r_{cb}') \right]^2. \tag{4.66}$$

Using the properties $[R\omega] = R[\omega]R^{\mathsf{T}}$, $[R\omega]^2 = R[\omega]^2 R^{\mathsf{T}}$, and $[u - v] = [u] - [v]$ for any rotation matrix R and $u, v, \omega \in \mathbb{R}^3$, $\mathcal{I}_b$ can be simplified as follows:

$$
\begin{aligned}
\mathcal{I}_b &= -\sum_i \mathrm{m}_i R_{bc} \left([r_i'] - [r_{cb}'] \right)^2 R_{bc}^{\mathsf{T}} \\
&= R_{cb}^{\mathsf{T}} \left(-\sum_i \mathrm{m}_i [r_i']^2 \right) R_{cb} - R_{cb}^{\mathsf{T}} \left(\sum_i \mathrm{m}_i \right) [r_{cb}']^2 R_{cb} \\
&\quad + R_{cb}^{\mathsf{T}} \left(\left(\sum_i \mathrm{m}_i [r_i'] \right) [r_{cb}'] + [r_{cb}'] \left(\sum_i \mathrm{m}_i [r_i'] \right) \right) R_{cb} \\
&= R_{cb}^{\mathsf{T}} \left(\mathcal{I}_c - \mathrm{m}[r_{cb}']^2 \right) R_{cb}.
\end{aligned}
\tag{4.67}
$$

Equation (4.67) follows from $\sum \mathrm{m}_i [r_i'] = 0$ (since the center of mass corresponds to the $\{c\}$ frame origin $r_i' = 0$). Once $\mathcal{I}_c$ is known, (4.67) allows us to compute the inertia matrix $\mathcal{I}_b$ with respect to arbitrary frames $\{b\}$ without having to re-evaluate integrals for the moments and products of inertia. For the special case when $R_{cb} = I$, formula (4.67) is called the **Parallel Axis Theorem**.

Returning to the moment equation about O in its original form, i.e.,

$$\mathrm{h}_o = \sum_i \mathrm{r}_i \times \mathrm{m}_i \dot{\mathrm{r}}_i, \tag{4.68}$$

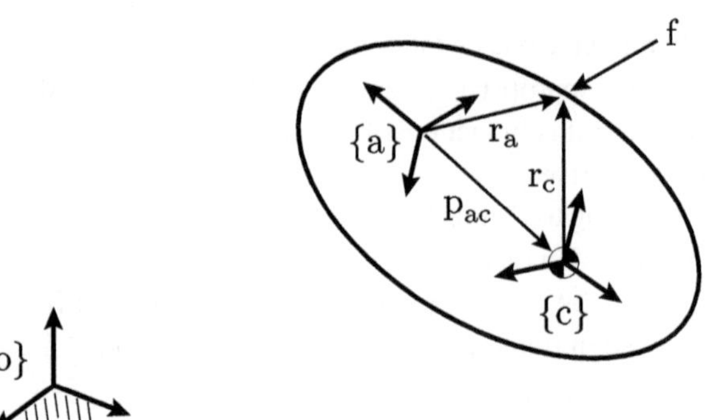

Figure 4.5 Transformation of moment equations for different choices of body frame: frames {c} and {a} are attached to different points on the rigid body.

taking time derivatives of both sides leads to

$$\dot{h}_o = \sum_i (\dot{r}_i \times m_i \dot{r}_i) + (r_i \times m_i \ddot{r}_i) = \sum_i r_i \times m_i \ddot{r}_i = \sum_i r_i \times f_i = m_o, \qquad (4.69)$$

where f_i is the external force applied at r_i, and m_o is the net external moment with respect to O applied to the rigid body.

We close this section by examining how the moment equations transform for different choices of body frame. Referring to Figure 4.5, frame {o} is the inertial frame, frame {c} is attached to the center of mass, and frame {a} is attached to some other arbitrary point on the rigid body. The vector from frame {a} to frame {c} is denoted p_{ac}. Observe that $r_a = r_c + p_{ac}$. A force f is applied to the rigid body as shown. The moment generated by f with respect to the origin of frame {c}, denoted m_c, is $m_c = r_c \times f$. The moment generated by the same force, this time with respect to the origin of frame {a}, is

$$m_a = r_a \times f = (r_c + p_{ac}) \times f = (r_c \times f) + (p_{ac} \times f) = m_c + (p_{ac} \times f). \qquad (4.70)$$

From the moment equation $m_c = \dot{h}_c$, the above can be written

$$m_a = \dot{h}_c + (p_{ac} \times f). \qquad (4.71)$$

If instead of $m_c = \dot{h}_c$ the moment equation is written with respect to a fixed point O, as $m_o = \dot{h}_o$, then (4.71) can also be written

$$m_a = \dot{h}_o + (p_{ao} \times f). \qquad (4.72)$$

Example 4.4   Consider once again the spherical pendulum example of Figure 4.3. This time we will use the moment equations with respect to fixed point O, $m_o = \dot{h}_o$. Attach a body frame {b} to the body at O, always in the same orientation as the center of mass frame {c} (i.e., frame {b} is attached to point O but moves with the body). Frames {b} and {c} always have the same orientation (i.e., $R_{cb} = I$) and therefore the same angular velocity components $\omega_b = \omega_c \in \mathbb{R}^3$:

$$\omega_b = (\dot{\theta}_2, \dot{\theta}_1 \sin \theta_2, \dot{\theta}_1 \cos \theta_2)^\top. \qquad (4.73)$$

Note also that $r_{cb} = (0, -a, 0)^\top$. The inertia matrix with respect to frame {b}, $\mathcal{I}_b$, can now be obtained from the Parallel Axis Theorem as

$$\mathcal{I}_b = \mathcal{I}_c - m[r_{cb}]^2 = \begin{bmatrix} \bar{\mu} & 0 & 0 \\ 0 & 0 & 0 \\ 0 & 0 & \bar{\mu} \end{bmatrix}, \quad \bar{\mu} = \mu + ma^2 = \frac{4}{3}ma^2. \tag{4.74}$$

Components of the angular momentum vector $h_o = h_x \hat{x}_b + h_y \hat{y}_b + h_z \hat{z}_b$ can now be obtained as

$$\begin{bmatrix} h_x \\ h_y \\ h_z \end{bmatrix} = \mathcal{I}_b \omega_b = \begin{bmatrix} \bar{\mu}\dot{\theta}_2 \\ 0 \\ \bar{\mu}\dot{\theta}_1 \cos \theta_2 \end{bmatrix}, \tag{4.75}$$

or $h_o = \bar{\mu}\dot{\theta}_2 \hat{x}_b + \bar{\mu}\dot{\theta}_1 \cos \theta_2 \hat{z}_b$. Then

$$\dot{h}_o = \bar{\mu}(\ddot{\theta}_2 + \dot{\theta}_1^2 \sin \theta_2 \cos \theta_2)\hat{x}_b + \bar{\mu}(\ddot{\theta}_1 \cos \theta_2 - 2\dot{\theta}_1\dot{\theta}_2 \sin \theta_2)\hat{z}_b. \tag{4.76}$$

What remains now is to determine $m_o = m_x \hat{x}_b + m_y \hat{y}_b + m_z \hat{z}_b$. This is straightforward:

$$(m_x, m_y, m_z)^\top = (-mga \cos \theta_2,\ 0,\ 0)^\top. \tag{4.77}$$

Matching terms from $m_o = \dot{h}_o$ we get (ignoring $0 = 0$)

$$-mga \cos \theta_2 = \bar{\mu}(\ddot{\theta}_2 + \dot{\theta}_1^2 \sin \theta_2 \cos \theta_2), \tag{4.78}$$

$$0 = \bar{\mu}(\ddot{\theta}_1 \cos \theta_2 - 2\dot{\theta}_1\dot{\theta}_2 \sin \theta_2). \tag{4.79}$$

The above can be further simplified to

$$\ddot{\theta}_1 = 2\dot{\theta}_1\dot{\theta}_2 \tan \theta_2, \tag{4.80}$$

$$\ddot{\theta}_2 = -\dot{\theta}_1^2 \sin \theta_2 \cos \theta_2 - \frac{3g}{4a} \cos \theta_2, \tag{4.81}$$

which agrees with the equations of motion derived earlier in Example 4.3. $\qquad \square$

4.4 Work and Energy

Formulas for the work and kinetic energy of a rigid body can be obtained by regarding a rigid body as a system of rigidly connected particles. Let particle i have mass m_i, position $r_i(t)$, and velocity $v_i(t) = \dot{r}_i(t)$, and be subject to an external force $f_i(t)$. The total work done by the external forces over some infinitesimal time interval dt is then

$$\delta W = \sum_i f_i \cdot v_i\, dt. \tag{4.82}$$

Now let r_c be the center of mass, and r_i' be the vector from the center of mass to particle i, so that $r_i = r_c + r_i'$. The velocity of particle i can then be written

$$v_i = v_c + \omega \times r_i', \tag{4.83}$$

where ω is the angular velocity of the rigid body. Substituting (4.83) into (4.82),

$$\delta W = \sum_i f_i \cdot \left(v_c + \omega \times r_i' \right) dt \tag{4.84}$$

$$= \left(\sum_i f_i \right) \cdot v_c\, dt + \sum_i f_i \cdot \left(\omega \times r_i' \right) dt. \tag{4.85}$$

Note that $\sum_i f_i = f$ is the net external force. Also, from the vector identity $a \cdot (b \times c) = c \cdot (a \times b) = b \cdot (c \times a)$, the second term simplifies to

$$\sum_i f_i \cdot (\omega \times r_i') = \omega \cdot \left(\sum_i r_i' \times f_i \right) = \omega \cdot m_c, \tag{4.86}$$

where $\sum_i r_i' \times f_i = m_c$ is the external moment with respect to the center of mass. Putting it all together,

$$\delta W = f \cdot v_c \, dt + m_c \cdot \omega \, dt. \tag{4.87}$$

The work done by the rigid body over some time interval $[0, T]$ is then

$$W = \int_0^T f \cdot v_c + m_c \cdot \omega \, dt \tag{4.88}$$

$$= \int_0^T f_c^\top v_c + m_c^\top \omega_c \, dt, \tag{4.89}$$

where (4.89) expresses the vector quantities of (4.88) in coordinates of the center of mass frame $\{c\}$.

The kinetic energy of a rigid body can also be obtained by regarding the rigid body as a system of rigidly connected particles:

$$\mathcal{K} = \sum_i \frac{1}{2} m_i \|v_i\|^2 \tag{4.90}$$

$$= \sum_i \frac{1}{2} m_i \|v_c + \omega \times r_i'\|^2 \tag{4.91}$$

$$= \sum_i \frac{1}{2} m_i \left(\|v_c\|^2 + 2 v_c \cdot (\omega \times r_i') + \|\omega \times r_i'\|^2 \right). \tag{4.92}$$

The first term is simply $\frac{1}{2} m \|v_c\|^2$, while the middle term simplifies to

$$\sum_i m_i v_c \cdot (\omega \times r_i') = v_c \cdot \omega \times \left(\sum_i m_i r_i' \right) = 0, \tag{4.93}$$

since from the definition of the center of mass $\sum_i m_i r_i' = 0$. The third term can be simplified using the vector identity $\|a \times b\|^2 = a \cdot (b \times (a \times b))$:[1]

$$\sum_i \frac{1}{2} m_i \|\omega \times r_i'\|^2 = \frac{1}{2} \omega \cdot \sum_i (r_i' \times m_i (\omega \times r_i')). \tag{4.94}$$

Recalling that the angular momentum of the rigid body with respect to its center of mass is given by $h_c = \sum_i r_i' \times m_i \dot{r}_i' = \sum_i r_i' \times m_i (\omega \times r_i')$, (4.94) simplifies to

$$\sum_i \frac{1}{2} m_i \|\omega \times r_i'\|^2 = \frac{1}{2} \omega \cdot h_c. \tag{4.95}$$

---

[1] In coordinates, $\|a \times b\|^2 = \|b \times a\|^2 = \|[b]a\|^2 = a^\top [b]^\top [b] a = a^\top [b][a]b = a \cdot (b \times (a \times b))$, where we make use of the identities $[b]^\top = -[b]$, $[b]a = -[a]b$.

Putting it all together,

$$\mathcal{K} = \frac{1}{2}m v_c \cdot v_c + \frac{1}{2}\omega \cdot h_c \tag{4.96}$$

$$= \frac{1}{2}m v_c^\top v_c + \frac{1}{2}\omega_c^\top I_c \omega_c, \tag{4.97}$$

where (4.97) is (4.96) expressed in coordinates of the center of mass frame $\{c\}$. Note that the first term corresponds to the translational kinetic energy of the rigid body, while the second term corresponds to the rotational kinetic energy of the rigid body.

The potential energy of a rigid body can also be obtained by regarding the rigid body as a system of particles; in this case the potential energy of the rigid body is the sum of the potential energies associated with each particle. As an example, in the case of a rigid body of mass m in a gravitational field, the potential energy is given by mgh, where g is the acceleration due to gravity and h is the height of the center of mass with respect to some chosen reference height. The potential energy of mechanical systems consisting of rigid bodies interconnected to each other by springs – such systems are widely used as models of vehicle suspension systems, for example – is covered in Chapter 7.

With the above notions of work and energy for rigid bodies, the same problem-solving techniques using the various conservation principles of work and energy, for example, $\mathcal{W} = \mathcal{K}_2 - \mathcal{K}_1, \mathcal{P}_1 + \mathcal{K}_1 = \mathcal{P}_2 + \mathcal{K}_2$, can also be applied to rigid bodies.

Example 4.5   Modeling a satellite as a rigid body, suppose the body frame is attached to its center of mass in such a way that its inertia matrix I_c is diagonal. Suppose further that the satellite undergoes only pure rotational motion about its center of mass, and that there are no external moments applied to the satellite. Under these assumptions, show that both the angular momentum and kinetic energy of the satellite are preserved, and that the components $\omega_c = (\omega_x, \omega_y, \omega_z)^\top$ of the angular velocity vector $\omega = \omega_x \hat{x}_c + \omega_y \hat{y}_c + \omega_z \hat{z}_c$ expressed in the center of mass frame are restricted to lie on the intersection of two ellipsoids.

Solution: Letting $I_c = \mathrm{diag}(\mu_x, \mu_y, \mu_z)$, the dynamic equations $I_c \dot{\omega}_c + \omega_c \times I_c \omega_c = 0$ for the satellite can be written

$$\mu_x \dot{\omega}_x + (\mu_z - \mu_y)\omega_y \omega_z = 0, \tag{4.98}$$

$$\mu_y \dot{\omega}_y + (\mu_x - \mu_z)\omega_z \omega_x = 0, \tag{4.99}$$

$$\mu_z \dot{\omega}_z + (\mu_y - \mu_x)\omega_x \omega_y = 0. \tag{4.100}$$

Expressing the rotational equations of motion in the form $m_c = \dot{h}_c$, where h_c is the angular momentum vector and m_c is the external moment, it naturally follows that if m_c is zero, then h_c is constant. We now verify this analytically. Letting $h_c \in \mathbb{R}^3$ be the frame $\{c\}$ representation of h_c, then $h_c = I_c \omega_c$, and

$$\|h_c\|^2 = \omega_c^\top I_c^\top I_c \omega_c = \mu_x^2 \omega_x^2 + \mu_y^2 \omega_y^2 + \mu_z^2 \omega_z^2. \tag{4.101}$$

Then

$$\frac{d}{dt}\|h_c\|^2 = 2(\mu_x^2 \omega_x \dot{\omega}_x + \mu_y^2 \omega_y \dot{\omega}_y + \mu_z^2 \omega_z \dot{\omega}_z). \tag{4.102}$$

Substituting for $\dot{\omega}_x$, $\dot{\omega}_y$, $\dot{\omega}_z$ from the equations of motion, it can then be verified that

$$\frac{d}{dt}\|h_c\|^2 = 2\left(\mu_x\mu_y - \mu_x\mu_z + \mu_y\mu_z - \mu_y\mu_x + \mu_z\mu_x - \mu_z\mu_y\right)\omega_x\omega_y\omega_z = 0, \quad (4.103)$$

or $\|h_c\|^2$ is equal to some constant nonnegative value c^2. Then from Equation (4.101), $\omega_c = (\omega_x, \omega_y, \omega_z)^\top$ is restricted to lie on the *angular momentum ellipsoid* defined by

$$\left(\frac{\mu_x}{c}\right)^2 \omega_x^2 + \left(\frac{\mu_y}{c}\right)^2 \omega_y^2 + \left(\frac{\mu_z}{c}\right)^2 \omega_z^2 = 1. \quad (4.104)$$

To show that kinetic energy is conserved, note that for pure rotational motion about the center of mass, the kinetic energy $\mathcal{K}$ of the satellite can be written

$$\mathcal{K} = \frac{1}{2}\omega_c^\top I_c\omega_c = \frac{1}{2}(\mu_x\omega_x^2 + \mu_y\omega_y^2 + \mu_z\omega_z^2). \quad (4.105)$$

Differentiating $\mathcal{K}$ with respect to time,

$$\begin{aligned}
\frac{d\mathcal{K}}{dt} &= \mu_x\omega_x\dot{\omega}_x + \mu_y\omega_y\dot{\omega}_y + \mu_z\omega_z\dot{\omega}_z \\
&= \omega_x(\mu_y - \mu_z)\omega_y\omega_z + \omega_y(\mu_z - \mu_x)\omega_z\omega_x + \omega_z(\mu_x - \mu_y)\omega_x\omega_y \\
&= 0.
\end{aligned} \quad (4.106)$$

Therefore $\mathcal{K} = d$ for some constant d, and ω_c is further restricted to lie on the *kinetic energy ellipsoid* defined by

$$\left(\frac{\mu_x}{2d}\right)\omega_x^2 + \left(\frac{\mu_y}{2d}\right)\omega_y^2 + \left(\frac{\mu_z}{2d}\right)\omega_z^2 = 1. \quad (4.107)$$

Equations (4.104) and (4.107) restrict ω_c to lie on the intersection of two ellipsoids, which is a curve called a *polhode*. $\qquad\square$

4.5 Summary

- A rigid body can be modeled as a system composed of an infinite number of particles. The equations of motion for a rigid body can then be derived by considering the equations of motion for a system of particles, and taking the limit as the number of particles increases to infinity.
- Letting f_i be the external force applied to particle i and $f = \sum_i f_i$ be the net external force on the rigid body, the equations of motion for the rigid body's center of mass are

$$f = m\ddot{r}_c, \quad (4.108)$$

where $m = \sum_i m_i$ is the total mass of the rigid body and $r_c = (\sum_i m_i r_i)/m$ is the position of the mass center. In terms of a reference frame {c} attached to the center of mass, the equations of motion can be explicitly written as

$$f_c = m(\dot{v}_c + \omega_c \times v_c), \quad (4.109)$$

where $f_c = (f_x, f_y, f_z)^\top$, $v_c = (v_x, v_y, v_z)^\top$, and $\omega_c = (\omega_x, \omega_y, \omega_z)^\top$ are, respectively, f, $\dot{r}_c$, and ω expressed in frame {c} coordinates, and $\dot{v}_c = (\dot{v}_x, \dot{v}_y, \dot{v}_z)^\top$.

- Given a rigid body with a body frame {b} attached to some point on the body (not necessarily its center of mass), let (x_i, y_i, z_i) be coordinates for particle i of the body with respect to frame {b}. The **inertia matrix** of the rigid body with respect to frame {b} is defined to be the 3×3 matrix

$$\mathcal{I}_b = \begin{bmatrix} I_{xx} & -I_{xy} & -I_{xz} \\ -I_{xy} & I_{yy} & -I_{yz} \\ -I_{xz} & -I_{yz} & I_{zz} \end{bmatrix}, \tag{4.110}$$

  where

$$I_{xx} = \Sigma_i \, \mathfrak{m}_i(y_i^2 + z_i^2), \quad I_{xy} = \Sigma_i \, \mathfrak{m}_i x_i y_i, \quad I_{xz} = \Sigma_i \, \mathfrak{m}_i x_i z_i,$$

$$I_{yy} = \Sigma_i \, \mathfrak{m}_i(x_i^2 + z_i^2), \quad I_{yz} = \Sigma_i \, \mathfrak{m}_i y_i z_i, \quad I_{zz} = \Sigma_i \, \mathfrak{m}_i(x_i^2 + y_i^2), \tag{4.111}$$

  the I_{xx}, I_{yy}, and I_{zz} terms are called the **moments of inertia** , while the I_{xy}, I_{yz}, and I_{xz} terms are called the **products of inertia**. In the limit as the number of particles goes to infinity, the summations for the moments and products of inertia can be replaced by integrals.
- Given a rigid body with two body frames {b} and {c}, with {c} attached to the center of mass, the inertia matrices $\mathcal{I}_b$ and $\mathcal{I}_c$ are related by

$$\mathcal{I}_b = R_{cb}^\top (\mathcal{I}_c - \mathfrak{m}[r_{cb}]^2)R_{cb}, \tag{4.112}$$

  where R_{cb} is the rotation matrix describing frame {b} relative to {c}, $r_{cb} \in \mathbb{R}^3$ is the vector from frame {c} to frame {b} expressed in frame {c} coordinates, and $[r_{cb}]$ is the 3×3 skew-symmetric representation of r_{cb}. When $R_{cb} = I$, Equation (4.112) is called the **Parallel Axis Theorem**.
- The **angular momentum** of a rigid body with respect to its center of mass is defined as

$$\mathfrak{h}_c = \sum_i \mathfrak{r}_i' \times \mathfrak{m} \mathfrak{v}_i, \tag{4.113}$$

  where $\mathfrak{v}_i$ is the velocity of particle i and $\mathfrak{r}_i'$ is the vector from the center of mass to particle i. $\mathfrak{h}_c$ simplifies to

$$\mathfrak{h}_c = \sum_i \mathfrak{r}_i' \times \mathfrak{m}\dot{\mathfrak{r}}_i' = \sum_i -\mathfrak{m}_i \left(\mathfrak{r}_i' \times (\mathfrak{r}_i' \times \omega) \right), \tag{4.114}$$

  where ω is the angular velocity of the rigid body. Letting $h_c \in \mathbb{R}^3$ be $\mathfrak{h}_c$ expressed in terms of the center of mass frame {c},

$$h_c = (h_x, h_y, h_z)^\top = \mathcal{I}_c \omega_c, \tag{4.115}$$

  where $\mathcal{I}_c$ is the 3×3 inertia matrix of the rigid body with respect to frame {c}, and $\omega_c \in \mathbb{R}^3$ is ω expressed in frame {c} coordinates.
- The moment equation with respect to the center of mass, $\mathfrak{m}_c = \dot{\mathfrak{h}}_c$, can be expressed in terms of the center of mass frame {c} coordinates as

$$m_c = \mathcal{I}_c \alpha_c + \omega_c \times \mathcal{I}_c \omega_c, \tag{4.116}$$

where $m_c \in \mathbb{R}^3$, $\omega_c \in \mathbb{R}^3$ are, respectively, m_c and ω expressed in frame $\{c\}$ coordinates, and $\alpha_c = \dot{\omega}_c \in \mathbb{R}^3$ is the angular acceleration expressed in frame $\{c\}$ coordinates.

- Given a rigid body with center of mass frame $\{c\}$ and another body frame $\{a\}$, and a force f applied to the body, the moment generated by f with respect to $\{a\}$ is given by

$$m_a = r_a \times f = m_c + (p_{ac} \times f), \tag{4.117}$$

where p_{ac} is the vector from frame $\{a\}$ to frame $\{c\}$ and m_c is the moment with respect to $\{c\}$.

- If a rigid body undergoes rotation about some fixed point O, the moment equation for the rigid body can also be expressed with respect to O as $m_o = \dot{h}_o$. If another body frame $\{a\}$ is attached to the body, the moment equation with respect to $\{a\}$ can then be written

$$m_a = \dot{h}_o + (p_{ao} \times f), \tag{4.118}$$

where p_{ao} is the vector from frame $\{a\}$ to frame $\{o\}$.

- Given a rigid body subject to an external force f and external moment m_c with respect to its center of mass, and whose center of mass frame $\{c\}$ moves with linear velocity v_c and angular velocity ω, the **work** $\mathcal{W}$ done by the body is given by

$$\mathcal{W} = \int_0^T f \cdot v_c + m_c \cdot \omega \, dt, \tag{4.119}$$

while the **kinetic energy** $\mathcal{K}$ of the body is given by

$$\mathcal{K} = \frac{1}{2} m v_c \cdot v_c + \frac{1}{2} h_c \cdot \omega$$
$$= \frac{1}{2} m v_c^\top v_c + \frac{1}{2} \omega_c^\top \mathcal{I}_c \omega_c. \tag{4.120}$$

- The **potential energy** of a rigid body is the sum of the potential energies associated with each particle constituting the rigid body. For the case of a rigid body of mass m in a gravitational field, the associated potential energy $\mathcal{P}$ is given by $\mathcal{P} = mgh$, where g is the acceleration due to gravity and h is the height of the center of mass with respect to some chosen reference height corresponding to $h = 0$.

- The conservation principles for work and energy, for example, $\mathcal{W} = \mathcal{K}_2 - \mathcal{K}_1$, $\mathcal{P}_1 + \mathcal{K}_1 = \mathcal{P}_2 + \mathcal{K}_2$, can also be applied to rigid bodies in the same manner as for particles.

4.6 Exercises

Exercise 4.1 For the rigid bodies with mass m shown in Figure 4.6, show that the inertia matrices with respect to their center of mass frames are given as follows.
(a) Rod of length l:

$$\mathrm{diag}\left(\frac{1}{12}ml^2, \, 0, \, \frac{1}{12}ml^2\right).$$

(b) Rectangular plate of width a and height b:

$$\mathrm{diag}\left(\frac{1}{12}ma^2,\ \frac{1}{12}mb^2,\ \frac{1}{12}m(a^2+b^2)\right).$$

(c) Solid disk of radius r:

$$\mathrm{diag}\left(\frac{1}{4}mr^2,\ \frac{1}{4}mr^2,\ \frac{1}{2}mr^2\right).$$

(d) Circular ring of radius r:

$$\mathrm{diag}\left(\frac{1}{2}mr^2,\ \frac{1}{2}mr^2,\ mr^2\right).$$

(e) Solid ball of radius r:

$$\mathrm{diag}\left(\frac{2}{5}mr^2,\ \frac{2}{5}mr^2,\ \frac{2}{5}mr^2\right).$$

(f) Solid cylinder of radius r and height h:

$$\mathrm{diag}\left(\frac{1}{12}m(3r^2+h^2),\ \frac{1}{12}m(3r^2+h^2),\ \frac{1}{2}mr^2\right).$$

Exercise 4.2   As shown in Figure 4.7, three identical rods of length l and mass m are connected as one rigid body of mass $3m$. Find the inertia matrix I_c of the rigid body with respect to the center of mass frame $\{c\}$. (*Hint*: Assign appropriate body frames to each rod.)

Exercise 4.3   Suppose the inertia matrix of a rigid body with respect to a body frame $\{c\}$ attached to the center of mass is

$$I_c = \begin{bmatrix} 1 & 0 & 0 \\ 0 & 5 & 0 \\ 0 & 0 & 2 \end{bmatrix}.$$

Another body frame $\{a\}$ is attached to the rigid body such that the orientation of frame $\{c\}$ relative to frame $\{a\}$ is

$$R_{ac} = \begin{bmatrix} \frac{1}{\sqrt{2}} & 0 & \frac{1}{\sqrt{2}} \\ 0 & 1 & 0 \\ -\frac{1}{\sqrt{2}} & 0 & \frac{1}{\sqrt{2}} \end{bmatrix}$$

and the vector from frame $\{a\}$ to frame $\{c\}$ expressed in frame $\{a\}$ coordinates is $p_{ac} = (1, 2, 1)^\mathsf{T}$. Find the inertia matrix I_a of the rigid body with respect to frame $\{a\}$. The mass of the rigid body is $m = 1$.

Exercise 4.4   (a) Let $I_c \in \mathbb{R}^{3\times3}$ be the inertia matrix of a rigid body with respect to its center of mass frame $\{c\}$, and let I_a be the inertia matrix of the same rigid body with respect to another frame $\{a\}$ attached at another point on the body. Find an expression for I_c in terms of I_a. You may assume the rotation matrix R_{ca} and the vector d_{ca} from

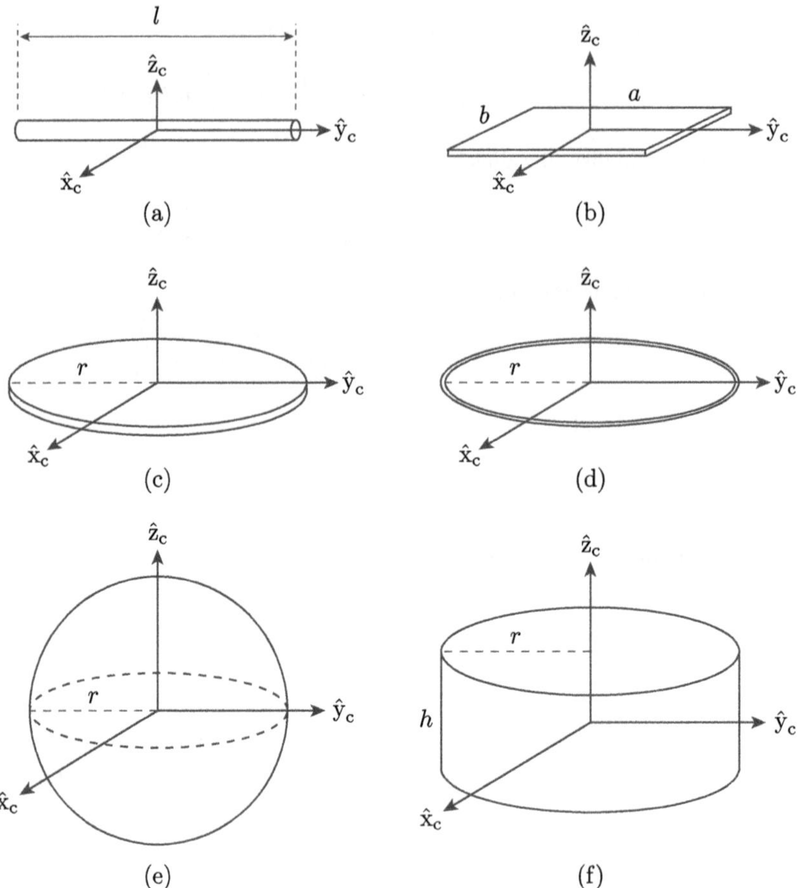

Figure 4.6 Rigid bodies of Exercise 4.1.

frame {c} to frame {a} (expressed in frame {c} coordinates) are known. The mass of the rigid body is $\mathfrak{m}$.

(b) Let $\mathcal{I}_b$ be the inertia matrix of the same rigid body with respect to another frame {b} that is not at the center of mass. Find an expression for $\mathcal{I}_b$ in terms of $\mathcal{I}_a$. You may assume the rotation matrix R_{ab} and the vectors d_{ab} from frame {a} to frame {b} and d_{ac} from frame {a} to frame {c} (expressed in frame {a} coordinates) are known.

Exercise 4.5 Let $\mathcal{I}_c \in \mathbb{R}^{3\times 3}$ be the inertia matrix of a rigid body with respect to its center of mass frame {c}. By definition $\mathcal{I}_c$ is always symmetric.

(a) Show that $\mathcal{I}_c$ is **positive-definite**; i.e., $\omega^\top \mathcal{I}_c \omega > 0$ for all nonzero $\omega \in \mathbb{R}^3$. (*Hint:* Use the fact that the kinetic energy of a rigid body in motion is always positive.)

(b) Let {b} be another body frame attached to the rigid body's center of mass but in a different orientation from {c}. Recall that the inertia matrix $\mathcal{I}_b \in \mathbb{R}^{3\times 3}$ is then related to $\mathcal{I}_c$ via

$$\mathcal{I}_b = R_{cb}^\top \mathcal{I}_c R_{cb},$$

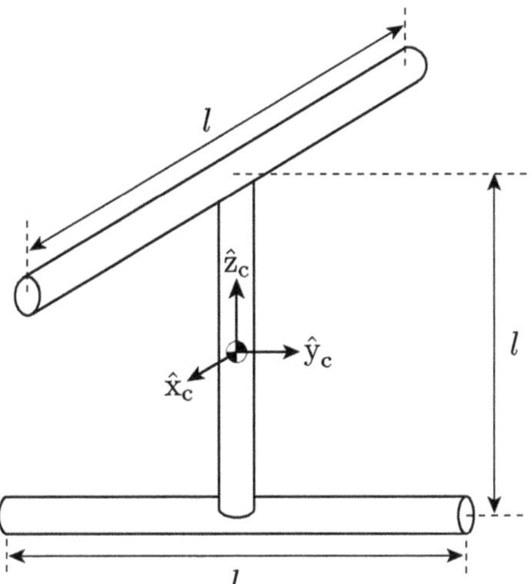

Figure 4.7 Three identical rods connected as a rigid body of Exercise 4.2.

where the rotation matrix R_{cb} represents the orientation of frame {b} relative to {c}. Show that I_b is also positive-definite.

(c) Assume that the eigenvalues of I_c are all real and distinct. Show that the {b} frame can always be chosen so that I_b is a diagonal matrix. (*Hint*: Use the fact that for any real-symmetric matrix A with distinct eigenvalues, its eigenvectors can always be chosen to be orthonormal to each other; i.e., each eigenvector is of unit length and orthogonal to each other.)

(d) More generally, any real symmetric matrix A can always be factorized as $A = R^T D R$, where D is diagonal and R is a rotation matrix. Verify this for the matrix

$$A = \begin{bmatrix} 0 & 1 & 1 \\ 1 & 0 & 1 \\ 1 & 1 & 0 \end{bmatrix}.$$

Exercise 4.6   We seek to find the kinetic energy of a rigid body rotating about some fixed point O with angular velocity ω. Referring to Figure 4.4, two body frames {c} and {b} are attached to the center of mass and to point O respectively. R_{cb} is the constant rotation matrix describing the orientation of frame {b} relative to frame {c}, and r_{bc} is the constant vector from frame {b} to frame {c}.

(a) Let ω_b and ω_c be ω expressed in frame {b} and frame {c} coordinates, respectively, and v_c be the linear velocity of the center of mass expressed in frame {c} coordinates. Show that

$$\omega_c = R_{cb}\omega_b,$$
$$v_c = [r'_{cb}]R_{cb}\omega_b,$$

where r'_{cb} is the vector from frame {c} to frame {b} expressed in frame {c} coordinates. (*Hint*: Given $u, v \in \mathbb{R}^3$, then $[u]v = [-v]u$.)

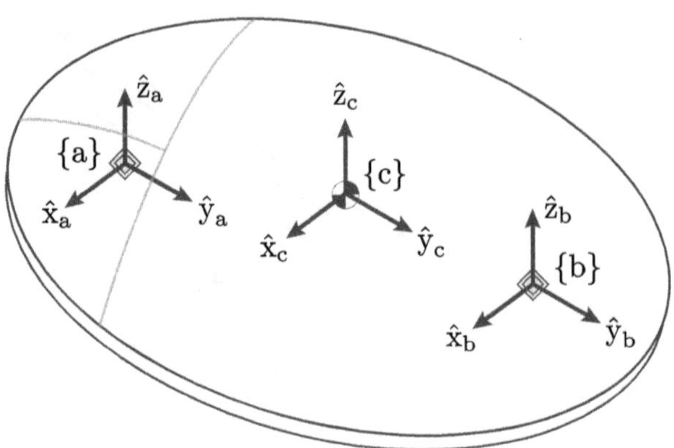

Figure 4.8 Computer mouse of Exercise 4.7.

(b) Let $\mathcal{I}_b$ be the inertia matrix of the rigid body with respect to frame {b}. Show that the kinetic energy $\mathcal{K}$ of the rigid body is

$$\mathcal{K} = \frac{1}{2}\omega_b^\top \mathcal{I}_b \omega_b.$$

Exercise 4.7   We seek to find the linear and angular velocities of the center of mass of a computer mouse, which moves on a plane, using two velocity sensors attached to the mouse. As shown in Figure 4.8, three body frames {a}, {b}, and {c} are attached to sensor 1, sensor 2, and the center of mass, respectively, where the three frames have the same orientation. The positions of the two sensors expressed in frame {c} coordinates are $p_{ca} = (p_{ax}, p_{ay})^\top$ and $p_{cb} = (p_{bx}, p_{by})^\top$ respectively. The velocities measured by the sensors expressed in each sensor frame are $v_a = (v_{ax}, v_{ay})^\top$ and $v_b = (v_{bx}, v_{by})^\top$ respectively. However, one sensor is broken, causing v_{ax} to be inaccurate.
(a) Express the angular velocity $\omega_c = \omega\hat{z}_c$ of the center of mass using the given parameters and accurate measurements.
(b) Express the linear velocity $v_c = v_x\hat{x}_c + v_y\hat{y}_c$ of the center of mass using the given parameters and accurate measurements.
(c) What conditions should be satisfied regarding the relative position of the two sensors to find ω_c and v_c?

Exercise 4.8   As shown in Figure 4.9, a rod of mass m and length l is connected to a rotating shaft by a pin. The shaft rotates at a constant rate ω and gravity g acts downward.
(a) Determine the angle $\theta > 0$ of the rod about the vertical axis. Assume ω is sufficiently large.
(b) Determine the kinetic energy $\mathcal{K}$ of the rod.

Exercise 4.9   As shown in Figure 4.10, a square plate of mass m and side length b is connected to a revolute joint with a horizontal rotation axis, while the whole system rotates about the vertical axis at a constant rate ω. A body frame {b} is assigned at the intersection of the two rotation axes ($\hat{y}_b$ is perpendicular to the plate), and angle θ is defined as shown in the figure. Gravity g acts downward.

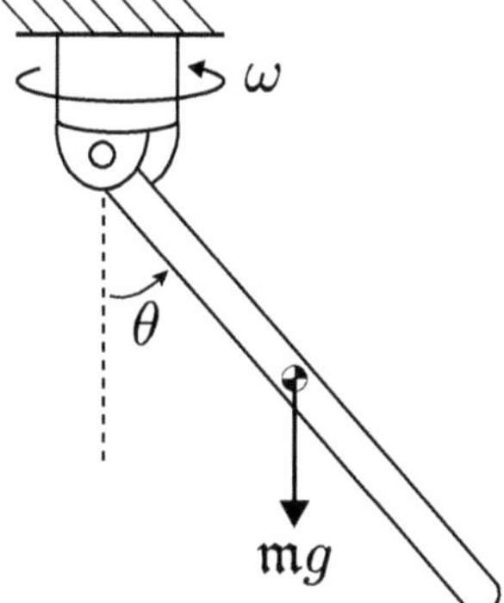

Figure 4.9 Rod connected to rotating shaft of Exercise 4.8.

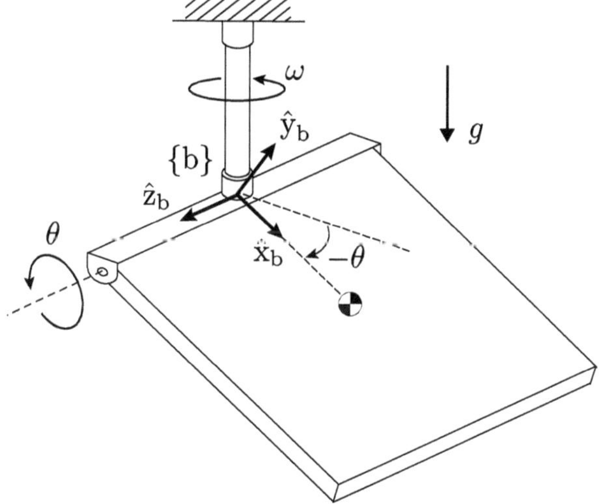

Figure 4.10 Rotating square plate of Exercise 4.9.

(a) Find the inertia matrix I_b of the plate with respect to frame $\{b\}$.

(b) Determine the angle $\theta > -\frac{\pi}{2}$ of the plate. Assume ω is sufficiently large.

Exercise 4.10   Consider the spherical pendulum of mass m and length L rotating about two orthogonal axes as shown in Figure 4.11. Two frames $\{b\}$ and $\{c\}$ are attached to the rod, where $\{b\}$ is assigned at the intersection of the two axes of rotation and $\{c\}$ at the rod center of mass. Joint angles θ_1 and θ_2 are defined as shown in the figure and gravity g acts downward.

(a) Find the inertia matrix I_b of the rod with respect to frame $\{b\}$.

(b) The angular momentum of the rod with respect to the intersection point of the two rotation axes is $h_x \hat{x}_b + h_y \hat{y}_b + h_z \hat{z}_b$. Find h_x, h_y, and h_z.

(c) Derive the equations of motion of the spherical pendulum.

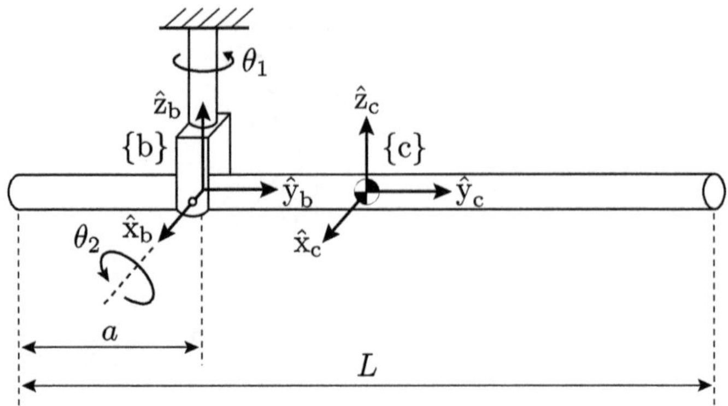

Figure 4.11 Spherical pendulum of Exercise 4.10.

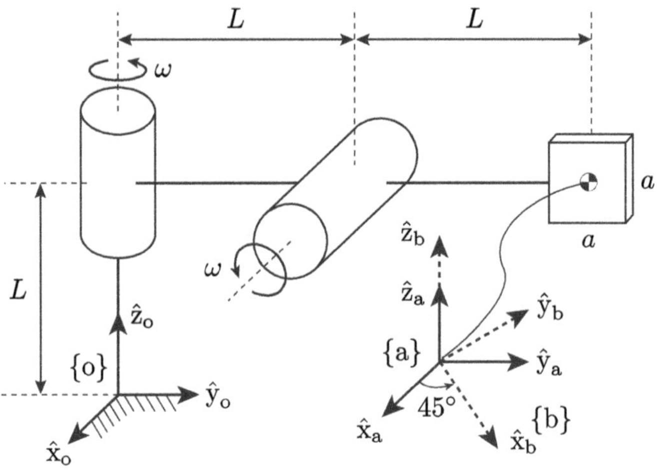

Figure 4.12 Square plate attached to robot arm of Exercise 4.11.

Exercise 4.11 As shown in Figure 4.12, a square plate of mass $\mathfrak{m}$ and side length a is attached to the tip of a robot arm. The length of the links is L and the joints rotate at a constant rate ω. Two body frames {a} and {b} with different orientations are attached to the plate's center of mass as shown.

(a) At the instant shown in the figure, find the angular velocity $\omega_a \in \mathbb{R}^3$ and linear velocity $v_a \in \mathbb{R}^3$ of the plate, both expressed in frame {a} coordinates.

(b) Determine the kinetic energy of the plate at the instant shown.

(c) Find the rotation matrix R_{ab} describing the orientation of frame {b} relative to frame {a} and the inertia matrix $\mathcal{I}_b$ of the plate with respect to frame {b}.

Exercise 4.12 As shown in Figure 4.13, a tilted disk of radius r_d and mass $\mathfrak{m}_d$ is rigidly attached to a thick rod of radius r_r, length l, and mass $\mathfrak{m}_r$, where the angle between the disk and rod is $45°$. The rod rotates at a constant rate ω. Find the kinetic energy of the system.

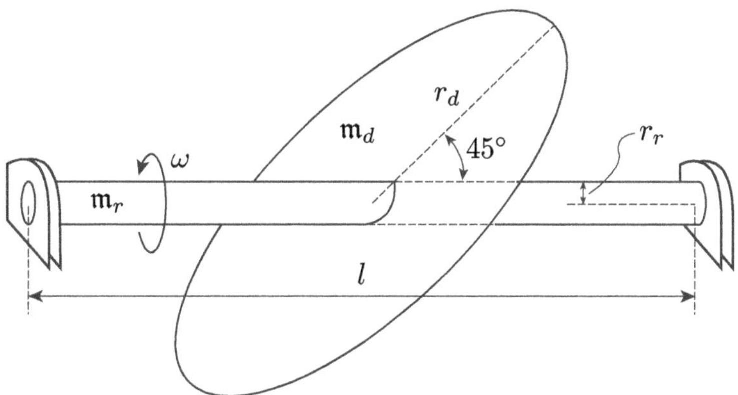

Figure 4.13 Rotating tilted disk of Exercise 4.12.

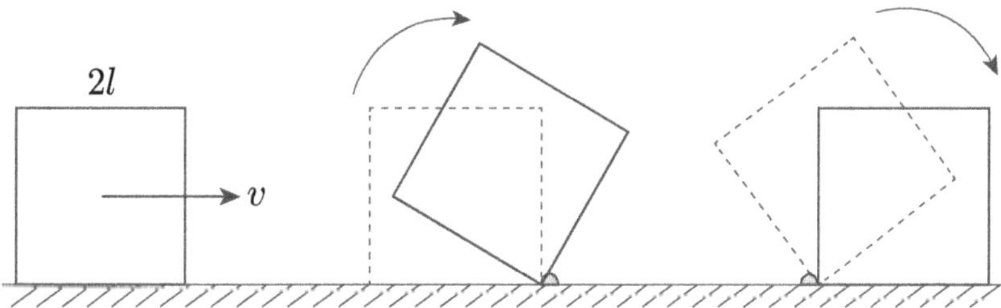

Figure 4.14 Sliding block of Exercise 4.13.

Exercise 4.13 As shown in Figure 4.14, a square block of mass m with side length $2l$ slides on a frictionless surface with linear velocity v. The moment of inertia with respect to the center of mass is given as I_c. Gravity g acts downward. The block meets a small bump of negligible size, then rolls over the bump and lands in the position shown in the picture. Find the minimum velocity v for the block to roll over the bump.

Exercise 4.14 As shown in Figure 4.15, a tube of mass m_t and length $2L$ is rotating about the vertical axis with angular velocity ω. A ball (particle) of mass m_b is dropped into the tube center. Neglecting friction, determine the angular velocity of the tube just after the ball has left the end of the tube. Assume the moment of inertia of the tube about the vertical axis is $I_{zz} = \frac{1}{12}m_t L^2$.

Exercise 4.15 A thin uniform rod AB of mass m and length L is connected to massless sliding joints A and B as shown in Figure 4.16. Angle θ and angular velocity ω are defined as shown in the figure.
(a) Find the velocity v_c of the rod's center of mass C in terms of θ, ω, and L. Express your answer in the fixed frame {o} coordinates.
(b) Derive the rod's kinetic energy in terms of ω, m, and L.
(c) At $t = 0$ the rod lies in the horizontal position ($\theta = 0$). Because of gravity g acting

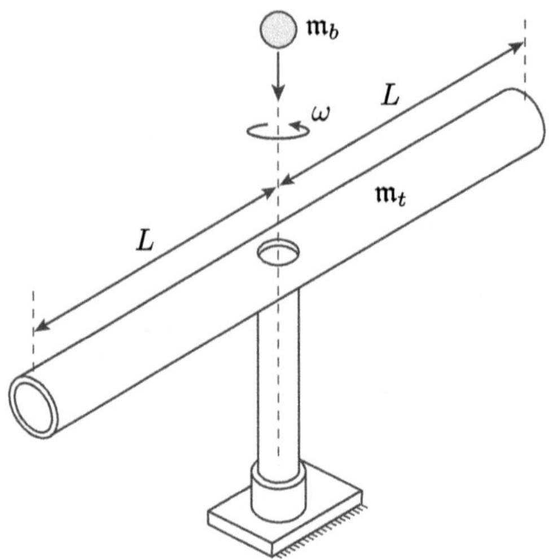

Figure 4.15 Ball dropped into rotating tube of Exercise 4.14.

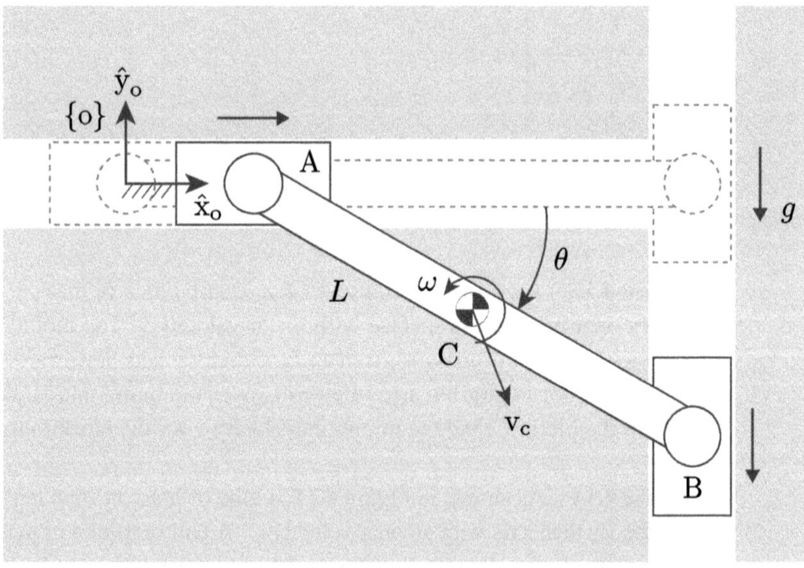

Figure 4.16 Rod connected to two orthogonal sliding joints of Exercise 4.15.

downward, joint B starts sliding down. Find the change in the potential energy of the rod as a function of θ. Also, express ω as a function of θ.

Exercise 4.16 A thin uniform disk of radius R and mass m is rigidly attached to a rotating vertical shaft, where the angle between the disk and shaft is θ (see Figure 4.17). A body frame {b} is attached to the disk at the center of mass C and the fixed frame {s} is also assigned at C. Assume the shaft is massless and there is no gravity.

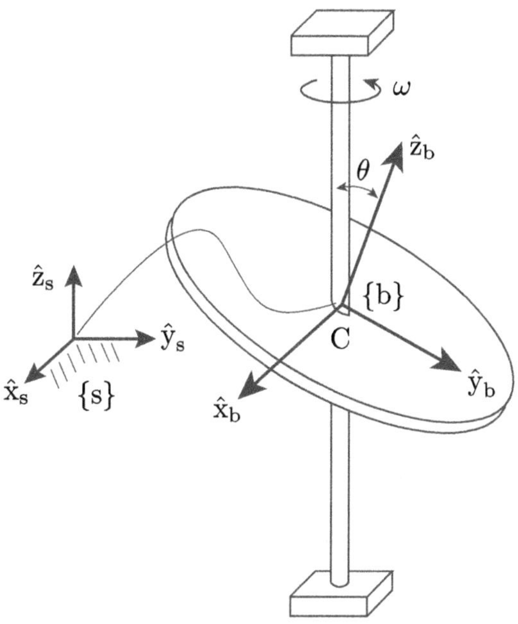

Figure 4.17 Vertically rotating tilted disk of Exercise 4.16.

(a) At the instant shown in the figure, $\hat{x}_s$ of frame {s} and $\hat{x}_b$ of frame {b} are aligned. Find the inertia matrix $\mathcal{I}_s$ of the disk with respect to frame {s}.

(b) Suppose the shaft rotates at a constant rate ω. At the instant shown in the figure, find the angular momentum h_c about C. Express your answer in the fixed frame {s} coordinates.

(c) Now suppose the shaft rotates with angular velocity ω and angular acceleration α. Find the torque at the shaft at the instant shown. Express your answer in the fixed frame {s} coordinates.

5 Multibody Dynamics: A Twist–Wrench Formulation

In this chapter we will derive the equations of motion for rigid body systems in terms of **twists** and **wrenches**. A twist is obtained by concatenating the pair of three-dimensional angular and linear velocity vectors into a single six-dimensional vector. A wrench is similarly obtained by concatenating the pair of three-dimensional moment and force vectors into a single six-dimensional vector.

Using twists and wrenches, the equations of motion for a rigid body can be unified into a single six-dimensional equation, rather than treated as two separate equations involving external forces and moments, respectively. To be more explicit, recall that the equations of motion for a rigid body expressed with respect to a body frame $\{c\}$ attached to the rigid body center of mass can be written

$$f_c = \mathfrak{m}(\dot{v}_c + \omega_c \times v_c), \tag{5.1}$$

$$m_c = I_c\,\dot{\omega}_c + \omega_c \times I_c\omega_c, \tag{5.2}$$

where $f_c \in \mathbb{R}^3$ is the external force, $m_c \in \mathbb{R}^3$ is the external moment, $\mathfrak{m}$ is the mass, $I_c \in \mathbb{R}^{3\times3}$ is the inertia matrix, $\omega_c \in \mathbb{R}^3$ and $v_c \in \mathbb{R}^3$ are the angular and linear velocities of body frame $\{c\}$, $\dot{v}_c \in \mathbb{R}^3$ is the time-derivative of v_c, and $\dot{\omega}_c \in \mathbb{R}^3$ is the angular acceleration. Equations (5.1)–(5.2) can be written in combined form as

$$\begin{bmatrix} m_c \\ f_c \end{bmatrix} = \begin{bmatrix} I_c & 0 \\ 0 & \mathfrak{m}\cdot\mathbb{1} \end{bmatrix}\begin{bmatrix} \dot{\omega}_c \\ \dot{v}_c \end{bmatrix} + \begin{bmatrix} [\omega_c] & [v_c] \\ 0 & [\omega_c] \end{bmatrix}\begin{bmatrix} I_c & 0 \\ 0 & \mathfrak{m}\cdot\mathbb{1} \end{bmatrix}\begin{bmatrix} \omega_c \\ v_c \end{bmatrix}, \tag{5.3}$$

where $\mathbb{1}$ denotes the 3×3 identity matrix.[1] Define the wrench $\mathcal{F}_c = (m_c, f_c)$, twist $\mathcal{V}_c = (\omega_c, v_c)$, and the following 6×6 matrix representation of $\mathcal{V}_c$:

$$[\mathcal{V}_c] = \begin{bmatrix} [\omega_c] & 0 \\ [v_c] & [\omega_c] \end{bmatrix}. \tag{5.4}$$

Equation (5.3) can now be written more compactly as

$$\mathcal{F}_c = \mathcal{G}_c\dot{\mathcal{V}}_c - [\mathcal{V}_c]^{\mathsf{T}}\mathcal{G}_c\mathcal{V}_c, \quad \mathcal{G}_c = \begin{bmatrix} I_c & 0 \\ 0 & \mathfrak{m}\cdot\mathbb{1} \end{bmatrix}. \tag{5.5}$$

Here $\mathcal{G}_c \in \mathbb{R}^{6\times6}$ is called the **spatial inertia matrix**, and can be thought of as a 6×6 generalization of the 3×3 inertia matrix I_c.

---

[1] Equation (5.3) makes use of the identity $[v_c]v_c = v_c \times v_c = 0$.

At this point it is fair to ask if the twist–wrench formulation offers any real advantage besides notational compactness. The answer is an unequivocal yes, particularly for mechanical systems involving multiple interacting rigid bodies (*multibody systems*). For such systems it is almost always necessary to transform forces, moments, velocities, accelerations and other quantities from the reference frame of one rigid body to another, especially between bodies that exert forces on each other (for example, connected by a joint or in contact with each other). The twist–wrench formalism offers a simple and concise set of rules for how such physical quantities transform under a change of reference frame.

As a preview of these transformation rules, attach reference frame {b} to another point on the rigid body, and suppose we wish to recast Equations (5.1)–(5.2) with respect to frame {b} rather than frame {c}. Drawing upon everything we have learned about transformation rules for linear and angular velocities and accelerations, forces, moments, and inertias, the equations of motion with respect to frame {b} now become, after considerable calculation,

$$f_b = \mathfrak{m}(\dot{v}_b + [\omega_b]v_b + [\dot{\omega}_b]r_b + [\omega_b]^2 r_b), \tag{5.6}$$

$$m_b = I_b\dot{\omega}_b + [\omega_b]I_b\omega_b + \mathfrak{m}[r_b]\dot{v}_b + \mathfrak{m}[r_b][\omega_b]v_b$$
$$+ \mathfrak{m}[r_b][\dot{\omega}_b]r_b + \mathfrak{m}[r_b][\omega_b]^2 r_b, \tag{5.7}$$

where $v_b \in \mathbb{R}^3$ and $\omega_b \in \mathbb{R}^3$ are, respectively, the linear and angular velocity of frame {b}, $r_b \in \mathbb{R}^3$ is the vector from the frame {b} origin to the center of mass, $f_b \in \mathbb{R}^3$ is the external force, $m_b \in \mathbb{R}^3$ is the external moment taken with respect to the frame {b} origin, and $I_b \in \mathbb{R}^{3\times3}$ is the 3×3 inertia matrix with respect to frame {b} (*Note*: All vectors with the subscript b are expressed in frame {b} coordinates).

Considerable effort is needed to obtain Equations (5.6)–(5.7) from the original (5.1)–(5.2). Using the twist–wrench formalism, a concise and systematic set of transformation rules can be derived for twists, wrenches, and spatial inertias, allowing for straightforward derivation of (5.6)–(5.7) from (5.1)–(5.2), and vice versa. Specifically, by defining a special 6×6 transformation matrix

$$\Phi_{bc} = \begin{bmatrix} R_{bc} & 0 \\ [p_{bc}] R_{bc} & R_{bc} \end{bmatrix}, \tag{5.8}$$

where R_{bc} is the rotation matrix describing the orientation of frame {c} relative to frame {b}, and $p_{bc} \in \mathbb{R}^3$ is the vector from {b} to {c} expressed in frame {b} coordinates, it can be shown that

$$\mathcal{V}_b = \Phi_{bc}\mathcal{V}_c, \tag{5.9}$$

$$\mathcal{F}_b = \Phi_{cb}^{\top}\mathcal{F}_c, \tag{5.10}$$

$$\mathcal{G}_b = \Phi_{cb}^{\top}\mathcal{G}_c\Phi_{cb}. \tag{5.11}$$

Φ_{bc} further satisfies $\Phi_{ab}\Phi_{bc} = \Phi_{ac}$ (and therefore $\Phi_{bc}^{-1} = \Phi_{cb}$) and $[\Phi_{cb}\mathcal{V}_b] = \Phi_{cb}[\mathcal{V}_b]\Phi_{cb}^{-1}$. With these properties it follows quite straightforwardly that

$$\mathcal{F}_b = \mathcal{G}_b\dot{\mathcal{V}}_b - [\mathcal{V}_b]^{\top}\mathcal{G}_b\mathcal{V}_b. \tag{5.12}$$

Note that (5.12) and (5.5) have the same structure; what this means is that the equations of motion can be written independent of any choice of body frame as

$$\mathcal{F} = \mathcal{G}\dot{\mathcal{V}} - [\mathcal{V}]^\top \mathcal{G}\mathcal{V}. \tag{5.13}$$

By observing the transformation rules (5.9)–(5.11), the equations of motion for a rigid body can be expressed with respect to any body frame, making the twist–wrench formalism a particularly powerful tool for multibody systems whose advantages go far beyond notational compactness and ease of bookkeeping. Later in this chapter we use the twist–wrench formalism to derive simple recursive algorithms for the dynamics of serially connected rigid bodies, or **open chains**. Open chains are ubiquitous – we have already encountered numerous examples in the previous chapters – and the corresponding recursive algorithms serve as a building block for the analysis of more complex multibody systems.

5.1   Velocity, Force, and Moment Transformation Formulas

In this section we derive the transformation formulas for the angular and linear velocities, and also forces and moments of a rigid body under a change of body reference frame. Before doing so we remind the reader of some notation and formulas from Chapter 2:

- Let $v_a, v_b \in \mathbb{R}^3$ be column vector representations of vector v with respect to frames $\{a\}$ and $\{b\}$, respectively. Then $v_a = R_{ab}v_b$ and $v_b = R_{ba}v_a$, where R_{ab} is the rotation matrix describing frame $\{b\}$ relative to frame $\{a\}$. Also recall that $R_{ab}R_{bc} = R_{ac}$ and $R_{ab}^{-1} = R_{ba}$.
- Given $\omega \in \mathbb{R}^3$, its 3×3 skew-symmetric matrix representation

$$[\omega] = \begin{bmatrix} 0 & -\omega_3 & \omega_2 \\ \omega_3 & 0 & -\omega_1 \\ -\omega_2 & \omega_1 & 0 \end{bmatrix} \tag{5.14}$$

  satisfies $[\omega] + [\omega]^\top = 0$ and $R[\omega]R^\top = [R\omega]$ for any rotation matrix R.
- Given $u, v \in \mathbb{R}^3$, then $u \times v = [u]v$, $[u \times v] = [u][v] - [v][u]$, and $[u]v = [-v]u$.

Angular Velocities

Referring to Figure 5.1, frame $\{o\}$ is the inertial frame, frame $\{c\}$ is attached to the center of mass, and frame $\{b\}$ is attached to some other arbitrary point on the rigid body. As the rigid body moves in space, frames $\{c\}$ and $\{b\}$ experience linear velocities v_c and v_b, respectively, and angular velocities ω_c and ω_b, respectively. Let p_{cb} be the vector from frame $\{c\}$ to frame $\{b\}$. If $\omega_c \in \mathbb{R}^3$ is the representation of ω_c in frame $\{c\}$ coordinates, while $\omega_b \in \mathbb{R}^3$ is the representation of ω_b in frame $\{b\}$ coordinates, then $[\omega_b] = R_{ob}^\top \dot{R}_{ob}$ and $[\omega_c] = R_{oc}^\top \dot{R}_{oc}$. Since $R_{ob} = R_{oc}R_{cb}$ with R_{cb} constant, it follows that

$$[\omega_b] = R_{ob}^\top \dot{R}_{ob} = R_{cb}^\top R_{oc}^\top (\dot{R}_{oc}R_{cb}) = R_{cb}^\top [\omega_c]R_{cb} = R_{bc}[\omega_c]R_{bc}^\top = [R_{bc}\omega_c], \tag{5.15}$$

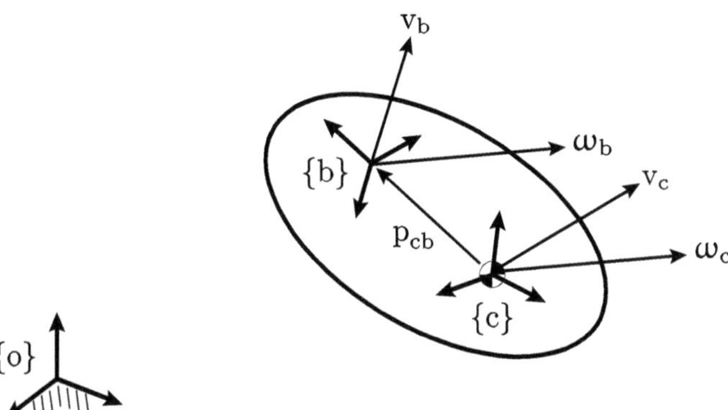

Figure 5.1 A rigid body with two reference frames attached: frame {c} is attached to the center of mass, while frame {b} is attached to some other arbitrary point on the rigid body.

or in vector form,

$$\omega_b = R_{bc}\omega_c, \quad \omega_c = R_{cb}\omega_b. \tag{5.16}$$

In effect, what we have just proven is that $\omega_c = \omega_b = \omega$, i.e., the angular velocity of a rigid body is the same regardless of where the body reference frame is attached to the rigid body.

Linear Velocities

Examining Figure 5.1, from what we have learned in Chapter 2 it should be apparent that v_b and v_c are related by $v_b = v_c + \omega \times p_{cb}$ (and although not needed here, the accelerations a_b and a_c are related by $a_b = a_c + \dot{\omega} \times p_{cb} + \omega \times (\omega \times p_{cb})$). Letting $v_b \in \mathbb{R}^3$ be v_b represented in frame {b} coordinates, and $v_c \in \mathbb{R}^3$ be v_c represented in frame {c} coordinates, we have

$$v_b = R_{bc}(v_c + [\omega_c]p_{cb}) = R_{bc}v_c + R_{bc}[\omega_c]p_{cb} = R_{bc}v_c + R_{bc}[-p_{cb}]\omega_c. \tag{5.17}$$

We remind the reader that $p_{cb} \in \mathbb{R}^3$ is the column vector representation of p_{cb} in frame {c} coordinates, and $p_{bc} \in \mathbb{R}^3$ is the column vector representation of p_{bc} in frame {b} coordinates. It then follows that $-p_{cb}$ is just p_{bc} expressed in frame {c} coordinates, i.e., $-p_{cb} = R_{cb}p_{bc}$. The above can therefore be rewritten

$$v_b = R_{bc}v_c + R_{bc}[R_{cb}p_{bc}]\omega_c = R_{bc}v_c + R_{bc}R_{cb}[p_{bc}]R_{cb}^{\mathsf{T}}\omega_c, \tag{5.18}$$

which further simplifies to

$$v_b = R_{bc}v_c + [p_{bc}]R_{bc}\omega_c. \tag{5.19}$$

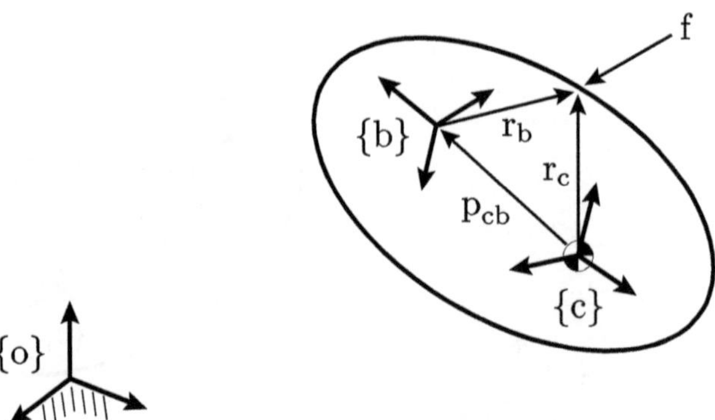

Figure 5.2 Transformation rule for forces and moments applied to a rigid body: frames {c} and {b} are attached to different points on the rigid body.

Forces and Moments

We now derive the transformation rule for applied forces and moments. Referring to Figure 5.2, let f be an external force applied to a rigid body. If $f_c \in \mathbb{R}^3$ is f expressed in frame {c} coordinates, and $f_b \in \mathbb{R}^3$ is f expressed in frame {b} coordinates, then f_b and f_c are related by

$$f_c = R_{cb} f_b, \quad f_b = R_{bc} f_c. \tag{5.20}$$

Let $m_b = r_b \times f$ be the moment generated by f with respect to the frame {b} origin, and $m_c = r_c \times f$ be the moment generated by f with respect to the frame {c} origin. Observing that $r_c = p_{cb} + r_b$, m_c can be written

$$m_c = (p_{cb} \times f) + (r_b \times f) = (p_{cb} \times f) + m_b. \tag{5.21}$$

Expressing the above in frame {c} coordinates,

$$\begin{aligned}
m_c &= [p_{cb}] f_c + R_{cb} m_b \\
&= [-R_{cb} p_{bc}](R_{cb} f_b) + R_{cb} m_b \\
&= -R_{cb}[p_{bc}] R_{cb}^{\mathsf{T}} R_{cb} f_b + R_{cb} m_b \\
&= -R_{cb}[p_{bc}] f_b + R_{cb} m_b.
\end{aligned} \tag{5.22}$$

Equations (5.16)–(5.22) are the transformation rules for linear and angular velocities, and also applied forces and moments, under a change of body reference frame. In Section 5.2 we will express these four transformation rules in terms of a pair of equations involving six-dimensional twists and wrenches.

5.2   Twists and Wrenches

Letting $\omega \in \mathbb{R}^3$ and $v \in \mathbb{R}^3$, respectively, be the column vector representations of ω and v with respect to the same reference frame, we now combine ω and v into a six-dimensional **twist** $\mathcal{V}$:

$$\mathcal{V} = \left[\begin{array}{c} \omega \\ v \end{array} \right] \in \mathbb{R}^6, \tag{5.23}$$

which we will also sometimes write more compactly as $\mathcal{V} = (\omega, v)$. $\mathcal{V} \in \mathbb{R}^6$ will also be written as the following 6×6 matrix $[\mathcal{V}]$:

$$[\mathcal{V}] = \left[\begin{array}{cc} [\omega] & 0 \\ [v] & [\omega] \end{array} \right]. \tag{5.24}$$

To express the reference frame transformation rules for $\mathcal{V}$ more succinctly, the following notation is introduced. Given a rotation matrix R and $p \in \mathbb{R}^3$, define the 6×6 matrix $\Phi(R, p)$ as follows:

$$\Phi(R, p) = \left[\begin{array}{cc} R & 0 \\ [p]\,R & R \end{array} \right]. \tag{5.25}$$

To simplify the notation we will in most cases suppress the arguments (R, p). We now list some properties of the matrix Φ.

Proposition 5.1   *Φ as defined in (5.25) satisfies the following properties:*

1. $\Phi^{-1} = \left[\begin{array}{cc} R^{\top} & 0 \\ R^{\top}[p]^{\top} & R^{\top} \end{array} \right].$

2. $\Phi[\mathcal{V}]\Phi^{-1} = [\Phi\mathcal{V}].$

3. Letting Φ_{ab} be defined as in (5.25) for some given rotation matrix R_{ab} and $p_{ab} \in \mathbb{R}^3$, and similarly for Φ_{bc} and Φ_{ac}, then $\Phi_{ab}\Phi_{bc} = \Phi_{ac}$.

4. $\Phi_{ab}^{-1} = \Phi_{ba}.$

Proof   (1) follows by verifying that the product of the given Φ and Φ^{-1} is the identity. To prove (2), we make use of the formula $[u][v] - [v][u] = [u \times v]$ for $u, v \in \mathbb{R}^3$. First, the right-hand side $[\Phi\mathcal{V}]$ is

$$[\Phi\mathcal{V}] = \left[\begin{array}{cc} [R\omega] & 0 \\ [p \times R\omega + Rv] & [R\omega] \end{array} \right]. \tag{5.26}$$

Expanding the left-hand side $\Phi[\mathcal{V}]\Phi^{-1}$, we get

$$\left[\begin{array}{cc} R[\omega]R^{\top} & 0 \\ [p]\,R[\omega]R^{\top} + R[v]R^{\top} + R[\omega]R^{\top}[p]^{\top} & R[\omega]R^{\top} \end{array} \right] \tag{5.27}$$

$$= \left[\begin{array}{cc} [R\omega] & 0 \\ [p]\,[R\omega] + [R\omega][p]^{\top} + [Rv] & [R\omega] \end{array} \right]. \tag{5.28}$$

Noting that $[p][R\omega] + [R\omega][p]^\top = [p][R\omega] - [R\omega][p] = [p \times R\omega]$, we get

$$\Phi[\mathcal{V}]\Phi^{-1} = \begin{bmatrix} [R\omega] & 0 \\ [p \times R\omega + Rv] & [R\omega] \end{bmatrix}. \tag{5.29}$$

Matching terms with the right-hand side $[\Phi\mathcal{V}]$, the result follows. To prove (3), expand $\Phi_{ab}\Phi_{bc}$ as follows:

$$\begin{bmatrix} R_{ab} & 0 \\ [p_{ab}]R_{ab} & R_{ab} \end{bmatrix}\begin{bmatrix} R_{bc} & 0 \\ [p_{bc}]R_{bc} & R_{bc} \end{bmatrix} = \begin{bmatrix} R_{ac} & 0 \\ [p_{ab}]R_{ac} + R_{ab}[p_{bc}]R_{bc} & R_{ac} \end{bmatrix}. \tag{5.30}$$

Rewrite the lower-left entry as $[p_{ab}]R_{ac} + R_{ab}[p_{bc}](R_{ab}^\top R_{ab})R_{bc} = [p_{ab}]R_{ac} + [R_{ab}p_{bc}]R_{ac} = [p_{ab} + R_{ab}p_{bc}]R_{ac}$. Since $p_{ab} + R_{ab}p_{bc} = p_{ac}$, it follows that

$$\Phi_{ab}\Phi_{bc} = \begin{bmatrix} R_{ac} & 0 \\ [p_{ac}]R_{ac} & R_{ac} \end{bmatrix} \tag{5.31}$$

as claimed. Finally, (4) follows by observing that $\Phi_{aa} = I$, so that $\Phi_{ab}\Phi_{ba} = I$, or $\Phi_{ab}^{-1} = \Phi_{ba}$ as claimed. $\qquad\square$

We are now ready to rewrite the transformation rules (5.16)–(5.22) in terms of twists and wrenches. First, Equations (5.16) and (5.19) – $\omega_b = R_{bc}\omega_c$ and $v_b = [p_{bc}]R_{bc}\omega_c + R_{bc}v_c$ – can be merged into a single equation involving the twists $\mathcal{V}_b = (\omega_b, v_b)$, $\mathcal{V}_c = (\omega_c, v_c)$, and Φ_{bc} straightforwardly as

$$\mathcal{V}_b = \Phi_{bc}\mathcal{V}_c. \tag{5.32}$$

Under the present assumption that $\{c\}$ and $\{b\}$ are both attached to the same rigid body, so that R_{bc} and p_{bc} (and consequently Φ_{bc}) are all constant, then

$$\dot{\mathcal{V}}_b = \Phi_{bc}\dot{\mathcal{V}}_c. \tag{5.33}$$

Equations (5.20) and (5.22), which we write in the slightly modified form

$$f_c = R_{cb}f_b = R_{bc}^\top f_b, \tag{5.34}$$

$$m_c = -R_{cb}[p_{bc}]f_b + R_{cb}m_b = R_{bc}^\top[p_{bc}]^\top f_b + R_{bc}^\top m_b, \tag{5.35}$$

can be merged into the following single equation:

$$\begin{bmatrix} m_c \\ f_c \end{bmatrix} = \begin{bmatrix} R_{bc}^\top & R_{bc}^\top[p_{bc}]^\top \\ 0 & R_{bc}^\top \end{bmatrix}\begin{bmatrix} m_b \\ f_b \end{bmatrix} = \begin{bmatrix} R_{bc} & 0 \\ [p_{bc}]R_{bc} & R_{bc} \end{bmatrix}^\top\begin{bmatrix} m_b \\ f_b \end{bmatrix}. \tag{5.36}$$

In terms of the wrenches $\mathcal{F}_c = (m_c, f_c)$, $\mathcal{F}_b = (m_b, f_b)$, and Φ_{bc}, the above can be written more compactly as

$$\mathcal{F}_c = \Phi_{bc}^\top\mathcal{F}_b. \tag{5.37}$$

In summary, the twist $\mathcal{V}_c$ transforms to $\mathcal{V}_b$ via Φ_{bc}, whereas the wrench $\mathcal{F}_b$ transforms to $\mathcal{F}_c$ via $\Phi_{bc}^\top$. One way to remember the order of the indices of Φ for the wrench transformation (5.37) is to regard the transpose operator as "switching" the order of b and c.

5.3   Spatial Inertia and Spatial Momentum

For this section let $\{c\}$ be the body frame attached to the center of mass of a rigid body. The 6×6 **spatial inertia matrix** is defined as follows:

$$\mathcal{G}_c = \begin{bmatrix} \mathcal{I}_c & 0 \\ 0 & \mathfrak{m} \cdot \mathbb{1} \end{bmatrix}, \tag{5.38}$$

where $\mathfrak{m}$ is the mass of the body and $\mathcal{I}_c$ is the 3×3 inertia matrix with respect to frame $\{c\}$.

A simple calculation verifies that in terms of the spatial inertia matrix, the kinetic energy $\mathcal{K}$ of the rigid body can be expressed as follows:

$$\mathcal{K} = \frac{1}{2}\mathfrak{m}v_c^\top v_c + \frac{1}{2}\omega_c^\top \mathcal{I}_c \omega_c$$
$$= \frac{1}{2}\begin{bmatrix} \omega_c^\top & v_c^\top \end{bmatrix}\begin{bmatrix} \mathcal{I}_c & 0 \\ 0 & \mathfrak{m} \cdot \mathbb{1} \end{bmatrix}\begin{bmatrix} \omega_c \\ v_c \end{bmatrix} = \frac{1}{2}\mathcal{V}_c^\top \mathcal{G}_c \mathcal{V}_c. \tag{5.39}$$

The kinetic energy of a rigid body should not depend on the choice of body frame, and we can use this fact to answer the question of how the spatial inertia matrix transforms under a change of body frame. Recalling that the twists $\mathcal{V}_c$ and $\mathcal{V}_b$ are related by $\mathcal{V}_c = \Phi_{cb}\mathcal{V}_b$, the kinetic energy $\mathcal{K}$ of the rigid body can be written

$$\mathcal{K} - \frac{1}{2}\mathcal{V}_c^\top \mathcal{G}_c \mathcal{V}_c = \frac{1}{2}\mathcal{V}_b^\top \Phi_{cb}^\top \mathcal{G}_c \Phi_{cb} \mathcal{V}_b. \tag{5.40}$$

Expressing the kinetic energy in terms of $\mathcal{V}_b$ as $\frac{1}{2}\mathcal{V}_b^\top \mathcal{G}_b \mathcal{V}_b$, it follows that

$$\mathcal{G}_b = \Phi_{cb}^\top \mathcal{G}_c \Phi_{cb}. \tag{5.41}$$

Equation (5.41) is the transformation rule for spatial inertia matrices under a change of body frame. Observe that the spatial inertia matrix is always symmetric and invertible, since both Φ_{cb} and $\mathcal{G}_c$ are always invertible.

Note that the Parallel Axis Theorem is a special case of Equation (5.41) when $R_{cb} = \mathbb{1}$ (the identity): from (5.41), $\mathcal{G}_c$ transforms to $\mathcal{G}_b$ as follows:

$$\mathcal{G}_b = \begin{bmatrix} \mathbb{1} & -[p_{cb}] \\ 0 & \mathbb{1} \end{bmatrix}\begin{bmatrix} \mathcal{I}_c & 0 \\ 0 & \mathfrak{m} \cdot \mathbb{1} \end{bmatrix}\begin{bmatrix} \mathbb{1} & 0 \\ [p_{cb}] & \mathbb{1} \end{bmatrix}$$
$$= \begin{bmatrix} \mathcal{I}_c - \mathfrak{m}[p_{cb}]^2 & -\mathfrak{m}[p_{cb}] \\ \mathfrak{m}[p_{cb}] & \mathfrak{m} \cdot \mathbb{1} \end{bmatrix}, \tag{5.42}$$

so that the 3×3 inertia matrix $\mathcal{I}_b$ with respect to frame $\{b\}$ is

$$\mathcal{I}_b = \mathcal{I}_c - \mathfrak{m}[p_{cb}]^2. \tag{5.43}$$

The spatial inertia matrix allows us to define the **spatial momentum** $\mathcal{H}_c \in \mathbb{R}^6$ by concatenating the linear and angular momenta:

$$\mathcal{H}_c = \mathcal{G}_c \mathcal{V}_c = \begin{bmatrix} \mathcal{I}_c & 0 \\ 0 & \mathfrak{m} \cdot \mathbb{1} \end{bmatrix}\begin{bmatrix} \omega_c \\ v_c \end{bmatrix} = \begin{bmatrix} \mathcal{I}_c \omega_c \\ \mathfrak{m}v_c \end{bmatrix}. \tag{5.44}$$

In terms of the spatial momentum, the kinetic energy can be written

$$\text{Kinetic Energy} = \frac{1}{2}\mathcal{H}_c^\top \mathcal{V}_c. \tag{5.45}$$

The transformation rule for spatial momentum can be verified straightforwardly from the corresponding rules for the spatial inertia and also twists. Noting that $\mathcal{V}_c = \Phi_{cb}\mathcal{V}_b$ and $\mathcal{G}_c = \Phi_{bc}^\top \mathcal{G}_b \Phi_{bc}$,

$$\mathcal{H}_c = \mathcal{G}_c \mathcal{V}_c = \Phi_{bc}^\top \mathcal{G}_b \Phi_{bc} \Phi_{cb} \mathcal{V}_b = \Phi_{bc}^\top \mathcal{G}_b \mathcal{V}_b = \Phi_{bc}^\top \mathcal{H}_b. \tag{5.46}$$

The spatial momentum obeys the same transformation rule for wrenches:

$$\mathcal{H}_c = \Phi_{bc}^\top \mathcal{H}_b, \quad \mathcal{H}_b = \Phi_{cb}^\top \mathcal{H}_c. \tag{5.47}$$

5.4   Twist–Wrench Formulation of the Equations of Motion

Using twists and wrenches, the rigid body equations of motion with respect to the center of mass frame (5.3) can be written

$$\mathcal{F}_c = \mathcal{G}_c \dot{\mathcal{V}}_c - [\mathcal{V}_c]^\top \mathcal{G}_c \mathcal{V}_c. \tag{5.48}$$

By applying the transformation rules for twists and wrenches, expressing (5.48) with respect to another body frame {b} is now straightforward. Using the fact that $\mathcal{F}_c = \Phi_{bc}^\top \mathcal{F}_b$ and $\mathcal{V}_c = \Phi_{cb}\mathcal{V}_b$, (5.48) can be written

$$\Phi_{bc}^\top \mathcal{F}_b = \mathcal{G}_c \Phi_{cb} \dot{\mathcal{V}}_b - [\Phi_{cb}\mathcal{V}_b]^\top \mathcal{G}_c \Phi_{cb} \mathcal{V}_b. \tag{5.49}$$

From the properties $\Phi_{bc}^{-1} = \Phi_{cb}$ and $[\Phi_{cb}\mathcal{V}_b] = \Phi_{cb}[\mathcal{V}_b]\Phi_{cb}^{-1}$, (5.49) simplifies to

$$\begin{aligned}
\mathcal{F}_b &= \Phi_{cb}^\top \mathcal{G}_c \Phi_{cb} \dot{\mathcal{V}}_b - \Phi_{cb}^\top [\Phi_{cb}\mathcal{V}_b]^\top \mathcal{G}_c \Phi_{cb} \mathcal{V}_b \\
&= \Phi_{cb}^\top \mathcal{G}_c \Phi_{cb} \dot{\mathcal{V}}_b - \Phi_{cb}^\top \Phi_{cb}^{-T} [\mathcal{V}_b]^\top \Phi_{cb}^\top \mathcal{G}_c \Phi_{cb} \mathcal{V}_b \\
&= \Phi_{cb}^\top \mathcal{G}_c \Phi_{cb} \dot{\mathcal{V}}_b - [\mathcal{V}_b]^\top \Phi_{cb}^\top \mathcal{G}_c \Phi_{cb} \mathcal{V}_b.
\end{aligned} \tag{5.50}$$

From the spatial inertia matrix transformation rule (5.41), the rigid body equations of motion with respect to frame {b} becomes

$$\mathcal{F}_b = \mathcal{G}_b \dot{\mathcal{V}}_b - [\mathcal{V}_b]^\top \mathcal{G}_b \mathcal{V}_b. \tag{5.51}$$

Comparing (5.51) and (5.48), the two equations have the same structure. Thus, by observing the transformation rules for twists, wrenches, and spatial inertias, the equations of motion can be written independent of any specific body frame as

$$\mathcal{F} = \mathcal{G}\dot{\mathcal{V}} - [\mathcal{V}]^\top \mathcal{G}\mathcal{V}. \tag{5.52}$$

Example 5.1 Consider once again the spherical pendulum example of Figure 4.3. Suppose a body frame {b} is attached to the body at point O as shown (keep in mind that frame {b} is attached to point O but moves with the body). Then

$$
R_{ob} = \begin{bmatrix} \cos\theta_1 & -\sin\theta_1 & 0 \\ \sin\theta_1 & \cos\theta_1 & 0 \\ 0 & 0 & 1 \end{bmatrix} \begin{bmatrix} 1 & 0 & 0 \\ 0 & \cos\theta_2 & -\sin\theta_2 \\ 0 & \sin\theta_2 & \cos\theta_2 \end{bmatrix}
$$

$$
= \begin{bmatrix} \cos\theta_1 & -\sin\theta_1\cos\theta_2 & \sin\theta_1\sin\theta_2 \\ \sin\theta_1 & \cos\theta_1\cos\theta_2 & -\cos\theta_1\sin\theta_2 \\ 0 & \sin\theta_2 & \cos\theta_2 \end{bmatrix}, \tag{5.53}
$$

and $p_{ob} = (0,0,0)^\top$. If the spatial inertia matrix $\mathcal{G}_c$ with respect to the center of mass frame {c} is known, then the corresponding spatial inertia $\mathcal{G}_b$ with respect to frame {b} is given by

$$
\mathcal{G}_b = \Phi_{cb}^\top \mathcal{G}_c \Phi_{cb}
$$

$$
= \begin{bmatrix} 1 & -[p_{cb}] \\ 0 & 1 \end{bmatrix} \begin{bmatrix} \mathcal{I}_c & 0 \\ 0 & m\cdot 1 \end{bmatrix} \begin{bmatrix} 1 & 0 \\ [p_{cb}] & 1 \end{bmatrix}
$$

$$
= \begin{bmatrix} \mathcal{I}_c - m[p_{cb}]^2 & -m[p_{cb}] \\ m[p_{cb}] & m\cdot 1 \end{bmatrix}
$$

$$
= \begin{bmatrix} \bar{\mu} & 0 & 0 & 0 & 0 & ma \\ 0 & 0 & 0 & 0 & 0 & 0 \\ 0 & 0 & \bar{\mu} & -ma & 0 & 0 \\ 0 & 0 & -ma & m & 0 & 0 \\ 0 & 0 & 0 & 0 & m & 0 \\ ma & 0 & 0 & 0 & 0 & m \end{bmatrix}, \tag{5.54}
$$

where $\bar{\mu} = \mu + ma^2 = \frac{4ma^2}{3}$, $\mathcal{I}_c = \mathrm{diag}(\mu, 0, \mu)$, and $p_{cb} = (0, -a, 0)^\top$.

We now calculate the link velocity and acceleration twists. From R_{ob} calculated above, the angular velocity of frame {b} expressed in frame {b} coordinates is obtained from $[\omega_b] = R_{ob}^\top \dot{R}_{ob}$:

$$
\omega_b = \begin{bmatrix} \dot{\theta}_2 \\ \dot{\theta}_1\sin\theta_2 \\ \dot{\theta}_1\cos\theta_2 \end{bmatrix}. \tag{5.55}
$$

Differentiating ω_b,

$$
\dot{\omega}_b = \begin{bmatrix} \ddot{\theta}_2 \\ \ddot{\theta}_1\sin\theta_2 + \dot{\theta}_1\dot{\theta}_2\cos\theta_2 \\ \ddot{\theta}_1\cos\theta_2 - \dot{\theta}_1\dot{\theta}_2\sin\theta_2 \end{bmatrix}. \tag{5.56}
$$

Also observe that the linear velocity $v_b = 0$, and therefore $\dot{v}_b = 0$. From the above we therefore get the following expressions for the velocity and acceleration twists of frame {b}:

$$
\mathcal{V}_b =
\begin{bmatrix}
\dot\theta_2 \\
\dot\theta_1 \sin\theta_2 \\
\dot\theta_1 \cos\theta_2 \\
0 \\
0 \\
0
\end{bmatrix}, \quad
\dot{\mathcal{V}}_b =
\begin{bmatrix}
\ddot\theta_2 \\
\ddot\theta_1 \sin\theta_2 + \dot\theta_1\dot\theta_2 \cos\theta_2 \\
\ddot\theta_1 \cos\theta_2 - \dot\theta_1\dot\theta_2 \sin\theta_2 \\
0 \\
0 \\
0
\end{bmatrix}.
\tag{5.57}
$$

We now calculate the wrench experienced by the body in frame {b} coordinates. The only forces and moments acting on the body are from the reaction forces at the joints, and gravity acting on the center of mass. Denoting the reaction force at the joint by $n = n_x\hat{x}_b + n_y\hat{y}_b + n_z\hat{z}_b$ and the gravitational force by $-mg(\sin\theta_2\hat{y}_b + \cos\theta_2\hat{z}_b)$, the resultant force is $n_x\hat{x}_b + (n_y - mg\sin\theta_2)\hat{y}_b + (n_z - mg\cos\theta_2)\hat{z}_b$. Since moments are calculated with respect to the frame {b} origin, the only force contributing to the moment is the gravitational force; the corresponding moment is $a\hat{y}_b \times (-mg(\sin\theta_2\hat{y}_b + \cos\theta_2\hat{z}_b)) = -mga\cos\theta_2\hat{x}_b$. Putting everything together, the net wrench on the body is

$$
\mathcal{F}_b =
\begin{bmatrix}
-mga\cos\theta_2 \\
0 \\
0 \\
n_x \\
n_y - mg\sin\theta_2 \\
n_z - mg\cos\theta_2
\end{bmatrix}.
\tag{5.58}
$$

Substituting the above into the dynamics equation $\mathcal{F}_b = \mathcal{G}_b\dot{\mathcal{V}}_b - [\mathcal{V}_b]^\top \mathcal{G}_b\mathcal{V}_b$, we get, after some calculation and rearrangement,

$$
0 = \bar\mu(\ddot\theta_2 + \dot\theta_1^2 \sin\theta_2 \cos\theta_2) + mga\cos\theta_2,
\tag{5.59}
$$

$$
0 = \bar\mu(\ddot\theta_1 \cos\theta_2 - 2\dot\theta_1\dot\theta_2 \sin\theta_2),
\tag{5.60}
$$

$$
n_x = ma(2\dot\theta_1\dot\theta_2 \sin\theta_2 - \ddot\theta_1 \cos\theta_2),
\tag{5.61}
$$

$$
n_y = -ma(\dot\theta_1^2 \cos^2\theta_2 + \dot\theta_2^2) + mg\sin\theta_2,
\tag{5.62}
$$

$$
n_z = ma(\ddot\theta_2 + \dot\theta_1^2 \sin\theta_2 \cos\theta_2) + mg\cos\theta_2.
\tag{5.63}
$$

The equations of motion for the spherical pendulum are therefore given by the two differential equations (5.59) and (5.60), which can be further simplified to

$$
\ddot\theta_1 = 2\dot\theta_1\dot\theta_2 \tan\theta_2,
\tag{5.64}
$$

$$
\ddot\theta_2 = -\dot\theta_1^2 \sin\theta_2 \cos\theta_2 - \frac{3g}{4a}\cos\theta_2.
\tag{5.65}
$$

$\square$

5.5   Dynamics of Open Chains

Using the twist–wrench formulation of the dynamics for a single rigid body, in this section we work out in detail the dynamic equations for an important class of multibody systems, open chains consisting of rigid bodies (or *links*) connected serially

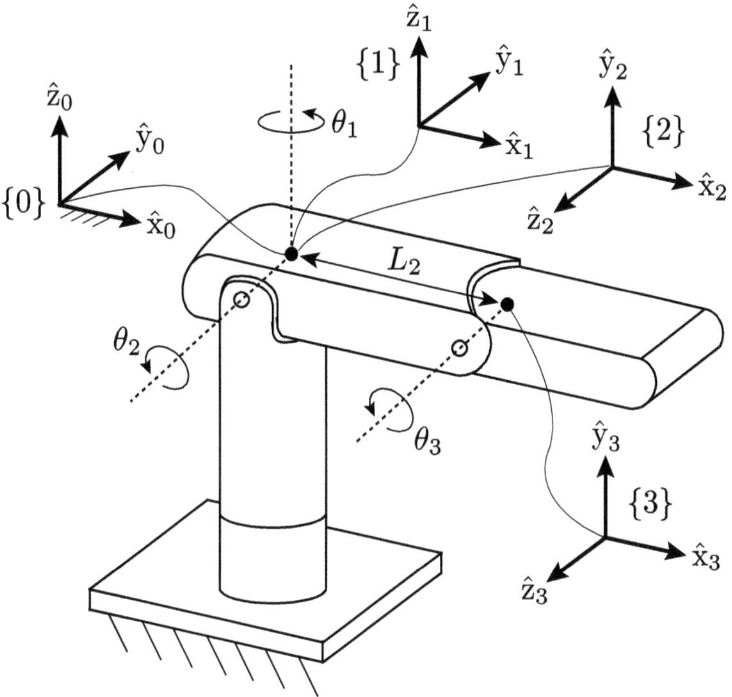

Figure 5.3 An elbow-type robot arm.

by mechanical joints. The first step in the dynamic analysis of more complex multibody systems with closed loops is to "cut" the loops, thereby obtaining a tree structure, and then deriving the dynamics for each open chain branch of the tree. Systems with closed loops are further addressed in the subsequent chapter on Lagrangian dynamics.

The present focus will be on computing the **inverse dynamics** of an open chain: assuming that the motion of all the links of the open chain are specified, the algorithm calculates the corresponding forces and torques at each of the joints (informally, the objective is to compute $f(t)$ from the given motion $r(t)$ in $f(t) = m\ddot{r}$). In turn it is possible to derive, in closed form, the equations of motion for the overall system from the algorithm.

5.5.1 Elbow-Type Robot Arm Dynamics

Rather than derive the algorithm in full generality, we first illustrate each step of the algorithm via a concrete example. The elbow-type robot arm of Figure 5.3 consists of three rigid links connected serially by revolute joints. The overall structure, and the frame assignments for each link, are shown in the figure. The 6×6 spatial inertia $\mathcal{G}_i$ for link i with respect to link reference frame $\{i\}$, $i = 1, 2, 3$, are assumed given. Starting from the base and moving outward, we first determine the link velocities and accelerations of each link frame. For link 1, note that

$$R_{01} = \begin{bmatrix} \cos\theta_1 & -\sin\theta_1 & 0 \\ \sin\theta_1 & \cos\theta_1 & 0 \\ 0 & 0 & 1 \end{bmatrix}, \tag{5.66}$$

and $[\omega_1] = R_{01}^\top \dot{R}_{01}$, which in vector form becomes

$$\omega_1 = (0, 0, \dot\theta_1)^\top. \tag{5.67}$$

ω_1 is the angular velocity of frame $\{1\}$, expressed in frame $\{1\}$ coordinates, and $\dot\omega_1 = (0, 0, \ddot\theta_1)^\top$. The linear velocity v_1 of frame $\{1\}$ is clearly zero, as is its derivative $\dot{v}_1$. Therefore the velocity and acceleration twists of frame $\{1\}$ are given by

$$\mathcal{V}_1 = \begin{bmatrix} \omega_1 \\ v_1 \end{bmatrix} = \begin{bmatrix} 0 \\ 0 \\ \dot\theta_1 \\ 0 \\ 0 \\ 0 \end{bmatrix}, \quad \dot{\mathcal{V}}_1 = \begin{bmatrix} \dot\omega_1 \\ \dot{v}_1 \end{bmatrix} = \begin{bmatrix} 0 \\ 0 \\ \ddot\theta_1 \\ 0 \\ 0 \\ 0 \end{bmatrix}. \tag{5.68}$$

We now proceed to link 2. Observe that

$$R_{02} = R_{01} R_{12} = \begin{bmatrix} \cos\theta_1 & -\sin\theta_1 & 0 \\ \sin\theta_1 & \cos\theta_1 & 0 \\ 0 & 0 & 1 \end{bmatrix} \begin{bmatrix} \cos\theta_2 & -\sin\theta_2 & 0 \\ 0 & 0 & -1 \\ \sin\theta_2 & \cos\theta_2 & 0 \end{bmatrix}$$

$$= \begin{bmatrix} \cos\theta_1 \cos\theta_2 & -\cos\theta_1 \sin\theta_2 & \sin\theta_1 \\ \sin\theta_1 \cos\theta_2 & -\sin\theta_1 \sin\theta_2 & -\cos\theta_1 \\ \sin\theta_2 & \cos\theta_2 & 0 \end{bmatrix}. \tag{5.69}$$

The angular velocity $\omega_2 \in \mathbb{R}^3$ of link frame $\{2\}$ is obtained from $[\omega_2] = R_{02}^\top \dot{R}_{02}$ as

$$\omega_2 = (\dot\theta_1 \sin\theta_2,\ \dot\theta_1 \cos\theta_2,\ \dot\theta_2)^\top, \tag{5.70}$$

from which

$$\dot\omega_2 = (\dot\theta_1 \dot\theta_2 \cos\theta_2 + \ddot\theta_1 \sin\theta_2,\ -\dot\theta_1 \dot\theta_2 \sin\theta_2 + \ddot\theta_1 \cos\theta_2,\ \ddot\theta_2)^\top. \tag{5.71}$$

Also by inspection we have $v_2 = \dot{v}_2 = 0$. The velocity and acceleration twists for link frame $\{2\}$ are therefore given by

$$\mathcal{V}_2 = \begin{bmatrix} \dot\theta_1 \sin\theta_2 \\ \dot\theta_1 \cos\theta_2 \\ \dot\theta_2 \\ 0 \\ 0 \\ 0 \end{bmatrix}, \quad \dot{\mathcal{V}}_2 = \begin{bmatrix} \dot\theta_1 \dot\theta_2 \cos\theta_2 + \ddot\theta_1 \sin\theta_2 \\ -\dot\theta_1 \dot\theta_2 \sin\theta_2 + \ddot\theta_1 \cos\theta_2 \\ \ddot\theta_2 \\ 0 \\ 0 \\ 0 \end{bmatrix}. \tag{5.72}$$

We now consider the third link. Since $R_{03} = R_{02}R_{23}$ with R_{02} obtained in (5.69) and

$$R_{23} = \begin{bmatrix} \cos\theta_3 & -\sin\theta_3 & 0 \\ \sin\theta_3 & \cos\theta_3 & 0 \\ 0 & 0 & 1 \end{bmatrix}, \tag{5.73}$$

ω_3 can be obtained as

$$[\omega_3] = R_{03}^{\top}\dot{R}_{03} = R_{23}^{\top}\left(R_{02}^{\top}\dot{R}_{02}\right)R_{23} + R_{23}^{\top}\dot{R}_{23}, \tag{5.74}$$

or in vector form as $\omega_3 = R_{32}\omega_2 + (0, 0, \dot{\theta}_3)^{\top}$:

$$\omega_3 = \begin{bmatrix} \dot{\theta}_1(\sin\theta_2\cos\theta_3 + \cos\theta_2\sin\theta_3) \\ \dot{\theta}_1(\cos\theta_2\cos\theta_3 - \sin\theta_2\sin\theta_3) \\ \dot{\theta}_2 + \dot{\theta}_3 \end{bmatrix} = \begin{bmatrix} \dot{\theta}_1\sin(\theta_2 + \theta_3) \\ \dot{\theta}_1\cos(\theta_2 + \theta_3) \\ \dot{\theta}_2 + \dot{\theta}_3 \end{bmatrix}. \tag{5.75}$$

Differentiating ω_3,

$$\dot{\omega}_3 = \begin{bmatrix} \ddot{\theta}_1\sin(\theta_2 + \theta_3) + \dot{\theta}_1(\dot{\theta}_2 + \dot{\theta}_3)\cos(\theta_2 + \theta_3) \\ \ddot{\theta}_1\cos(\theta_2 + \theta_3) - \dot{\theta}_1(\dot{\theta}_2 + \dot{\theta}_3)\sin(\theta_2 + \theta_3) \\ \ddot{\theta}_2 + \ddot{\theta}_3 \end{bmatrix}. \tag{5.76}$$

We now calculate $v_3 \in \mathbb{R}^3$, the linear velocity of frame $\{3\}$ expressed in frame $\{3\}$ coordinates. Let v_3 be the linear velocity of frame $\{3\}$; then $v_3 = \omega_2 \times L_2\hat{x}_2$, where L_2 is the length of the second link. Since v_3 expressed in frame $\{2\}$ coordinates is

$$\begin{bmatrix} \dot{\theta}_1\sin\theta_2 \\ \dot{\theta}_1\cos\theta_2 \\ \dot{\theta}_2 \end{bmatrix} \times \begin{bmatrix} L_2 \\ 0 \\ 0 \end{bmatrix} = \begin{bmatrix} 0 \\ L_2\dot{\theta}_2 \\ -L_2\dot{\theta}_1\cos\theta_2 \end{bmatrix}, \tag{5.77}$$

v_3 expressed in frame $\{3\}$ coordinates can be obtained by multiplying the above by R_{32}, i.e.,

$$v_3 = \begin{bmatrix} \cos\theta_3 & \sin\theta_3 & 0 \\ -\sin\theta_3 & \cos\theta_3 & 0 \\ 0 & 0 & 1 \end{bmatrix} \begin{bmatrix} 0 \\ L_2\dot{\theta}_2 \\ -L_2\dot{\theta}_1\cos\theta_2 \end{bmatrix} = \begin{bmatrix} L_2\dot{\theta}_2\sin\theta_3 \\ L_2\dot{\theta}_2\cos\theta_3 \\ -L_2\dot{\theta}_1\cos\theta_2 \end{bmatrix}. \tag{5.78}$$

Differentiating v_3,

$$\dot{v}_3 = \begin{bmatrix} L_2(\ddot{\theta}_2\sin\theta_3 + \dot{\theta}_2\dot{\theta}_3\cos\theta_3) \\ L_2(\ddot{\theta}_2\cos\theta_3 - \dot{\theta}_2\dot{\theta}_3\sin\theta_3) \\ -L_2(\ddot{\theta}_1\cos\theta_2 - \dot{\theta}_1\dot{\theta}_2\sin\theta_2) \end{bmatrix}. \tag{5.79}$$

The velocity twist $\mathcal{V}_3 = (\omega_3, v_3)$ and its derivative $\dot{\mathcal{V}}_3$ can then be calculated from the obtained values for ω_3 and v_3. This completes the forward iteration for the link velocities and accelerations.

We now iterate from the tip link to the base link and calculate the joint forces and torques. For the third link we have

$$\mathcal{F}_3 = \mathcal{G}_3\dot{\mathcal{V}}_3 - [\mathcal{V}_3]^{\top}\mathcal{G}_3\mathcal{V}_3, \tag{5.80}$$

where $\mathcal{F}_3 = (m_3, f_3)$ is the wrench exerted on the third link by the second link. Then $\tau_3 \in \mathbb{R}$ is the $\hat{z}$-component of the moment m_3. For the second link,

$$\mathcal{F}_2 = \mathcal{G}_2 \dot{\mathcal{V}}_2 - [\mathcal{V}_2]^{\top} \mathcal{G}_2 \mathcal{V}_2 + \Phi_{32}^{\top} \mathcal{F}_3, \tag{5.81}$$

where $\mathcal{F}_2 = (m_2, f_2)$ is the wrench exerted on the second link by the first link. Then $\tau_2 \in \mathbb{R}$ is the $\hat{z}$-component of the moment m_2. Finally for link 1,

$$\mathcal{F}_1 = \mathcal{G}_1 \dot{\mathcal{V}}_1 - [\mathcal{V}_1]^{\top} \mathcal{G}_1 \mathcal{V}_1 + \Phi_{21}^{\top} \mathcal{F}_2, \tag{5.82}$$

where $\mathcal{F}_1 = (m_1, f_1)$ is the wrench exerted on the first link by ground (regarded as link 0). Then $\tau_1 \in \mathbb{R}$ is the $\hat{z}$-component of the moment m_1.

In the derivations above we have not accounted for gravity. Gravity can be accounted for in one of two ways. The first and more obvious way is to include the wrench corresponding to gravity when calculating $\mathcal{F}_i$ for the equation

$$\mathcal{F}_i = \mathcal{G}_i \dot{\mathcal{V}}_i - [\mathcal{V}_i]^{\top} \mathcal{G}_i \mathcal{V}_i + \Phi_{i+1,i}^{\top} \mathcal{F}_{i+1}. \tag{5.83}$$

A second way that is particularly effective for open chains is to assume the base frame $\{0\}$, or ground, accelerates with acceleration twist

$$\dot{\mathcal{V}}_0 = \begin{bmatrix} \dot{\omega}_0 \\ \dot{v}_0 \end{bmatrix} = (0, 0, 0, 0, 0, -g)^{\top}. \tag{5.84}$$

5.5.2   Recursive Inverse Dynamics Algorithm

The inverse dynamics procedure worked out for the elbow-type robot arm can be generalized to arbitrary open chain structures. In this section we present a recursive algorithm for doing so.

- **Preliminaries**: Choose an inertial reference frame $\{0\}$, and assign link reference frames $\{i\}$ to each link $i = 1, \ldots, n$. Spatial inertia matrices $\mathcal{G}_i$ with respect to each link frame $\{i\}$ are assumed given. Analytic expressions for the rotation matrices $R_{i-1,i}$ and vectors $p_{i-1,i}$ are also assumed available. The following are also required: (i) $\mathcal{V}_0 = (\omega_0, v_0)$, the velocity twist of the base (inertial) frame $\{0\}$ expressed in frame $\{0\}$ coordinates; (ii) $\dot{\mathcal{V}}_0 = (\dot{\omega}_0, \dot{v}_0)$, the acceleration twist of the base frame $\{0\}$ expressed in frame $\{0\}$ coordinates; (iii) $\mathcal{F}_{\text{tip}} = (m_{\text{tip}}, f_{\text{tip}})$, the external wrench applied to link n, expressed in frame $\{n\}$ coordinates.
- **Initialization**: $\mathcal{V}_0 = \text{given}$, $\dot{\mathcal{V}}_0 = \text{given}$, $\mathcal{F}_{n+1} = \mathcal{F}_{\text{tip}}$.
- **Forward Iteration**: for $i = 1$ to n do

$$[\omega_i'] = R_{i-1,i}^{\top} \dot{R}_{i-1,i}, \tag{5.85}$$

$$v_i' = R_{i-1,i}^{\top} \dot{p}_{i-1,i}, \tag{5.86}$$

$$\mathcal{V}_i' = (\omega_i', v_i'), \tag{5.87}$$

$$\mathcal{V}_i = \Phi_{i,i-1} \mathcal{V}_{i-1} + \mathcal{V}_i', \tag{5.88}$$

$$\dot{\mathcal{V}}_i = \Phi_{i,i-1} \dot{\mathcal{V}}_{i-1} + \dot{\mathcal{V}}_i' - [\mathcal{V}_i'] \Phi_{i,i-1} \mathcal{V}_{i-1}. \tag{5.89}$$

- **Backward Iteration**: for $i = n$ to 1 do

$$\mathcal{F}_i = \Phi_{i+1,i}^\top \mathcal{F}_{i+1} + \mathcal{G}_i \dot{\mathcal{V}}_i - [\mathcal{V}_i]^\top \mathcal{G}_i \mathcal{V}_i. \tag{5.90}$$

The above formulation does not explicitly take into account gravity, but as described in the earlier elbow arm example, the forces and moments due to gravity can be inserted into the backward iteration Equation (5.90). A simpler way to account for gravity is to set the base frame to have an acceleration twist $\dot{\mathcal{V}}_0 = (0, g)$, where $g \in \mathbb{R}^3$ is the gravitational acceleration vector expressed in the base frame $\{0\}$ coordinates. In this case it is important to remember that the link acceleration calculated within the recursive algorithm is not its true acceleration, but rather its true acceleration minus g.

5.6   Summary

- Let $\{b\}$ be an arbitrary body frame attached to a moving rigid body. Suppose frame $\{b\}$ moves with linear velocity v and angular velocity ω, and let $v_b, \omega_b \in \mathbb{R}^3$, respectively, be v and ω expressed in frame $\{b\}$ coordinates. The **twist** $\mathcal{V}_b$ is defined as

$$\mathcal{V}_b = \begin{bmatrix} \omega_b \\ v_b \end{bmatrix} \in \mathbb{R}^6, \tag{5.91}$$

also written more compactly as $\mathcal{V}_b = (\omega_b, v_b)$.
- Given a twist $\mathcal{V} = (\omega, v)$, its 6×6 matrix representation is denoted

$$[\mathcal{V}] = \begin{bmatrix} [\omega] & 0 \\ [v] & [\omega] \end{bmatrix}. \tag{5.92}$$

- Suppose the same rigid body is subject to a force f and moment m_b with respect to the origin of frame $\{b\}$, and let $f_b, m_b \in \mathbb{R}^3$ be f and m_b expressed in frame $\{b\}$ coordinates, respectively. The **wrench** $\mathcal{F}_b$ is defined as

$$\mathcal{F}_b = \begin{bmatrix} m_b \\ f_b \end{bmatrix} \in \mathbb{R}^6, \tag{5.93}$$

also written more compactly as $\mathcal{F}_b = (m_b, f_b)$.
- Given some rotation matrix R and vector $p \in \mathbb{R}^3$, the 6×6 matrix $\Phi(R, p) = \Phi$ is defined as

$$\Phi = \begin{bmatrix} R & 0 \\ [p]R & R \end{bmatrix}. \tag{5.94}$$

Φ is always invertible, with its inverse Φ^{-1} given by

$$\Phi^{-1} = \begin{bmatrix} R^\top & 0 \\ R^\top [p]^\top & R^\top \end{bmatrix}. \tag{5.95}$$

Φ also satisfies $\Phi[\mathcal{V}]\Phi^{-1} = [\Phi\mathcal{V}]$ for any twist $\mathcal{V} = (\omega, v)$.

- Given two frames {a} and {b}, let R_{ab} be the rotation matrix describing frame {b} relative to frame {a}, and $p_{ab} \in \mathbb{R}^3$ be the vector from frame {a} to frame {b} expressed in frame {a} coordinates. Then

$$\Phi_{ab} = \begin{bmatrix} R_{ab} & 0 \\ [p_{ab}] R_{ab} & R_{ab} \end{bmatrix} \tag{5.96}$$

  satisfies $\Phi_{ab}\Phi_{bc} = \Phi_{ac}$ and $\Phi_{ab}^{-1} = \Phi_{ba}$.

- Let {a} and {b} be two body frames attached to the same moving body and subject to an external force and moment. The twists $\mathcal{V}_a$ and $\mathcal{V}_b$ are then related by

$$\mathcal{V}_a = \Phi_{ab}\mathcal{V}_b, \quad \mathcal{V}_b = \Phi_{ba}\mathcal{V}_a, \tag{5.97}$$

  while the wrenches $\mathcal{F}_a$ and $\mathcal{F}_b$ are related by

$$\mathcal{F}_a = \Phi_{ba}^{\mathsf{T}}\mathcal{F}_b, \quad \mathcal{F}_b = \Phi_{ab}^{\mathsf{T}}\mathcal{F}_a. \tag{5.98}$$

- Let {c} be the body frame attached to the center of mass of a rigid body. The 6×6 **spatial inertia matrix** $\mathcal{G}_c$ is defined as

$$\mathcal{G}_c = \begin{bmatrix} \mathcal{I}_c & 0 \\ 0 & \mathfrak{m} \cdot \mathbb{1} \end{bmatrix}, \tag{5.99}$$

  where $\mathfrak{m}$ is the mass of the body and $\mathcal{I}_c$ is the 3×3 inertia matrix with respect to frame {c}. If $\mathcal{V}_c$ is the twist associated with the center of mass frame {c}, the **kinetic energy** $\mathcal{K}$ of the rigid body is then given by

$$\mathcal{K} = \frac{1}{2}\mathcal{V}_c^{\mathsf{T}}\mathcal{G}_c\mathcal{V}_c. \tag{5.100}$$

- Let {b} be another body frame attached to the rigid body distinct from the center of mass frame {c}. The spatial inertia $\mathcal{G}_b$ with respect to frame {b} is then defined as

$$\mathcal{G}_b = \Phi_{cb}^{\mathsf{T}}\mathcal{G}_c\Phi_{cb}. \tag{5.101}$$

  If {a} is another body frame distinct from {b}, the spatial inertias $\mathcal{G}_a$ and $\mathcal{G}_b$ are related by

$$\mathcal{G}_a = \Phi_{ba}^{\mathsf{T}}\mathcal{G}_b\Phi_{ba}, \quad \mathcal{G}_b = \Phi_{ab}^{\mathsf{T}}\mathcal{G}_a\Phi_{ab}. \tag{5.102}$$

- The **spatial momentum** vector $\mathcal{H}_c \in \mathbb{R}^6$ with respect to the center of mass frame {c} is defined as

$$\mathcal{H}_c = \mathcal{G}_c\mathcal{V}_c = \begin{bmatrix} \mathcal{I}_c & 0 \\ 0 & \mathfrak{m} \cdot \mathbb{1} \end{bmatrix}\begin{bmatrix} \omega_c \\ v_c \end{bmatrix} = \begin{bmatrix} \mathcal{I}_c\omega_c \\ \mathfrak{m}v_c \end{bmatrix}. \tag{5.103}$$

  The spatial momentum $\mathcal{H}_b$ with respect to an arbitrary body frame {b} is then defined as $\mathcal{H}_b = \mathcal{G}_b\mathcal{V}_b$. The spatial momentum vector obeys the same transformation rule for wrenches:

$$\mathcal{H}_a = \Phi_{ba}^{\mathsf{T}}\mathcal{H}_b, \quad \mathcal{H}_b = \Phi_{ab}^{\mathsf{T}}\mathcal{H}_a. \tag{5.104}$$

  In terms of the spatial momentum, the kinetic energy can be written $\mathcal{K} = \frac{1}{2}\mathcal{H}_b^{\mathsf{T}}\mathcal{V}_b$ for any body frame {b}.

- The rigid body equations of motion with respect to any body frame can be written

$$\mathcal{F} = \mathcal{G}\dot{\mathcal{V}} - [\mathcal{V}]^{\top}\mathcal{G}\mathcal{V}, \tag{5.105}$$

  where $\mathcal{G}$ is the spatial inertia, $\mathcal{V}$ is the twist and $\dot{\mathcal{V}}$ its derivative, and $\mathcal{F}$ is the external applied wrench, all expressed with respect to the chosen body frame.

- Given a multibody system consisting of rigid bodies connected by mechanical joints, in the **inverse dynamics** problem the goal is to determine the forces and torques that must be applied at the joints in order to generate a desired motion trajectory of the multibody system. In the **forward dynamics** problem, the goal is to determine the subsequent motion of the multibody system from the external forces and torques applied at the joints.

- An **open chain** consists of rigid bodies (links) connected serially by mechanical joints. Recursive algorithms for computing the inverse dynamics of open chains can be systematically derived using the twist–wrench formalism.

5.7   Exercises

Exercise 5.1   Let $\mathcal{V}_1 = (\omega_1, v_1)$ and $\mathcal{V}_2 = (\omega_2, v_2)$ be two twists, and

$$[\mathcal{V}_1]\mathcal{V}_2 - \begin{bmatrix} [\omega_1] & 0 \\ [v_1] & [\omega_1] \end{bmatrix} \begin{bmatrix} \omega_2 \\ v_2 \end{bmatrix} = \begin{bmatrix} \omega_3 \\ v_3 \end{bmatrix} = \mathcal{V}_3$$

Verify that

$$[\mathcal{V}_3] = [\mathcal{V}_1][\mathcal{V}_2] - [\mathcal{V}_2][\mathcal{V}_1].$$

Exercise 5.2   Recall that the transformation formula for wrenches is given by

$$\mathcal{F}_b = \Phi_{ab}^{\top}\mathcal{F}_a,$$

where $\mathcal{F}_a = (m_a, f_a)$, $\mathcal{F}_b = (m_b, f_b)$, and

$$\Phi_{ab} = \begin{bmatrix} R_{ab} & 0 \\ [p_{ab}] R_{ab} & R_{ab} \end{bmatrix}.$$

Someone claims that by switching the order of the forces and moments, i.e., defining $\mathcal{F}'_a = (f_a, m_a)$ and $\mathcal{F}'_b = (f_b, m_b)$, the wrench transformation formula can be written in the more natural and intuitive way as

$$\mathcal{F}'_b = \Phi_{ba}\mathcal{F}'_a.$$

Is this claim correct? Explain your answer.

Exercise 5.3   Suppose the spatial inertia matrix $\mathcal{G}_c \in \mathbb{R}^{6\times6}$ of a rigid body of mass $\mathfrak{m}$ with respect to a body frame $\{c\}$ attached to the center of mass is

$$\mathcal{G}_c = \begin{bmatrix} \mathcal{I}_c & 0 \\ 0 & \mathfrak{m}\cdot\mathbb{1} \end{bmatrix}, \quad \mathcal{I}_c = \begin{bmatrix} 1 & 0 & 0 \\ 0 & 5 & 0 \\ 0 & 0 & 2 \end{bmatrix}.$$

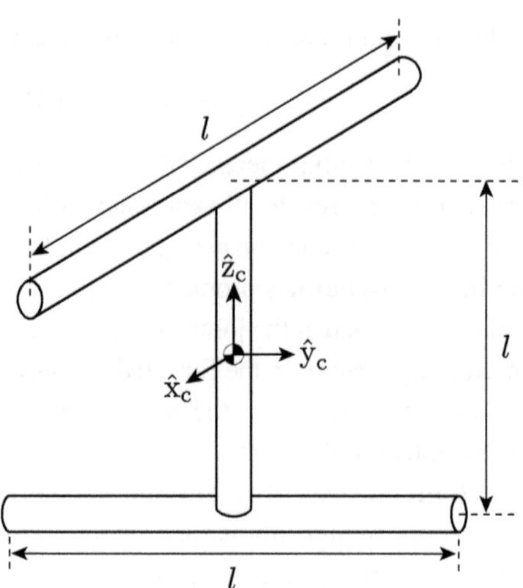

Figure 5.4 Three identical rods connected as a rigid body of Exercise 5.4.

Another body frame {a} is attached to the rigid body such that the orientation of frame {c} relative to frame {a} is

$$
R_{ac} = \begin{bmatrix} \frac{1}{\sqrt{2}} & 0 & \frac{1}{\sqrt{2}} \\ 0 & 1 & 0 \\ -\frac{1}{\sqrt{2}} & 0 & \frac{1}{\sqrt{2}} \end{bmatrix}
$$

and the vector from frame {a} to frame {c} expressed in frame {a} coordinates is $p_{ac} = (1, 2, 1)^{\top}$. Find the spatial inertia matrix $\mathcal{G}_a \in \mathbb{R}^{6 \times 6}$ of the rigid body with respect to frame {a}.

Exercise 5.4   As shown in Figure 5.4, three identical rods of length l and mass $\mathfrak{m}$ are connected as one rigid body of mass $3\mathfrak{m}$. Find the spatial inertia matrix $\mathcal{G}_c \in \mathbb{R}^{6 \times 6}$ of the rigid body with respect to the center of mass frame {c}.

Exercise 5.5   A body frame {c} is attached to the center of mass of a rigid body. The mass of the rigid body is $\mathfrak{m}$ and the inertia matrix with respect to frame {c} is $\mathcal{I}_c \in \mathbb{R}^{3 \times 3}$. Frame {c} moves with angular velocity ω and linear velocity v, where $\omega_c \in \mathbb{R}^3$ and $v_c \in \mathbb{R}^3$ are ω and v expressed in frame {c} coordinates respectively.
(a) Let {s} be the fixed frame, and $\omega_s \in \mathbb{R}^3$ and $v_s \in \mathbb{R}^3$ be ω and v expressed in {s} frame coordinates respectively. Express ω_s and v_s in terms of ω_c and v_c. Define whatever other vectors and matrices you need for your answer.
(b) Suppose frame {b} is attached to a different point on the rigid body. Frame {b} then moves with angular velocity ω' and linear velocity v', where $\omega_b \in \mathbb{R}^3$ and $v_b \in \mathbb{R}^3$ are ω' and v' expressed in {b} frame coordinates respectively. Express ω_b and v_b in terms of ω_c and v_c. Define whatever other vectors and matrices you need for your answer.
(c) Express ω_b and v_b in terms of ω_s and v_s from (a). Define whatever other vectors and matrices you need for your answer.

(d) Given ω_b and v_b, find the kinetic energy of the rigid body. Define whatever other vectors and matrices you need for your answer.

(e) Now suppose frames {b} and {c} have the same origin, and $\mathcal{I}_b$ and $\mathcal{I}_c$ are given as

$$\mathcal{I}_b = \begin{bmatrix} 1 & 0 & 0 \\ 0 & 2 & 0 \\ 0 & 0 & 3 \end{bmatrix}, \quad \mathcal{I}_c = \begin{bmatrix} \frac{3}{2} & -\frac{1}{2} & 0 \\ -\frac{1}{2} & \frac{3}{2} & 0 \\ 0 & 0 & 3 \end{bmatrix}.$$

Find the rotation matrix R_{bc}.

(f) Suppose a force f is applied to the origin of frame {b}. Let $f_b \in \mathbb{R}^3$ be f expressed in {b} frame coordinates. The following values are given:

$$f_b = \begin{bmatrix} 1 \\ 1 \\ 1 \end{bmatrix}, \quad \omega_b = \begin{bmatrix} 0 \\ 0 \\ 1 \end{bmatrix}, \quad v_b = \begin{bmatrix} 1 \\ 0 \\ 0 \end{bmatrix}.$$

Assuming $m = 1$ and using the $\mathcal{I}_b$ given in (e), find the angular acceleration and linear acceleration of frame {b}, and express them in {b} frame coordinates.

Exercise 5.6   A thin uniform disk of radius R and mass m_d is rigidly attached to a rotating vertical shaft, where the angle between the disk and shaft is β and angle θ is defined as shown in Figure 5.5. The shaft is modeled as a thick rod of radius r, length l, and mass m_r. The disk and the shaft have the same center of mass C. Two body frames {b} and {c} are attached to the system at C. Assume there is no gravity.
(a) Find the spatial inertia matrix $\mathcal{G}_b$ of the system with respect to frame {b}.
(b) Find the twist $\mathcal{V}_b$ and its derivative $\dot{\mathcal{V}}_b$ of frame {b}.
(c) An external wrench $\mathcal{F}_b = (m_b, f_b)$ is applied at the system. Derive the equations of motion using the twist–wrench formulation.

Exercise 5.7   As shown in Figure 5.6, a square plate of mass m and side length b is connected to a revolute joint with a horizontal rotation axis, while the whole system rotates about the vertical axis. A body frame {b} is assigned at the intersection of the two rotation axes ($\hat{y}_b$ is perpendicular to the plate), and angles θ_1 and θ_2 are defined as shown in the figure. Gravity g acts downward.
(a) Derive the equations of motion using the twist–wrench formulation.
(b) Now suppose $\dot{\theta}_1 = \omega$, where ω is constant. Assuming ω is sufficiently large, determine the angle $\theta_2 > -\frac{\pi}{2}$ using the equations of motion derived from (a). Compare your answer with Exercise 4.9 of Chapter 4.

Exercise 5.8   The robot shown in Figure 5.7 consists of a rigid link of mass m connected to the ground by a screw joint, where the screw joint has a pitch of η m/rad (i.e., for every 1 radian counterclockwise rotation of θ, the joint moves linearly in the fixed frame $\hat{z}_0$-axis direction by η meters). The center of mass frame {c} is attached to the link center as shown, and the inertia matrix with respect to frame {c} is given as $\mathrm{diag}(I_{xx}, I_{yy}, I_{zz})$. Let {a} be another body frame attached to the upper surface of the link as shown in the figure. We further define the following:

- ω_c, v_c: angular and linear velocity vectors of frame {c} respectively.

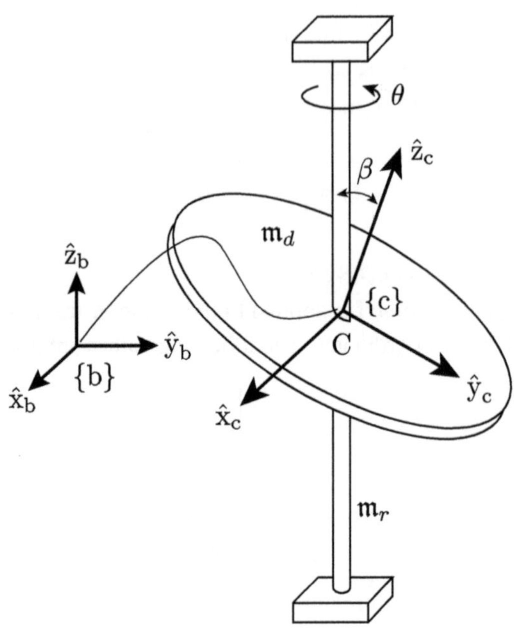

Figure 5.5 Vertically rotating tilted disk of Exercise 5.6.

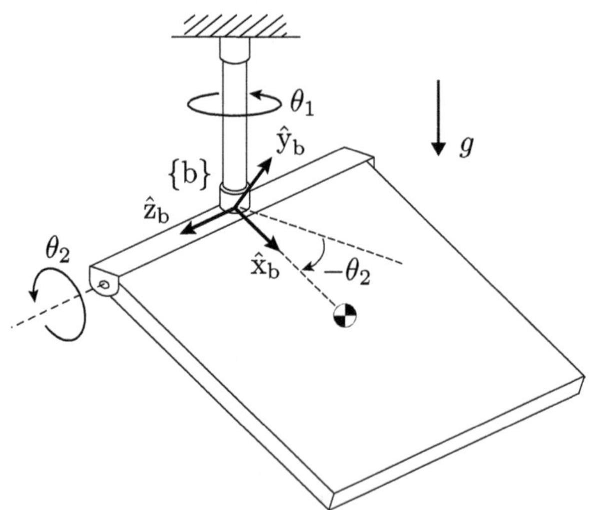

Figure 5.6 Rotating square plate of Exercise 5.7.

- $\omega_c, v_c \in \mathbb{R}^3$: ω_c and v_c expressed in frame $\{c\}$ coordinates respectively.
- ω_a, v_a: angular and linear velocity vectors of frame $\{a\}$ respectively.
- $\omega_a, v_a \in \mathbb{R}^3$: ω_a and v_a expressed in frame $\{a\}$ coordinates respectively.

Gravity g acts downward.

(a) Express ω_a and v_a in terms of the given parameters and variables.

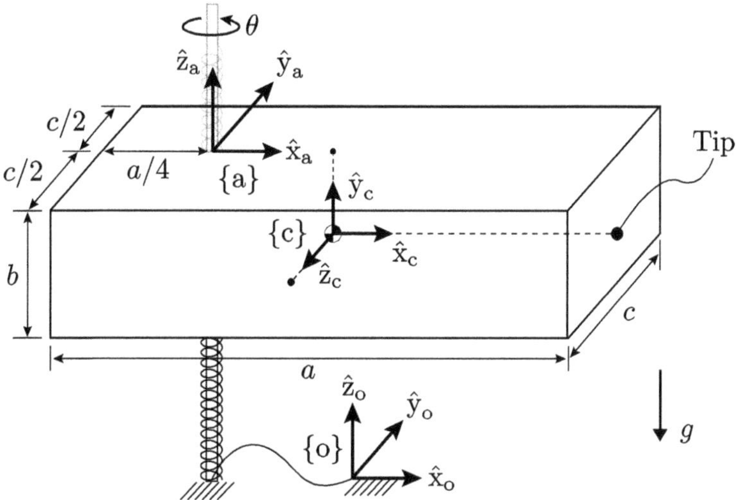

Figure 5.7 Screw robot of Exercise 5.8.

(b) Express ω_c and v_c in terms of the given parameters and variables.

(c) Find the kinetic energy $\mathcal{K}(\theta, \dot{\theta})$ and potential energy $\mathcal{P}(\theta)$ of the link.

(d) Let $\tau \in \mathbb{R}$ be an external torque applied at the screw joint. What is the work done by the external torque when θ has changed by $\Delta\theta$?

(e) Derive the equations of motion as a second-order differential equation of the form $\tau = \alpha\ddot{\theta} + \beta$ (α, β constant) using the conservation principles of work and energy.

(f) Now suppose an external force $f = -p\hat{y}_a$ (p constant) is continuously applied to the tip of the link. Determine the reaction force $f_r = f_x\hat{x}_a + f_y\hat{y}_a + f_z\hat{z}_a$ at frame {a} origin as a function of $(\theta, \dot{\theta}, \ddot{\theta})$.

Exercise 5.9   Figure 5.8 shows the "squirrel barrel" ride in its zero position ($\theta_1 = \theta_2 = 0$, the seat frame {3} is in its highest position when $\theta_2 = 0$). The fixed frame {0}, moving frames {1}, {2}, {3}, and all dimensions and mass-inertia parameters are as shown in Figure 5.8. Assume there is no gravity, and take the masses and inertias of any unlabeled bodies to be zero.

(a) For arbitrary θ_1 and θ_2, find the position of frame {3} with respect to frame {0}, expressed in frame {0} coordinates.

(b) Assuming $\theta_1 = \theta_2 = 0$, find the linear velocity of the seat frame {3}, expressed in frame {3} coordinates.

(c) Assume $\dot{\theta}_1$ and $\dot{\theta}_2$ are constant. Find $\mathcal{V}_2$, the velocity twist of frame {2} expressed in frame {2} coordinates.

(d) Now assume $\dot{\theta}_1$ is constant and $\theta_2 = \dot{\theta}_2 = 0$. Compute the wrench applied at the base of axis θ_1 with respect to frame {1} when $\theta_1 = 0$.

Exercise 5.10   Two identical rods of mass m and length L are connected as shown in Figure 5.9, forming a rotational inverted pendulum (this is also known as the Furuta

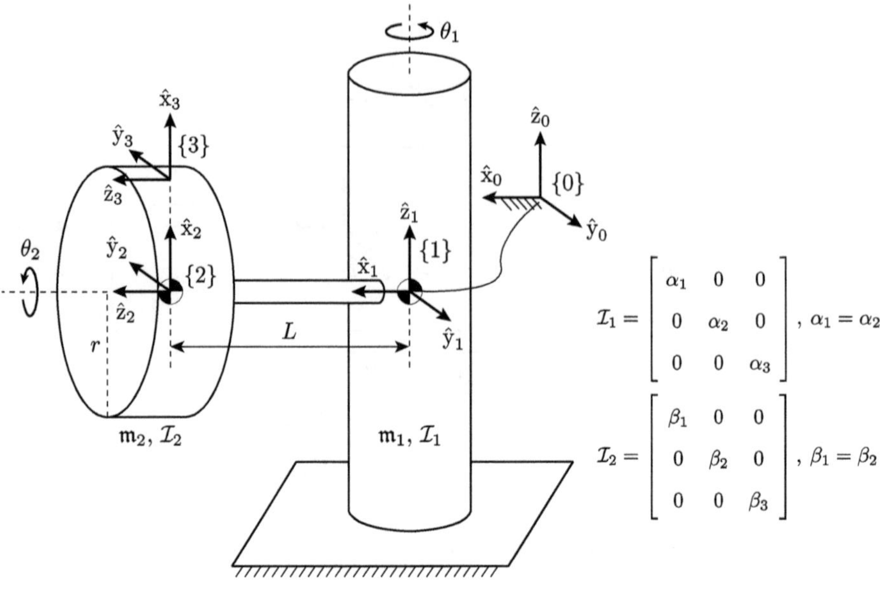

Figure 5.8 Squirrel barrel ride of Exercise 5.9.

pendulum). Angles θ_1 and θ_2 are defined as shown, and the figure shows the pendulum when $\theta_1 = \theta_2 = 0$. The fixed frame $\{0\}$ is assigned at the intersection of joint 1 and rod 1. Body frames $\{1\}$ and $\{2\}$ are attached to rod 1 and rod 2, respectively, as shown in the figure. Assume $L = 1$.

(a) Find the rotation matrix R_{02} and the coordinates $p_0 \in \mathbb{R}^3$ of point P expressed in frame $\{0\}$ coordinates.

(b) Find the angular and linear velocities of frame $\{2\}$ expressed in frame $\{2\}$ coordinates.

(c) Suppose an external force f is applied to point P of link 2. Let $f_0 \in \mathbb{R}^3$ be f expressed in frame $\{0\}$ coordinates, and $m_0 \in \mathbb{R}^3$ be the moment generated by f with respect to the origin of frame $\{0\}$, expressed in frame $\{0\}$ coordinates. Find $f_2 \in \mathbb{R}^3$, the force f expressed in frame $\{2\}$ coordinates. Also find $m_2 \in \mathbb{R}^3$, the moment generated by f with respect to the origin of frame $\{2\}$, expressed in frame $\{2\}$ coordinates.

(d) Suppose the pendulum is at rest at some θ_1 and θ_2. If an external force f is applied to point P, the pendulum will move. What are the input joint torques τ_1 and τ_2 that keep the pendulum at rest? (*Hint*: Consider a conservation of power argument, in which the power generated by the joint torques is equal to the power generated by the external force f. Expressing things in terms of frame $\{2\}$ coordinates will be convenient.)

Exercise 5.11   The quadrotor of Figure 5.10 can be modeled as a rigid body with four rotors. Let $\{s\}$ be the fixed frame, and $\{b\}$ be the body frame attached to the quadrotor's center of mass (each rotor axis points in the $\hat{z}_b$-direction). The mass of the quadrotor is m and $I_b \in \mathbb{R}^{3\times3}$ is the inertia matrix of the quadrotor with respect to the center of mass frame $\{b\}$. The orientation R_{sb} and position p_{sb} of the center of mass frame as

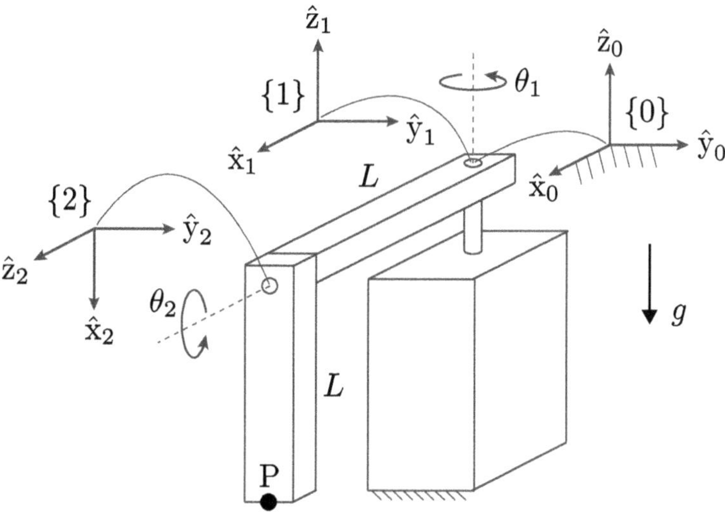

Figure 5.9 Rotational inverted pendulum when $\theta_1 = \theta_2 = 0$ of Exercise 5.10.

seen from the fixed frame are known (both are expressed in {s} frame coordinates).
You may ignore gravity.

(a) Suppose the quadrotor's center of mass moves with linear velocity $v_s \in \mathbb{R}^3$ and angular velocity $\omega_s \in \mathbb{R}^3$, where v_s and ω_s are both expressed in {s} frame coordinates. Derive the kinetic energy of the quadrotor in terms of v_s and ω_s.

(b) Suppose an external force $f_s \in \mathbb{R}^3$ and external moment $m_s \in \mathbb{R}^3$ are applied to the quadrotor body, where f_s and m_s are both expressed in {s} frame coordinates (m_s is the moment about the {s} frame origin). Find the linear acceleration $a_b \in \mathbb{R}^3$ and angular acceleration $\alpha_b \in \mathbb{R}^3$ of the quadrotor frame {b}, where a_b and α_b are both expressed in {b} frame coordinates.

(c) The four rotors apply a force $f_r \in \mathbb{R}^3$ and moment $m_r \in \mathbb{R}^3$ on the quadrotor body (f_r and m_r are both expressed in {b} frame coordinates):

$$f_r = \left(0,\ 0,\ k_1(\dot\theta_1^2 + \dot\theta_2^2 + \dot\theta_3^2 + \dot\theta_4^2)\right)^\top,$$

$$m_r = \left(k_1 d(\dot\theta_4^2 - \dot\theta_2^2),\ k_1 d(\dot\theta_1^2 - \dot\theta_3^2),\ k_1 d(\dot\theta_1^2 - \dot\theta_2^2 + \dot\theta_3^2 - \dot\theta_4^2)\right)^\top,$$

where $\dot\theta_i$ denotes the rate of rotation of rotor i, d is the distance from each rotor axis to the center of mass, and k_1 and k_2 are constants determined by factors such as propeller geometry, air density, air velocity. To simplify the calculation, assume $k_1 = k_2 = d = \mathfrak{m} = 1$ and $\mathcal{I}_b$ is the 3×3 identity matrix I. Suppose that at $t = 0$, the quadrotor starts at the origin of {s} and $R_{sb}(0) = I$. Set the rotor inputs $\dot\theta_i = 1$, $i = 1, \ldots, 4$. Describe in detail the motion of the quadrotor by analyzing the equations of motion. How does the motion depend on the initial linear velocity $v_b(0)$ and initial angular velocity $\omega_b(0)$ (both are expressed in {b} frame coordinates)? To help you

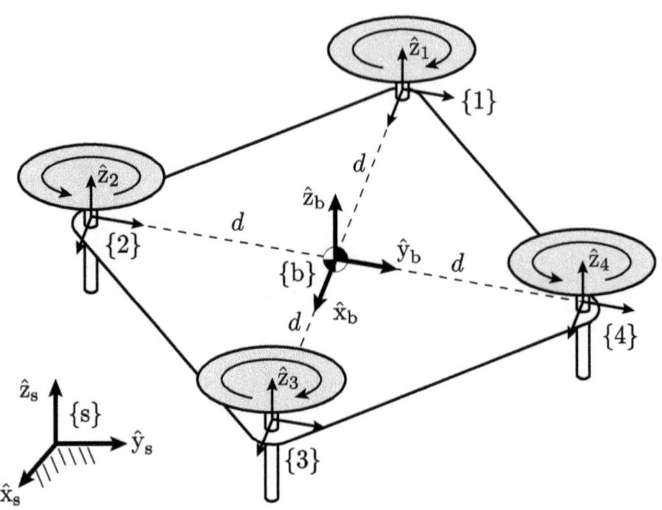

Figure 5.10 Quadrotor of Exercise 5.11.

think about this exercise, you can try different values of $v_b(0)$ and $\omega_b(0)$ (for example, try $(0, 0, 0)^\top$, $(1, 0, 0)^\top$, and $(0, 0, 1)^\top$).

6 Lagrangian Dynamics

Up to this point in our study of dynamics, we have relied on Newton's laws of motion as the foundation. For each rigid body in a system, we draw free-body diagrams and apply Newton's second law to express the resulting equations of motion while taking into account external forces and moments. These equations are then manipulated and simplified to obtain a minimal set of differential equations that describe the motion of the entire system.

In this chapter, we take a different approach to deriving the equations of motion – one based on **Lagrangian dynamics**. Here, the equations are derived from Hamilton's **principle of stationary action**, which states that a mechanical system evolves in a way that extremizes the action – defined as the integral of the difference between kinetic and potential energy over time. In other words, the system follows a trajectory for which the action is a minimum, maximum, or saddle point.

This method is both elegant and powerful. In many cases, it allows us to derive the equations of motion without explicitly calculating constraint forces between bodies. We begin by illustrating this approach with a motivating example.

Example 6.1 The rod shown in Figure 6.1 is connected to a revolute joint, with gravity acting downward. The first step in the Lagrangian dynamics formalism is to determine the **degrees of freedom** of the system; intuitively, this is the number of independent coordinates needed to represent the state of the system. In the case of the rod, knowing the joint value θ is enough to completely specify the position and orientation of the rod, so the degrees of freedom is one. The coordinate $\theta \in [0, 2\pi]$ then acts as the **generalized coordinates** for the system.

The next step is to derive an expression for the **Lagrangian** of the system in terms of the generalized coordinate θ and its time derivative $\dot{\theta}$. The Lagrangian $\mathcal{L}$ is the difference between the kinetic energy and the potential energy:

$$\mathcal{L}(\theta, \dot{\theta}) = \mathcal{K}(\theta, \dot{\theta}) - \mathcal{P}(\theta), \tag{6.1}$$

where $\mathcal{K}(\theta, \dot{\theta})$ is the kinetic energy and $\mathcal{P}(\theta)$ is the potential energy. For the rod, the moments of inertia with respect to the frame indicated in the figure are $I_{xx} = I_{yy} = \frac{1}{3}mL^2$, $I_{zz} = 0$. Therefore

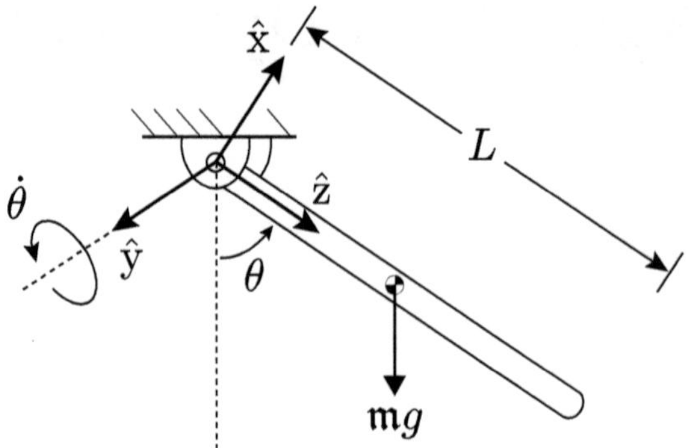

Figure 6.1 A rod connected to the ceiling by a revolute joint.

$$\mathcal{K}(\theta, \dot{\theta}) = \frac{1}{6}mL^2\dot{\theta}^2, \tag{6.2}$$

$$\mathcal{P}(\theta) = -\frac{mgL}{2}\cos\theta, \tag{6.3}$$

and the Lagrangian is

$$\mathcal{L}(\theta, \dot{\theta}) = \frac{1}{6}mL^2\dot{\theta}^2 + \frac{mgL}{2}\cos\theta. \tag{6.4}$$

The **Euler–Lagrange equations** are now applied. For this system they are of the form

$$\tau = \frac{d}{dt}\frac{\partial\mathcal{L}}{\partial\dot{\theta}} - \frac{\partial\mathcal{L}}{\partial\theta}, \tag{6.5}$$

where $\tau \in \mathbb{R}$ is the torque about joint θ, so that $\int \tau^\top \dot{\theta}\, dt$ represents the work of the system, or equivalently, $\tau^\top \dot{\theta}$ represents the power. Substituting the expression (6.4) for $\mathcal{L}$ into (6.5),

$$\tau = \frac{1}{3}mL^2\ddot{\theta} + \frac{mgL}{2}\sin\theta. \tag{6.6}$$

These are the equations of motion for the rod. If no input torques are applied at the joint, then τ is set to zero. $\quad\square$

For systems with low to moderate degrees of freedom, the Lagrangian method is quite effective and efficient. However, for more complex systems with many degrees of freedom, the calculations can become cumbersome, and additional analysis may be needed to determine quantities such as the reaction forces between connected bodies. We now discuss in more detail each of the steps in the Lagrangian dynamics formalism, beginning with an introduction to some basic results in optimization and the calculus of variations.

6.1 Optimization and the Calculus of Variations

6.1.1 Minimization of Functions

Let $f(x)$, $f : \mathbb{R} \to \mathbb{R}$ be a differentiable function of the scalar variable x that we seek to minimize; $f(x)$ is called the **objective function**. Let x^* be a minimizer of $f(x)$. Clearly x^* satisfies $\frac{df}{dx}(x^*) = 0$. This result generalizes to multidimensional functions $f : \mathbb{R}^n \to \mathbb{R}$ in the following way:

Proposition 6.1 *Let $f(x)$, $f : \mathbb{R}^n \to \mathbb{R}$, be a differentiable function of $x = (x_1, \ldots, x_n)^\top$ (i.e., all partial derivatives $\frac{\partial f}{\partial x_i}$ are well defined everywhere), and let $x^* \in \mathbb{R}^n$ be a minimizer (local or global). Then x^* satisfies*

$$\frac{\partial f}{\partial x}(x^*) = \left[\begin{array}{ccc} \frac{\partial f}{\partial x_1}(x^*) & \cdots & \frac{\partial f}{\partial x_n}(x^*) \end{array} \right] = 0. \tag{6.7}$$

The above is a **first-order necessary condition** (FONC) for x^* to be a minimizer of $f(x)$. Note that this is not a sufficient condition; any x^* satisfying the above can also be a maximizer or a saddle point. The following notation will also be used:

$$\nabla f(x) = \left[\begin{array}{c} \frac{\partial f}{\partial x_1} \\ \vdots \\ \frac{\partial f}{\partial x_n} \end{array} \right] = \frac{\partial f}{\partial x}^\top . \tag{6.8}$$

$\nabla f(x)$ is referred to as the **gradient** of f at x. At a point x, $\nabla f(x)$ indicates the direction of steepest ascent of f (or equivalently, $-\nabla f(x)$ indicates the direction of steepest descent of f). Any point x^* that satisfies $\nabla f(x^*) = 0$ is a **critical point** of f.

Example 6.2 Solutions to the linear equation $Ax = b$, where $A \in \mathbb{R}^{m \times n}$ and $b \in \mathbb{R}^m$ are given and $x \in \mathbb{R}^n$ is to be determined, may or may not exist depending on the relative sizes of m and n and also the rank of A. If $m > n$, then the number of equations m exceeds the number of variables n, and in this case a solution will in general not exist. One can, however, seek a best-fit solution $x \in \mathbb{R}^n$ in the sense of minimizing the difference between Ax and b:

$$\min_x \frac{1}{2}\|Ax - b\|^2 = \frac{1}{2}(Ax - b)^\top (Ax - b) = \frac{1}{2}\left(x^\top A^\top Ax - 2x^\top A^\top b + b^\top b\right). \tag{6.9}$$

Since $b^\top b$ is constant this term can be ignored in the minimization. The objective function to be minimized is therefore

$$f(x) = \frac{1}{2}x^\top A^\top Ax - x^\top A^\top b. \tag{6.10}$$

To obtain the gradient of (6.10), consider a simpler version of the objective function $h(x) = \frac{1}{2}x^\top Qx$, where $x \in \mathbb{R}^2$ and $Q \in \mathbb{R}^{2 \times 2}$:

$$h(x) = \frac{1}{2}x^\top Qx = \frac{1}{2}\left(q_{11}x_1^2 + q_{12}x_1x_2 + q_{21}x_2x_1 + q_{22}x_2^2\right). \tag{6.11}$$

Its gradient $\nabla h(x)$ is

$$\nabla h(x) = \begin{bmatrix} \frac{\partial h}{\partial x_1} \\ \frac{\partial h}{\partial x_2} \end{bmatrix} = \begin{bmatrix} q_{11}x_1 + \frac{1}{2}(q_{12} + q_{21})x_2 \\ \frac{1}{2}(q_{21} + q_{12})x_1 + q_{22}x_2 \end{bmatrix}$$
$$= \begin{bmatrix} q_{11} & \frac{1}{2}(q_{12} + q_{21}) \\ \frac{1}{2}(q_{21} + q_{12}) & q_{22} \end{bmatrix} \begin{bmatrix} x_1 \\ x_2 \end{bmatrix} = \frac{1}{2}(Q + Q^\top)x. \tag{6.12}$$

It is not difficult to see that the above result also holds when Q is $n \times n$. Also note that if Q is symmetric, then $\nabla h(x) = Qx$.

With this result, and observing that $A^\top A$ is symmetric, the corresponding first-order necessary condition for the objective function (6.10) is $A^\top Ax - A^\top b = 0$, or

$$A^\top Ax = A^\top b. \tag{6.13}$$

Since A is $m \times n$ with $m \geq n$, $A^\top A$ is an $n \times n$ square matrix. $A^\top A$ is moreover invertible if A is of maximal rank n, in which case the unique solution x^* is given by

$$x^* = (A^\top A)^{-1}A^\top b. \tag{6.14}$$

The x^* above is referred to as the **least-squares solution** to $Ax = b$. $\qquad\square$

We now add the following twist: minimize the same objective function $f(x)$ as before, but the minimizer x^* must now satisfy the constraint $g(x^*) = 0$ for some given $g: \mathbb{R}^n \to \mathbb{R}^m$. The function

$$g(x) = \begin{bmatrix} g_1(x_1, \ldots, x_n) \\ \vdots \\ g_m(x_1, \ldots, x_n) \end{bmatrix} \tag{6.15}$$

is an equality constraint. Note that if $m > n$, then the equation $g(x) = 0$ may be overconstrained (i.e., the number of independent equations exceeds the number of variables) and a solution may not exist. Therefore to make the optimization problem meaningful, typically $m \leq n$.

Let x^* be a minimizer of the above equality-constrained minimization problem. Then x^* must satisfy the following first-order necessary condition:

Proposition 6.2 *Let $f(x)$, $f: \mathbb{R}^n \to \mathbb{R}$ be a differentiable objective function to be minimized, and let $g(x)$, $g: \mathbb{R}^n \to \mathbb{R}^m$, $m \leq n$, be a differentiable function. The x^* that minimizes $f(x)$ while satisfying $g(x^*) = 0$ must then satisfy the following first-order necessary condition: there exists a $\lambda \in \mathbb{R}^m$ such that*

$$\frac{\partial f}{\partial x}(x^*) + \lambda^\top \frac{\partial g}{\partial x}(x^*) = 0, \tag{6.16}$$

where

$$\frac{\partial g}{\partial x} = \begin{bmatrix} \frac{\partial g_1}{\partial x_1} & \cdots & \frac{\partial g_1}{\partial x_n} \\ \vdots & \ddots & \vdots \\ \frac{\partial g_m}{\partial x_1} & \cdots & \frac{\partial g_m}{\partial x_n} \end{bmatrix} \in \mathbb{R}^{m \times n}. \tag{6.17}$$

λ is referred to as the **Lagrange multiplier**. An equivalent statement of this first-order necessary condition is to define a function $H(x, \lambda) = f(x) + \lambda^\top g(x)$. Treating $H(x, \lambda)$ as an unconstrained objective function to be minimized with respect to x and λ, the corresponding first-order necessary conditions for $H(x, \lambda)$ are then

$$\frac{\partial H}{\partial x}(x^*, \lambda) = \frac{\partial f}{\partial x}(x^*) + \lambda^\top \frac{\partial g}{\partial x}(x^*) = 0, \tag{6.18}$$

$$\frac{\partial H}{\partial \lambda}(x^*, \lambda) = g^\top(x^*) = 0. \tag{6.19}$$

Note that the second equation is just the original constraint $g(x^*) = 0$. Equations (6.18) and (6.19), respectively, consist of n and m equations, in terms of the $n + m$ unknowns $x \in \mathbb{R}^n$ and $\lambda \in \mathbb{R}^m$. For those interested in an application of the fundamental theorem of linear algebra, a proof of the first-order necessary condition for the equality-constrained optimization problem is provided in the Appendix.

Example 6.3   Now suppose $m < n$ in $Ax = b$, where $A \in \mathbb{R}^{m \times n}$ and $b \in \mathbb{R}^m$ are given as before. In this case the number of variables exceeds the number of constraints, so that an infinite number of solutions x^* will in general exist. One way to choose x^* is to find the **minimum norm solution**:

$$\min_{x \in \mathbb{R}^n} \frac{1}{2}\|x\|^2, \tag{6.20}$$

subject to the equality constraint $Ax = b$. Using Lagrange multipliers, the corresponding first-order necessary conditions can be written

$$A^\top \lambda + x = 0, \tag{6.21}$$

or $-A^\top \lambda = x$. Premultiplying both sides of this equation by A,

$$-AA^\top \lambda = Ax = b, \tag{6.22}$$

where in the last equality we make use of the given constraint $Ax = b$. Observe that $AA^\top$ is $m \times m$, with $m \le n$. If $AA^\top$ is invertible, or equivalently, the rank of A is m, then $\lambda = -(AA^\top)^{-1}b$, and the solution x^* is given by

$$x^* = -A^\top \lambda = A^\top (AA^\top)^{-1}b. \tag{6.23}$$

$\square$

Example 6.4   Consider the problem of finding the point on the line represented by the equation $x_1 + x_2 = 1$ that is closest to the origin. This problem can be formulated as minimum norm problem in $\mathbb{R}^2$ with objective function $f(x) = \frac{1}{2}(x_1^2 + x_2^2)$ and equality constraint $g(x) = x_1 + x_2 - 1 = 0$. Setting $H(x, \lambda) = \frac{1}{2}(x_1^2 + x_2^2) + \lambda(x_1 + x_2 - 1)$, the first-order necessary conditions are

$$\frac{\partial H}{\partial x} = \begin{bmatrix} x_1 + \lambda & x_2 + \lambda \end{bmatrix} = \begin{bmatrix} 0 & 0 \end{bmatrix}, \tag{6.24}$$

$$\frac{\partial H}{\partial \lambda} = x_1 + x_2 - 1 = 0. \tag{6.25}$$

The above can be arranged into the linear equation

$$
\begin{bmatrix} 1 & 0 & 1 \\ 0 & 1 & 1 \\ 1 & 1 & 0 \end{bmatrix} \begin{bmatrix} x_1 \\ x_2 \\ \lambda \end{bmatrix} = \begin{bmatrix} 0 \\ 0 \\ 1 \end{bmatrix}, \tag{6.26}
$$

whose solution is $x_1 = \frac{1}{2}$, $x_2 = \frac{1}{2}$, $\lambda = -\frac{1}{2}$. $\square$

Note that we could have also used the closed-form solution given by (6.23), but this would require inverting $AA^\top$. This example shows that the inverse does not have to be explicitly calculated; in most cases it is enough to solve the linear equation (6.22) for x, which is computationally much easier than inverting a matrix.

6.1.2   Calculus of Variations

Scalar Case

Now consider the problem of finding a differentiable scalar trajectory $x(t)$, $t \in [t_0, t_f]$, that minimizes the integral

$$
\int_{t_0}^{t_f} \mathcal{L}(x(t), \dot{x}(t), t) \, dt, \tag{6.27}
$$

subject to the boundary conditions $x(t_0) = x_0$, $x(t_f) = x_f$. Suppose $x^*(t)$ is the minimizer, and let

$$
x(t) = x^*(t) + \epsilon \eta(t), \tag{6.28}
$$

where ϵ is a scalar parameter and $\eta(t)$, $t \in [t_0, t_f]$ is a scalar **admissible variation** in the sense that $x(t)$ satisfies all given requirements and boundary conditions. For the given problem, $\eta(t)$ must be differentiable, and the boundary conditions $x(t_0) = x_0$, $x(t_f) = x_f$ require that $\eta(t_0) = \eta(t_f) = 0$.

Define the function $f(\epsilon)$, $f : \mathbb{R} \to \mathbb{R}$, as follows:

$$
f(\epsilon) = \int_{t_0}^{t_f} \mathcal{L}(x^* + \epsilon\eta, \dot{x}^* + \epsilon\dot{\eta}, t) \, dt. \tag{6.29}
$$

For what value of ϵ does $f(\epsilon)$ achieve its minimum? Since it was assumed that $x^*(t)$ is the minimizer, clearly f is minimized at $\epsilon = 0$; this further implies that

$$
\begin{aligned}
\left. \frac{\partial f}{\partial \epsilon} \right|_{\epsilon=0} &= \int_{t_0}^{t_f} \left. \left(\frac{\partial \mathcal{L}}{\partial x} \frac{\partial x}{\partial \epsilon} + \frac{\partial \mathcal{L}}{\partial \dot{x}} \frac{\partial \dot{x}}{\partial \epsilon} \right) \right|_{\epsilon=0} dt \\
&= \int_{t_0}^{t_f} \left(\frac{\partial \mathcal{L}}{\partial x}(x^*, \dot{x}^*, t)\eta + \frac{\partial \mathcal{L}}{\partial \dot{x}}(x^*, \dot{x}^*, t)\dot{\eta} \right) dt \\
&= 0.
\end{aligned} \tag{6.30}
$$

The second term $\int_{t_0}^{t_f} \frac{\partial \mathcal{L}}{\partial \dot{x}}(x^*, \dot{x}^*, t)\dot{\eta} \, dt$ can be integrated by parts:

$$
\int_{t_0}^{t_f} \frac{\partial \mathcal{L}}{\partial \dot{x}}(x^*, \dot{x}^*, t)\dot{\eta} \, dt = \frac{\partial \mathcal{L}}{\partial \dot{x}}(x^*, \dot{x}^*, t)\eta(t) \Big|_{t_0}^{t_f} - \int_{t_0}^{t_f} \eta(t) \frac{d}{dt} \left(\frac{\partial \mathcal{L}}{\partial \dot{x}}(x^*, \dot{x}^*, t) \right) dt. \tag{6.31}
$$

Since $\eta(t_0) = \eta(t_f) = 0$, the first term above is zero, and we therefore have

$$\frac{\partial f}{\partial \epsilon}\bigg|_{\epsilon=0} = \int_{t_0}^{t_f} \left(\frac{\partial \mathcal{L}}{\partial x}(x^*, \dot{x}^*, t) - \frac{d}{dt}\frac{\partial \mathcal{L}}{\partial \dot{x}}(x^*, \dot{x}^*, t) \right) \eta \, dt = 0. \tag{6.32}$$

We now make use of the following result from calculus:

Lemma 6.1 *Let $x(t)$ be a given continuous function of t on the interval $[t_0, t_f]$. Suppose $\int_{t_0}^{t_f} x(t)\eta(t) \, dt = 0$ for all continuous $\eta(t)$, $t \in [t_0, t_f]$. Then $x(t) = 0$ for all $t \in [t_0, t_f]$.*

Applying this lemma to Equation (6.32), we can conclude that

$$\frac{d}{dt}\frac{\partial \mathcal{L}}{\partial \dot{x}}(x^*, \dot{x}^*, t) - \frac{\partial \mathcal{L}}{\partial x}(x^*, \dot{x}^*, t) = 0. \tag{6.33}$$

The above **Euler–Lagrange equations** are first-order necessary conditions for $x^*(t)$ to be a minimizer in the calculus of variations problem.

Example 6.5 The Euler–Lagrange equations for the integral

$$\min_{x(t)} \frac{1}{2} \int_{t_0}^{t_f} (x^2 + \dot{x}^2) \, dt \tag{6.34}$$

are, with $\mathcal{L}(x, \dot{x}) = \frac{1}{2}(x^2 + \dot{x}^2)$,

$$\ddot{x} = x. \tag{6.35}$$

$\square$

Example 6.6 Now consider minimizing the integral

$$\int_0^2 x^2(1 - \dot{x})^2 dt, \tag{6.36}$$

with boundary conditions $x(0) = 0$, $x(2) = 1$. Setting $\mathcal{L} = x^2(1 - \dot{x})^2$, we have

$$\frac{\partial \mathcal{L}}{\partial x} = 2x(1 - \dot{x})^2, \tag{6.37}$$

$$\frac{\partial \mathcal{L}}{\partial \dot{x}} = -2x^2(1 - \dot{x}), \tag{6.38}$$

$$\frac{d}{dt}\frac{\partial \mathcal{L}}{\partial \dot{x}} = -4x\dot{x}(1 - \dot{x}) + 2x^2\ddot{x}, \tag{6.39}$$

and the Euler–Lagrange equations simplify to

$$x(x\ddot{x} + \dot{x}^2 - 1) = 0. \tag{6.40}$$

The above is a nonlinear second-order differential equation with split boundary conditions $x(0) = 0$ and $x(2) = 1$, which can in principle be solved to find a differentiable $x^*(t)$ that potentially minimizes the integral (remember that the Euler–Lagrange equations are only a necessary condition; further verification is necessary to confirm that the solution is a minimizer). On the other hand, if we allow for nondifferentiable but continuous $x(t)$, then observe that the integral is always nonnegative (the integrand involves a squared term and hence is nonnegative). The minimum possible value for this integral is zero, and therefore any $x(t)$ for which the integral is zero can be considered a (global) minimizer. Observe that the following $x(t)$ makes the integral zero:

$$x(t) = \begin{cases} 0, & t \in [0, 1], \\ t - 1, & t \in [1, 2]. \end{cases} \tag{6.41}$$

Note that this $x(t)$ is continuous, but fails to be differentiable at $t = 1$. □

As can be seen from the previous example, solutions to the Euler–Lagrange equations can vary depending on whether or not piecewise differentiable trajectories are admissible (but even then, the Euler–Lagrange equations still hold over the intervals in which the trajectory is differentiable). We will not delve into the details of piecewise differentiable solutions, since our main purpose is to show that with an appropriate formulation of the Lagrangian $\mathcal{L}$, the equations of motion of a mechanical system can be identified with the corresponding Euler–Lagrange equations.

Multidimensional Case

Up to now we have considered only scalar $x(t)$, and assumed that the boundary conditions $x(t_0)$ and $x(t_f)$ are specified. We now consider multidimensional $x(t)$, and also relax the boundary conditions slightly. The first-order necessary conditions are also derived using a more direct argument based on *first variations*, but the resulting Euler–Lagrange equations will be identical to the scalar case (but in multi-dimensional form). Let $x(t)$, $x \colon [t_0, t_f] \to \mathbb{R}^n$ be a differentiable n-dimensional trajectory with elements $x_i(t)$, $i = 1, \ldots, n$. As before, we seek to find the $x(t)$ that minimizes the integral

$$J(x, \dot{x}, t) = \int_{t_0}^{t_f} \mathcal{L}(x(t), \dot{x}(t), t)\, dt. \tag{6.42}$$

The boundary conditions are also relaxed: $x(t_0) = x_0$ as before, but t_f and $x(t_f)$ are now left free.

The **first variation** of J, denoted δJ, is given by

$$\delta J = \int_{t_0}^{t_f} \left(\frac{\partial \mathcal{L}}{\partial x} \delta x + \frac{\partial \mathcal{L}}{\partial \dot{x}} \delta \dot{x} \right) dt + \mathcal{L}(x(t_f), \dot{x}(t_f), t_f)\, \delta t_f, \tag{6.43}$$

where $\delta x(t) \in \mathbb{R}^n$ appearing in the integral is the admissible variation in $x(t)$ and $\delta \dot{x}$ its time derivative, the $\frac{\partial \mathcal{L}}{\partial x}$ and $\frac{\partial \mathcal{L}}{\partial \dot{x}}$ are the row vectors

$$\frac{\partial \mathcal{L}}{\partial x} = \begin{bmatrix} \frac{\partial \mathcal{L}}{\partial x_1} & \cdots & \frac{\partial \mathcal{L}}{\partial x_n} \end{bmatrix}, \tag{6.44}$$

$$\frac{\partial \mathcal{L}}{\partial \dot{x}} = \begin{bmatrix} \frac{\partial \mathcal{L}}{\partial \dot{x}_1} & \cdots & \frac{\partial \mathcal{L}}{\partial \dot{x}_n} \end{bmatrix}, \tag{6.45}$$

and the last term $\mathcal{L}(x(t_f), \dot{x}(t_f), t_f)\, \delta t_f$ in (6.43) follows by taking the first variation of J with respect to t_f (which is now a free variable); for this, we use the following fundamental theorem for ordinary integrals:

$$\frac{d}{dt} \int_a^t f(s)\, ds = f(t) \tag{6.46}$$

for any differentiable scalar function $f(t)$. Integrating the second term inside the integral, $\frac{\partial \mathcal{L}}{\partial \dot{x}} \delta \dot{x}$, by parts as before,

$$\delta J = \int_{t_0}^{t_f} \left(\frac{\partial \mathcal{L}}{\partial x} - \frac{d}{dt}\frac{\partial \mathcal{L}}{\partial \dot{x}} \right) \delta x \; dt + \frac{\partial \mathcal{L}}{\partial \dot{x}}(x(t_f),\dot{x}(t_f),t_f) \; \delta x(t_f)$$
$$+ \; \mathcal{L}(x(t_f),\dot{x}(t_f),t_f) \; \delta t_f. \tag{6.47}$$

The first-order necessary condition for optimality then requires that δJ in (6.47) be zero for all admissible variations of $\delta x(t)$ as well as $\delta x(t_f)$ and δt_f. From this optimality condition we have

$$\frac{d}{dt}\frac{\partial \mathcal{L}}{\partial \dot{x}} - \frac{\partial \mathcal{L}}{\partial x} = 0, \tag{6.48}$$

which is exactly the same form as the Euler–Lagrange equations obtained for the scalar $x(t)$ case. The boundary condition at $t = t_0$ is $x(t_0) = x_0$, while those at $t = t_f$ are extracted from the condition

$$\frac{\partial \mathcal{L}}{\partial \dot{x}}(x(t_f),\dot{x}(t_f),t_f) \; \delta x(t_f) + \mathcal{L}(x(t_f),\dot{x}(t_f),t_f) \; \delta t_f = 0. \tag{6.49}$$

As an example of how to use (6.49) to obtain the boundary condition for (6.48) at $t = t_f$, if, for example, both t_f and $x(t_f)$ were fixed, then $\delta x(t_f) = \delta t_f = 0$ in (6.49), and the boundary condition at t_f is simply $x(t_f) = x_f$. If, however, t_f were fixed and $x(t_f)$ free, then the corresponding boundary condition at t_f would be $\frac{\partial \mathcal{L}}{\partial \dot{x}}(x(t_f),\dot{x}(t_f),t_f) = 0$.

6.2   Lagrangian Dynamics

In Lagrangian dynamics, the equations of motion of a mechanical system can be obtained as the Euler–Lagrange equations to the Lagrangian $\mathcal{L} = \mathcal{K} - \mathcal{P}$, where $\mathcal{K}$ denotes the kinetic energy and $\mathcal{P}$ the potential energy. The first step in Lagrangian dynamics is to determine a set of **generalized coordinates** x for the system. Loosely speaking, these are a set of independent coordinates that completely specify the configuration of the system. For example, in the case of a system of rigid bodies, these would be independent coordinates for the position and orientation of every rigid body. The Lagrangian $\mathcal{L}(x, \dot{x}, t)$ is then obtained, and the equations of motion are derived from the Euler–Lagrange equations

$$\frac{d}{dt}\frac{\partial \mathcal{L}}{\partial \dot{x}} - \frac{\partial \mathcal{L}}{\partial x} = F^{\mathsf{T}}, \tag{6.50}$$

where $F \in \mathbb{R}^n$ are the **generalized forces**. The generalized forces include nonconservative forces, as well as other forces (including conservative forces) that are not already reflected in the choice of Lagrangian $\mathcal{L}$. In the following examples we illustrate why the generalized forces are needed in (6.50), and the physical intuition behind them.

Example 6.7   Consider a particle of mass $\mathfrak{m}$ moving in the plane. Once a fixed frame has been established, choose $x \in \mathbb{R}^2$ to be the generalized coordinates for the particle. The kinetic energy is $\frac{1}{2}\mathfrak{m}(\dot{x}_1^2 + \dot{x}_2^2)$, while the potential energy is left in the general form $\mathcal{P}(x)$ (in the case of gravity, $P(x) = \mathfrak{m}gx_2$, for example). The Lagrangian $\mathcal{L} = \mathcal{K} - \mathcal{P} = \frac{\mathfrak{m}}{2}\dot{x}^{\top}\dot{x} - \mathcal{P}(x)$, and the Euler–Lagrange equations $\frac{d}{dt}\frac{\partial \mathcal{L}}{\partial \dot{x}} - \frac{\partial \mathcal{L}}{\partial x} = 0$ lead to

$$\mathfrak{m}\ddot{x} = -\frac{\partial \mathcal{P}}{\partial x}^{\top}(x). \tag{6.51}$$

The term $-\frac{\partial \mathcal{P}}{\partial x}^{\top}(x) = f_c \in \mathbb{R}^2$ is a conservative force, and we obtain the familiar $f_c = \mathfrak{m}a$ for a particle. Any nonconservative force $f_{nc} \in \mathbb{R}^2$ applied to the particle therefore needs to be added to the right-hand side of the equations of motion, i.e.,

$$\frac{d}{dt}\frac{\partial \mathcal{L}}{\partial \dot{x}} - \frac{\partial \mathcal{L}}{\partial x} = f_{nc}^{\top}. \tag{6.52}$$

Additional conservative forces applied to the particle can be accounted for by adding the corresponding potential function to $\mathcal{P}(x)$, or by adding the force term directly to the right-hand side of (6.52).   □

Example 6.8   Now consider a particle of mass $\mathfrak{m}$ moving in the $\{x_1, x_2\}$ plane, connected to the origin of the fixed frame by a massless rod of length L; the rod rotates about the origin with angle θ. The generalized coordinate for this system can be taken to be the angle θ: once θ is specified, the position $(x_1, x_2)^{\top}$ of the particle is given by $x_1 = L\cos\theta$, $x_2 = L\sin\theta$. The kinetic energy is $\mathcal{K} = \frac{\mathfrak{m}}{2}\dot{x}^{\top}\dot{x} = \frac{\mathfrak{m}}{2}L^2\dot{\theta}^2$, while the potential energy term is of the form $\mathcal{P}(x(\theta))$. (In the case of gravity, for example, $\mathcal{P}(x(\theta)) = \mathfrak{m}gL\sin\theta$.)

The Lagrangian is then

$$\mathcal{L} = \mathcal{K} - \mathcal{P} = \frac{\mathfrak{m}L^2}{2}\dot{\theta}^2 - \mathcal{P}(x(\theta)), \tag{6.53}$$

and the Euler–Lagrange equation $\frac{d}{dt}\frac{\partial \mathcal{L}}{\partial \dot{\theta}} - \frac{\partial \mathcal{L}}{\partial \theta} = 0$ becomes

$$\mathfrak{m}L^2\ddot{\theta} = -\frac{\partial \mathcal{P}}{\partial \theta}. \tag{6.54}$$

To gain an intuitive understanding of what $\frac{\partial \mathcal{P}}{\partial \theta}$ is physically, recall that the potential $\mathcal{P}$ has units of energy and that the time-derivative of energy has units of power. Differentiating $\mathcal{P}(\theta)$ with respect to t, we get

$$\frac{d}{dt}\mathcal{P}(\theta) = \frac{\partial \mathcal{P}}{\partial \theta}\dot{\theta}. \tag{6.55}$$

For this particular example, since $\dot{\theta}$ is angular velocity, it follows that $\frac{\partial \mathcal{P}}{\partial \theta}$ must therefore correspond to a torque about θ in order for $\frac{\partial \mathcal{P}}{\partial \theta}\dot{\theta}$ to have units of power. Any other external torques τ_{ext} applied to joint θ are added to the right-hand side of the equations of motion as

$$\frac{d}{dt}\frac{\partial \mathcal{L}}{\partial \dot{\theta}} - \frac{\partial \mathcal{L}}{\partial \theta} = \tau_{\text{ext}}. \tag{6.56}$$

□

Based on the above examples, we now summarize the Lagrangian approach to deriving the dynamics equations for a general mechanical system. Once the **degrees of freedom** n of the system has been determined and an appropriate set of **generalized coordinates** $q \in \mathbb{R}^n$ have been chosen, the total kinetic energy $\mathcal{K}(q, \dot{q})$ and total

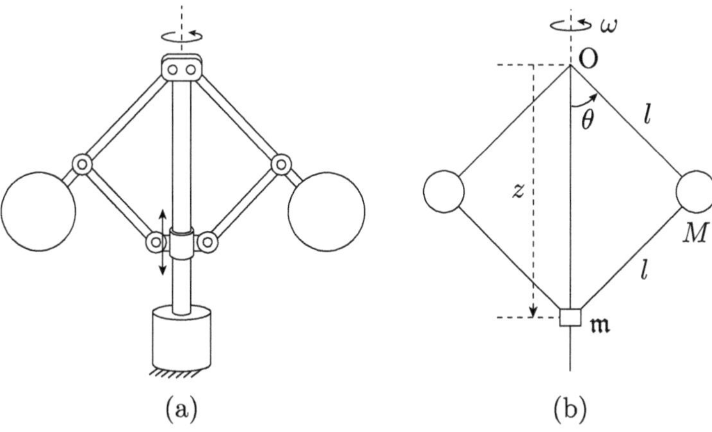

Figure 6.2 (a) The flyball governor; (b) Model.

potential energy $\mathcal{P}(q)$ are determined. Once $\mathcal{K}$ and $\mathcal{P}$ have been determined, the Lagrangian $\mathcal{L}$ is set to $\mathcal{L}(q,\dot{q}) = \mathcal{K}(q,\dot{q}) - \mathcal{P}(q)$. The equations of motion are then given by

$$\frac{d}{dt}\frac{\partial\mathcal{L}}{\partial\dot{q}} - \frac{\partial\mathcal{L}}{\partial q} = F^{\mathsf{T}}, \tag{6.57}$$

where $F \in \mathbb{R}^n$ are the **generalized forces**, and include all forces not captured in the Lagrangian, for example, nonconservative forces, or other conservative forces whose potentials have not been accounted for in the potential energy term $\mathcal{P}$. The physical meaning of F can be understood from the fact that $F^{\mathsf{T}}\dot{q}$ should have the physical units of power.

Example 6.9   Returning to the spherical pendulum from Example 5.1 in Chapter 5, choose the joint angles θ_1, θ_2 to be the generalized coordinates. The kinetic energy $\mathcal{K}$ and potential energy $\mathcal{P}$ are, respectively,

$$\mathcal{K} = \frac{1}{2}mv_c^{\mathsf{T}}v_c + \frac{1}{2}\omega_c^{\mathsf{T}}I_c\omega_c = \frac{\bar{\mu}}{2}(\dot{\theta}_2^2 + \dot{\theta}_1^2\cos^2\theta_2), \tag{6.58}$$

$$\mathcal{P} = mga\sin\theta_2, \tag{6.59}$$

where $\bar{\mu} = 4ma^2/3$. The corresponding Lagrangian $\mathcal{L} = \mathcal{K} - \mathcal{P}$ is

$$\mathcal{L} = \frac{\bar{\mu}}{2}(\dot{\theta}_2^2 + \dot{\theta}_1^2\cos^2\theta_2) - mga\sin\theta_2. \tag{6.60}$$

This time let us allow for the possibility of external torques τ_1 and τ_2 applied at θ_1 and θ_2 respectively. The Euler–Lagrange equations are, after some calculation,

$$\tau_1 = \bar{\mu}(\ddot{\theta}_1\cos^2\theta_2 - 2\dot{\theta}_1\dot{\theta}_2\sin\theta_2\cos\theta_2), \tag{6.61}$$

$$\tau_2 = \bar{\mu}(\ddot{\theta}_2 + \dot{\theta}_1^2\sin\theta_2\cos\theta_2) + mga\cos\theta_2. \tag{6.62}$$

Setting $\tau_1 = \tau_2 = 0$ results in the same set of equations as derived in Chapter 5.   □

Example 6.10 The flyball governor of Figure 6.2(a), also known as a centrifugal governor or centrifugal regulator, is a classical mechanism designed to maintain constant speed of an engine by regulating the flow of fuel. Early versions of the flyball governor were used in windmill grindstones, steam engines, spring-loaded record players, and clocks.

Figure 6.2(b) shows a simple model of the flyball governor. The system consists of two balls connected to a vertical shaft by the pantograph-like mechanism shown in the figure. The vertical shaft is connected to the engine (for example, by a belt not shown in the figure), so that as the engine rotates, the balls also rotate about the shaft at the same rate. As the engine rotates more quickly, the balls rotate more quickly and also begin to rise vertically. As the balls rise higher, a valve supplying fuel to the engine begins to close, limiting the flow of the fuel and thereby slowing the engine rate of rotation. As the engine slows, the balls in turn begin to lower due to gravity, which in turn further opens the valve and increases the fuel fed to the engine. In this way the flyball governor is able to maintain near-constant speed of the engine.

To construct a dynamic model of the flyball governor, model the two balls as point masses each of mass M. Assume that the connecting rods of the pantograph mechanism are massless, and that the mass of the mechanism is concentrated in the collar sliding vertically along the shaft; the collar is modeled as a particle of mass m.

The flyball governor has one degree of freedom, which we parameterize by the angle θ between the vertical shaft and the connecting rod of length l (see Figure 6.2(b)). Choosing θ as the generalized coordinate for the system, the position z of the collar on the vertical shaft is determined from kinematic constraints of the mechanism as

$$z = 2l \cos \theta. \tag{6.63}$$

Assume the vertical shaft rotates at rate ω. The total kinetic energy $\mathcal{K}$ of the system is then given by

$$\mathcal{K} = \frac{1}{2}Ml^2(\dot{\theta}^2 + \omega^2 \sin^2 \theta) \times 2 + \frac{1}{2}m\dot{z}^2$$
$$= Ml^2(\dot{\theta}^2 + \omega^2 \sin^2 \theta) + 2ml^2\dot{\theta}^2 \sin^2 \theta. \tag{6.64}$$

The first term $Ml^2(\dot{\theta}^2 + \omega^2 \sin^2 \theta)$ is the sum of the kinetic energies of the two rotating balls, while the second term $2ml^2\dot{\theta}^2 \sin^2 \theta$ is the kinetic energy of the collar of mass m (Equation (6.63) is used to express $\dot{z}$ in terms of θ and $\dot{\theta}$). The total potential energy $\mathcal{P}$ of the system is obtained as the sum of the potential energies of the two balls and the collar:

$$\mathcal{P} = -Mgl \cos \theta \times 2 - mgz = -2(M + m)gl \cos \theta, \tag{6.65}$$

where g is the gravitational acceleration acting downwards. The corresponding Lagrangian $\mathcal{L} = \mathcal{K} - \mathcal{P}$ is

$$\mathcal{L} = Ml^2(\dot{\theta}^2 + \omega^2 \sin^2 \theta) + 2ml^2\dot{\theta}^2 \sin^2 \theta + 2(M + m)gl \cos \theta. \tag{6.66}$$

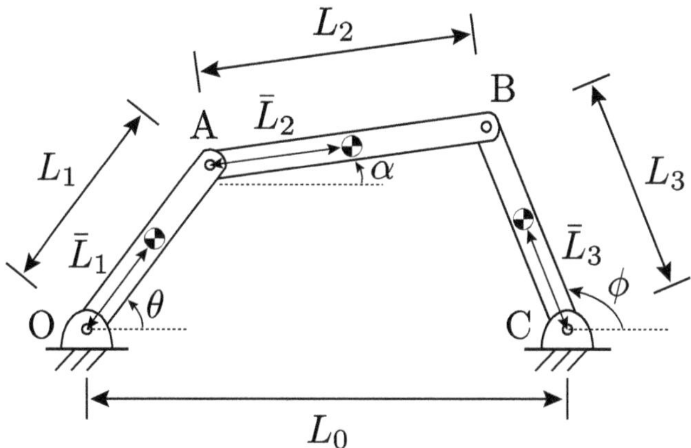

Figure 6.3 A planar four-bar linkage.

The equations of motion are now obtained from the Euler–Lagrange equations $\frac{d}{dt}\frac{\partial \mathcal{L}}{\partial \dot{\theta}} - \frac{\partial \mathcal{L}}{\partial \theta} = 0$, which after some calculation becomes

$$\ddot{\theta} = \frac{(-(M+m)g + (M\omega^2 - 2m\dot{\theta}^2)l\cos\theta)\sin\theta}{l(M + 2m\sin^2\theta)}. \tag{6.67}$$

If $m \ll M$, or equivalently $\frac{m}{M} \approx 0$, Equation (6.67) simplifies to

$$\ddot{\theta} \approx \left(-\frac{g}{l} + \omega^2\cos\theta\right)\sin\theta. \tag{6.68}$$

Steady-state solutions to (6.68) can now be found by setting $\dot{\theta} = \ddot{\theta} = 0$. One solution is $\bar{\theta}_1 = 0$, which corresponds to overlapping arms. Provided the condition $\omega \geq \sqrt{\frac{g}{l}}$ is satisfied, a second steady-state solution is given by

$$\bar{\theta}_2 = \cos^{-1}\frac{g}{l\omega^2}. \tag{6.69}$$

$\square$

Example 6.11   The four-bar linkage of Figure 6.3 is a planar closed chain mechanism consisting of three links arranged in a closed loop (the "fourth" linkage is a stationary one connecting joints O and C, and can be regarded as part of ground). The four-bar linkage is a one-degree-of-freedom mechanism: knowing the joint angle θ, it is then possible to determine the remaining angles α and ϕ, and therefore the position and orientation of each of the three links.

We will choose angle θ as the generalized coordinate for the system, but in the intermediate steps of the derivation also use angles α and ϕ to express the kinetic and potential energies of each of the links. The total kinetic energy is obtained as the sum of the kinetic energies for each of the three links:

$$\mathcal{K} = \frac{1}{2}\left(m_1\|\bar{v}_1\|^2 + I_1\dot{\theta}^2\right) + \frac{1}{2}\left(m_2\|\bar{v}_2\|^2 + I_2\dot{\alpha}^2\right) + \frac{1}{2}\left(m_3\|\bar{v}_3\|^2 + I_3\dot{\phi}^2\right), \tag{6.70}$$

where m_i is the mass of link i, I_i is the moment of inertia of link i with respect to its center of mass (all rotations are about the axis normal to the plane of motion), and

$\bar{v}_i \in \mathbb{R}^2$ is the velocity of the center of mass of link i expressed with respect to the inertial frame attached at O. The link velocity norms are given by

$$\|\bar{v}_1\|^2 = \bar{L}_1^2 \dot{\theta}^2, \tag{6.71}$$

$$\|\bar{v}_2\|^2 = L_1^2 \dot{\theta}^2 + \bar{L}_2^2 \dot{\alpha}^2 + 2 L_1 \bar{L}_2 \cos(\theta - \alpha) \dot{\theta} \dot{\alpha}, \tag{6.72}$$

$$\|\bar{v}_3\|^2 = \bar{L}_3^2 \dot{\phi}^2. \tag{6.73}$$

The total potential energy is also obtained as the sum of the potential energies for each of the three links:

$$\mathcal{P} = \mathfrak{m}_1 g \bar{L}_1 \sin\theta + \mathfrak{m}_2 g (L_1 \sin\theta + \bar{L}_2 \sin\alpha) + \mathfrak{m}_3 g \bar{L}_3 \sin\phi, \tag{6.74}$$

where g is the gravitational acceleration (gravity acts downward). The Lagrangian $\mathcal{L} = \mathcal{K} - \mathcal{P}$ is then dependent on the variables $(\theta, \alpha, \phi, \dot{\theta}, \dot{\alpha}, \dot{\phi})$.

Since θ is chosen to be the generalized coordinate (i.e., the independent variable), the angles α and ϕ need to be determined as a function of θ. These can be obtained from the *loop closure equations*:

$$L_1 \cos\theta + L_2 \cos\alpha = L_0 + L_3 \cos\phi, \tag{6.75}$$

$$L_1 \sin\theta + L_2 \sin\alpha = L_3 \sin\phi. \tag{6.76}$$

By taking the sum of the squares of (6.75) and (6.76) after leaving only $L_2 \cos\alpha$ and $L_2 \sin\alpha$ on the left-hand sides and solving for $t = \tan\frac{\phi}{2}$ (then $\sin\phi = \frac{2t}{1+t^2}$, $\cos\phi = \frac{1-t^2}{1+t^2}$), the angles ϕ and α can be obtained as functions of θ:

$$\phi = 2 \tan^{-1} \left(\frac{-k_1 \pm \sqrt{k_1^2 + k_2^2 - k_3^2}}{k_3 - k_2} \right), \tag{6.77}$$

$$\alpha = \tan^{-1} \left(\frac{-L_1 \sin\theta + L_3 \sin\phi}{L_0 - L_1 \cos\theta + L_3 \cos\phi} \right), \tag{6.78}$$

where $k_1(\theta) = -2 L_1 L_3 \sin\theta$, $k_2(\theta) = 2 L_3 (L_0 - L_1 \cos\theta)$, $k_3(\theta) = L_0^2 + L_1^2 - L_2^2 + L_3^2 - 2 L_0 L_1 \cos\theta$. For this example analytic equations can be derived, but in the general case it may be necessary to resort to numerical solutions.

The Euler–Lagrange equations are then

$$\frac{d}{dt} \frac{\partial \mathcal{L}}{\partial \dot{\theta}} - \frac{\partial \mathcal{L}}{\partial \theta} = \tau_{\text{ext}}, \tag{6.79}$$

where τ_{ext} denotes the external torque applied at joint θ. Note that because α and ϕ are dependent on θ, evaluation of the second term $\frac{\partial \mathcal{L}}{\partial \theta}$ requires the partial derivatives $\frac{\partial \mathcal{L}}{\partial \alpha} \frac{\partial \alpha}{\partial \theta}$ and $\frac{\partial \mathcal{L}}{\partial \phi} \frac{\partial \phi}{\partial \theta}$. The terms $\frac{\partial \alpha}{\partial \theta}$ and $\frac{\partial \phi}{\partial \theta}$ can be obtained by taking time derivatives of the loop closure equations (6.75), (6.76):

$$\begin{bmatrix} L_2 \sin\alpha & -L_3 \sin\phi \\ -L_2 \cos\alpha & L_3 \cos\phi \end{bmatrix} \begin{bmatrix} \dot{\alpha} \\ \dot{\phi} \end{bmatrix} = \begin{bmatrix} -L_1 \sin\theta \\ L_1 \cos\theta \end{bmatrix} \dot{\theta}, \tag{6.80}$$

or

$$\begin{bmatrix} \dot\alpha \\ \dot\phi \end{bmatrix} = \begin{bmatrix} L_2 \sin\alpha & -L_3 \sin\phi \\ -L_2 \cos\alpha & L_3 \cos\phi \end{bmatrix}^{-1} \begin{bmatrix} -L_1 \sin\theta \\ L_1 \cos\theta \end{bmatrix} \dot\theta = \begin{bmatrix} S_1 \\ S_2 \end{bmatrix} \dot\theta, \qquad (6.81)$$

where $S_1 = \frac{L_1 \sin(\phi-\theta)}{L_2 \sin(\alpha-\phi)}$ and $S_2 = \frac{L_1 \sin(\alpha-\theta)}{L_3 \sin(\alpha-\phi)}$. Since $\dot\alpha = \frac{\partial\alpha}{\partial\theta}\dot\theta$ and $\dot\phi = \frac{\partial\phi}{\partial\theta}\dot\theta$, expressions for $\frac{\partial\alpha}{\partial\theta}$ and $\frac{\partial\phi}{\partial\theta}$ can be obtained directly from (6.81). These expressions are also required in the evaluation of the first term $\frac{d}{dt}\frac{\partial\mathcal{L}}{\partial\dot\theta}$ of the Euler–Lagrange equations.

The equations of motion are then, after some calculation,

$$2\left(J_1 + J_2 S_1^2 + J_3 S_2^2 + K S_1 \cos(\theta - \alpha)\right)\ddot\theta$$
$$+ \left[2J_2 S_1\left(\frac{\partial S_1}{\partial\theta} + S_1 \frac{\partial S_1}{\partial\alpha} + S_2 \frac{\partial S_1}{\partial\phi}\right) + 2J_3 S_2 \left(\frac{\partial S_2}{\partial\theta} + S_1 \frac{\partial S_2}{\partial\alpha} + S_2 \frac{\partial S_2}{\partial\phi}\right)\right.$$
$$+ \left. K\left(\left(\frac{\partial S_1}{\partial\theta} + S_1 \frac{\partial S_1}{\partial\alpha} + S_2 \frac{\partial S_1}{\partial\phi}\right)\cos(\theta - \alpha) + (S_1^2 - S_1)\sin(\theta - \alpha)\right)\right]\dot\theta$$
$$+ (m_1 \bar{L}_1 + m_2 L_1)g\cos\theta + S_1 m_2 g \bar{L}_2 \cos\alpha + S_2 m_3 g \bar{L}_3 \cos\phi = \tau_{\text{ext}}, \qquad (6.82)$$

where $J_1 = \frac{1}{2}(m_1 \bar{L}_1^2 + I_1 + m_2 L_1^2)$, $J_2 = \frac{1}{2}(m_2 \bar{L}_2^2 + I_2)$, $J_3 = \frac{1}{2}(m_3 \bar{L}_3^2 + I_3)$, and $K = m_2 L_1 \bar{L}_2$. As can be seen from this example, the calculations can get quite involved in a Lagrangian dynamics formulation of the equations of motion, even for one-degree-of-freedom systems. $\qquad\square$

6.3 Summary

- Let $f(x)$, $f\colon \mathbb{R}^n \to \mathbb{R}$, be a differentiable function of $x = (x_1, \ldots, x_n)^\top$ that we seek to minimize; $f(x)$ is called the **objective function**. Let x^* be a minimizer of $f(x)$. Then x^* satisfies

$$\frac{\partial f}{\partial x}(x^*) = \begin{bmatrix} \frac{\partial f}{\partial x_1}(x^*) & \cdots & \frac{\partial f}{\partial x_n}(x^*) \end{bmatrix} = 0. \qquad (6.83)$$

The above is a **first-order necessary condition** (FONC) for x^* to be a minimizer of $f(x)$.

- For a function $f(x)$, $f\colon \mathbb{R}^n \to \mathbb{R}$, the **gradient** of f at a point $x \in \mathbb{R}^n$

$$\nabla f(x) = \begin{bmatrix} \frac{\partial f}{\partial x_1} \\ \vdots \\ \frac{\partial f}{\partial x_n} \end{bmatrix} = \frac{\partial f}{\partial x}^\top \qquad (6.84)$$

indicates the direction of steepest ascent of f (or equivalently, $-\nabla f(x)$ indicates the direction of steepest descent of f). Any point x^* that satisfies $\nabla f(x^*) = 0$ is a **critical point** of f.

- Let $f(x)$, $f\colon \mathbb{R}^n \to \mathbb{R}$ be a differentiable objective function to be minimized, and let $g(x)$, $g\colon \mathbb{R}^n \to \mathbb{R}^m$, $m \le n$, be a differentiable function of the form

$$g(x) = \begin{bmatrix} g_1(x_1, \ldots, x_n) \\ \vdots \\ g_m(x_1, \ldots, x_n) \end{bmatrix}. \tag{6.85}$$

The x^* that minimizes $f(x)$ while satisfying the equality constraint $g(x^*) = 0$ must then satisfy the following first-order necessary condition: there exists a $\lambda \in \mathbb{R}^m$ such that

$$\frac{\partial f}{\partial x}(x^*) + \lambda^\top \frac{\partial g}{\partial x}(x^*) = 0, \tag{6.86}$$

where

$$\frac{\partial g}{\partial x} = \begin{bmatrix} \frac{\partial g_1}{\partial x_1} & \cdots & \frac{\partial g_1}{\partial x_n} \\ \vdots & \ddots & \vdots \\ \frac{\partial g_m}{\partial x_1} & \cdots & \frac{\partial g_m}{\partial x_n} \end{bmatrix} \in \mathbb{R}^{m \times n}. \tag{6.87}$$

λ is referred to as the **Lagrange multiplier**. An equivalent statement of this first-order necessary condition can be derived by defining a function $H(x, \lambda) = f(x) + \lambda^\top g(x)$. Treating $H(x, \lambda)$ as an unconstrained objective function to be minimized with respect to x and λ, the corresponding first-order necessary conditions for $H(x, \lambda)$ are then

$$\frac{\partial H}{\partial x}(x^*, \lambda) = \frac{\partial f}{\partial x}(x^*) + \lambda^\top \frac{\partial g}{\partial x}(x^*) = 0, \tag{6.88}$$

$$\frac{\partial H}{\partial \lambda}(x^*, \lambda) = g^\top(x^*) = 0. \tag{6.89}$$

The second equation is just the original equality constraint $g(x^*) = 0$.

- Let $x(t)$, $t \in [t_0, t_f]$, be a differentiable scalar trajectory. The $x^*(t)$ that minimizes the integral

$$\int_{t_0}^{t_f} \mathcal{L}(x(t), \dot{x}(t), t)\, dt, \tag{6.90}$$

subject to the boundary conditions $x(t_0) = x_0$, $x(t_f) = x_f$ must satisfy

$$\frac{d}{dt}\frac{\partial \mathcal{L}}{\partial \dot{x}}(x^*, \dot{x}^*, t) - \frac{\partial \mathcal{L}}{\partial x}(x^*, \dot{x}^*, t) = 0. \tag{6.91}$$

The above **Euler–Lagrange equations** are first-order necessary conditions for $x^*(t)$ to be a minimizer in the scalar calculus of variations problem.

- Let $x(t)$, $x\colon [t_0, t_f] \to \mathbb{R}^n$ be a differentiable n-dimensional trajectory with elements $x_i(t)$, $i = 1, \ldots, n$. The $x^*(t)$ that minimizes the integral

$$\int_{t_0}^{t_f} \mathcal{L}(x(t), \dot{x}(t), t)\, dt, \tag{6.92}$$

subject to the boundary conditions $x(t_0) = x_0$ (t_f and $x(t_f)$ are left free) must satisfy the Euler–Lagrange equations

$$\frac{d}{dt}\frac{\partial \mathcal{L}}{\partial \dot{x}}(x^*, \dot{x}^*, t) - \frac{\partial \mathcal{L}}{\partial x}(x^*, \dot{x}^*, t) = 0, \tag{6.93}$$

where the $\frac{\partial \mathcal{L}}{\partial x}$ and $\frac{\partial \mathcal{L}}{\partial \dot{x}}$ are the row vectors

$$\frac{\partial \mathcal{L}}{\partial x} = \begin{bmatrix} \frac{\partial \mathcal{L}}{\partial x_1} & \cdots & \frac{\partial \mathcal{L}}{\partial x_n} \end{bmatrix}, \tag{6.94}$$

$$\frac{\partial \mathcal{L}}{\partial \dot{x}} = \begin{bmatrix} \frac{\partial \mathcal{L}}{\partial \dot{x}_1} & \cdots & \frac{\partial \mathcal{L}}{\partial \dot{x}_n} \end{bmatrix}. \tag{6.95}$$

- The **generalized coordinates** $q \in \mathbb{R}^n$ for a system are a set of independent coordinates that completely specify the configuration of the system, and the number of generalized coordinates n is the **degrees of freedom** of the system.
- In **Lagrangian dynamics**, the equations of motion of a mechanical system are obtained as the Euler–Lagrange equations to the **Lagrangian** of the system

$$\mathcal{L}(q, \dot{q}) = \mathcal{K}(q, \dot{q}) - \mathcal{P}(q), \tag{6.96}$$

where $q \in \mathbb{R}^n$ is an appropriate set of generalized coordinates for the system, $\mathcal{K}(q, \dot{q})$ is the total kinetic energy, and $\mathcal{P}(q)$ is the total potential energy. The equations of motion are then given by

$$\frac{d}{dt} \frac{\partial \mathcal{L}}{\partial \dot{q}} - \frac{\partial \mathcal{L}}{\partial q} = F^\top, \tag{6.97}$$

where $F \in \mathbb{R}^n$ are the **generalized forces**, which include all forces not captured in the Lagrangian $\mathcal{L}$.

6.4 Exercises

Exercise 6.1 Find the critical points of the following objective functions $f(x)$, $f : \mathbb{R} \to \mathbb{R}$.

(a) $f(x) = x^2 + 2x$.
(b) $f(x) = x^2 e^{-x^2}$.
(c) $f(x) = x^4 + 4x^3 + 6x^2 + 4x$.
(d) $f(x) = x + \sin x$.

Exercise 6.2 Find the critical point of the following objective function $f(x)$, $f : \mathbb{R}^2 \to \mathbb{R}$:

$$f(x) = x_1^2 + x_1 x_2 + \frac{1}{2} x_2^2 - x_1 - x_2.$$

What is the value of $f(x)$ at the critical point?

Exercise 6.3 Consider the following objective function $f(x)$, $f : \mathbb{R}^2 \to \mathbb{R}$:

$$f(x) = \frac{1}{2} x^\top A x + x^\top b + 6, \quad A = \begin{bmatrix} 1 & 2 \\ 4 & 7 \end{bmatrix}, \quad b = \begin{bmatrix} 3 \\ 5 \end{bmatrix}.$$

(a) Find the gradient $\nabla f(x)$ of f at point $(1, 1)^\top$.
(b) Find the point x^* that satisfies the first-order necessary conditions for a minimum.

Exercise 6.4 Minimize the objective function $f(x) = x_1^2 + x_2^2 + x_3^2$ subject to an equality constraint $g(x) = x_1^2 + \frac{x_2^2}{4} + \frac{x_3^2}{9} - 1 = 0$. From this minimum norm problem in $\mathbb{R}^3$, you can find the point on the ellipsoid represented by the equation $g(x) = 0$ that is closest to the origin.

Exercise 6.5 Find the distance from a point $p = (p_x, p_y, p_z)^\top$ to a plane represented by the equation $ax + by + cz + d = 0$. You need to find the point on the plane that is closest to p.

Exercise 6.6 Write down the first-order necessary conditions (but do not solve) for minimizing the following objective function $f(x)$, $f : \mathbb{R}^3 \to \mathbb{R}$,

$$f(x) = x_1^3 + 2(x_1 - x_2)^2 + 3x_1^2 - x_3^3,$$

subject to two equality constraints

$$g_1(x) = x_1 + 5x_2 - 10 = 0,$$
$$g_2(x) = (x_2 - 2)^2 - x_3 = 0.$$

Exercise 6.7 (a) Solve the following minimization problem

$$\min_{x \in \mathbb{R}^n} f(x) = \frac{1}{2} x^\top Q x,$$

subject to an equality constraint $1 - x^\top x = 0$, where $Q \in \mathbb{R}^{n \times n}$ is symmetric and positive-definite. Find the minimum value of the objective function and the corresponding minimizer in terms of the eigenvalues and eigenvectors of Q.

(b) Consider the planar robot arm of Figure 6.4. The tip position $x = (x_1, x_2)^\top$ is related to the joint position $\theta = (\theta_1, \theta_2)^\top$ by

$$x_1 = L_1 \cos\theta_1 + L_2 \cos(\theta_1 + \theta_2),$$
$$x_2 = L_1 \sin\theta_1 + L_2 \sin(\theta_1 + \theta_2).$$

The tip velocity $\dot{x} = (\dot{x}_1, \dot{x}_2)^\top$ is then related to the joint velocity $\dot{\theta} = (\dot{\theta}_1, \dot{\theta}_2)^\top$ by

$$\begin{bmatrix} \dot{x}_1 \\ \dot{x}_2 \end{bmatrix} = \begin{bmatrix} -L_1 \sin\theta_1 - L_2 \sin(\theta_1 + \theta_2) & -L_2 \sin(\theta_1 + \theta_2) \\ L_1 \cos\theta_1 + L_2 \cos(\theta_1 + \theta_2) & L_2 \cos(\theta_1 + \theta_2) \end{bmatrix} \begin{bmatrix} \dot{\theta}_1 \\ \dot{\theta}_2 \end{bmatrix},$$

or $\dot{x} = J(\theta)\dot{\theta}$. Suppose $L_1 = \sqrt{6}$ m, $L_2 = 2$ m, $\theta_1 = 0$ and $\theta_2 = 90°$. Find the unit magnitude joint velocity $\dot{\theta}$ (i.e., $\dot{\theta}_1^2 + \dot{\theta}_2^2 = 1$) that produces the tip velocity of maximum speed.

Exercise 6.8 (a) Find a differentiable scalar trajectory $x(t), t \in [0, \pi/2]$ that minimizes the integral

$$\int_0^{\pi/2} (\dot{x}^2 - x^2)\, dt$$

with boundary conditions $x(0) = 0$, $x(\pi/2) = 1$.

(b) Find a differentiable scalar trajectory $x(t), t \in [0, 2]$ that minimizes the integral

$$\int_0^2 (t\dot{x} + \dot{x}^2)\, dt$$

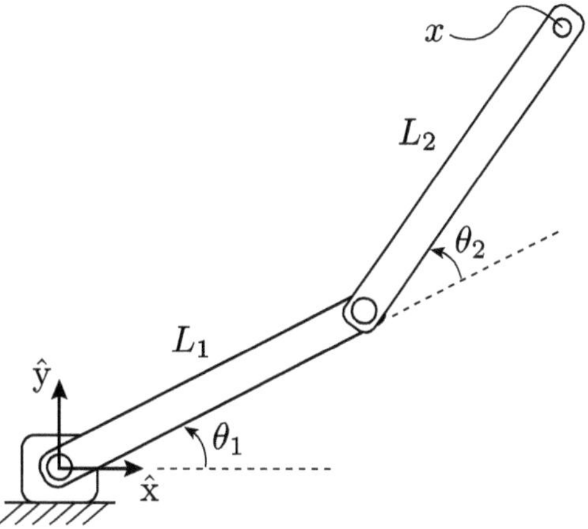

Figure 6.4 Planar robot arm of Exercise 6.7.

with boundary condition $x(0) = 1$. What should the value of $x(2)$ be? (*Hint*: Let $x(2) = c$ and solve the Euler–Lagrange equation to find the optimal value of the integral as a function of c.)

Exercise 6.9   We want to find the differentiable trajectory $x(t)$, $x \colon [t_0, t_f] \to \mathbb{R}^n$ that minimizes the integral

$$J(x, \dot{x}, \ddot{x}, t) = \int_{t_0}^{t_f} \mathcal{L}(x(t), \dot{x}(t), \ddot{x}(t), t) \, dt,$$

where $t_0, x(t_0), \dot{x}(t_0), t_f, x(t_f), \dot{x}(t_f)$ are fixed. Derive the Euler–Lagrange equations and boundary conditions.

Exercise 6.10   A massless torus of radius r is spinning about the vertical axis at a constant angular speed ω as shown in Figure 6.5. A particle of mass m is sliding around the torus, where θ is the angle about the vertical axis and gravity g acts downward. Choosing θ as the generalized coordinate, derive the equations of motion of the system using the Euler–Lagrange equations.

Exercise 6.11   As shown in Figure 6.6, a mass m attached to a string of length l swings freely under gravity, forming a spherical pendulum. Angles θ and ϕ are defined as shown in the figure and gravity g acts downward.
(a) Show that the Lagrangian of the system is

$$\mathcal{L}(\theta, \phi, \dot{\theta}, \dot{\phi}) = \frac{1}{2} m l^2 (\dot{\theta}^2 + \dot{\phi}^2 \sin^2 \theta) + m g l \cos \theta,$$

where the two angles (θ, ϕ) are the generalized coordinates.
(b) From the Euler–Lagrange equations, show that

$$\dot{\phi} \sin^2 \theta = c,$$

where c is constant.

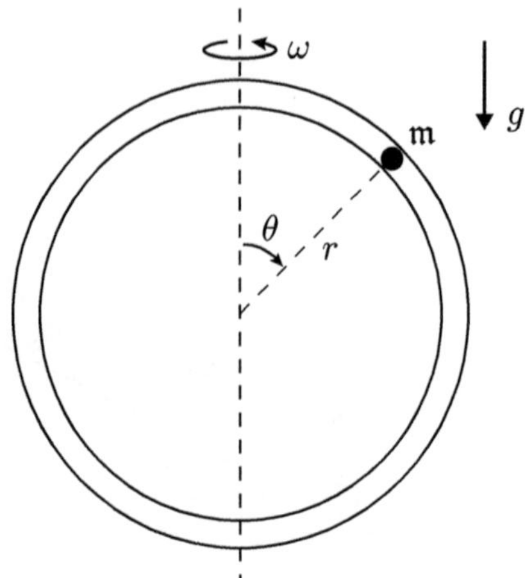

Figure 6.5 Spinning torus and particle of Exercise 6.10.

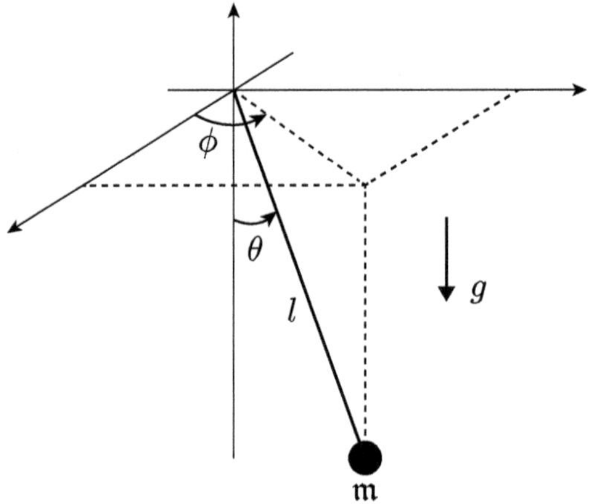

Figure 6.6 Spherical pendulum of Exercise 6.11.

(c) Show that the equations of motion are given by

$$\ddot{\theta} = -\frac{g}{l}\sin\theta + \dot{\phi}^2 \sin\theta\cos\theta,$$
$$\ddot{\phi} = -2\dot{\theta}\dot{\phi}\cot\theta.$$

Exercise 6.12 As shown in Figure 6.7, the spherical pendulum of Exercise 6.11 is now attached to a massless rod that can move vertically (up and down), where the length of the rod is z. From the Euler–Lagrange equations, show that the equations of motion are given by

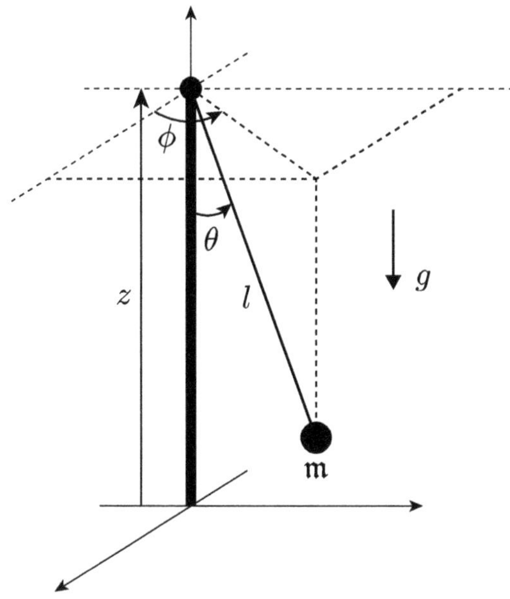

Figure 6.7 Spherical pendulum on vertical rod of Exercise 6.12.

$$\ddot{\theta} = (\dot{\theta}^2 + \dot{\phi}^2)\tan\theta,$$
$$\ddot{\phi} = -2\dot{\theta}\dot{\phi}\cot\theta,$$
$$\ddot{z} = -l(\dot{\theta}^2 + \dot{\phi}^2\sin^2\theta)\sec\theta - g.$$

Exercise 6.13   Figure 6.8 shows a sliding pendulum. A slider of mass m_1 moves horizontally along a frictionless guide and a ball of mass m_2 is connected to the slider by a massless cord of length l. The fixed frame $\{\hat{x}, \hat{y}\}$ is attached to the guide, and the position x of the slider and the angle θ of the cord are defined as shown in the figure. Gravity g acts downward. Choosing x and θ as the generalized coordinates, use the Euler–Lagrange equations to derive the equations of motion of the system. Model the slider and the ball as particles.

Exercise 6.14   As shown in Figure 6.9, a particle of mass m is sliding inside a frictionless cone of angle β. The fixed frame $\{\hat{x}, \hat{y}, \hat{z}\}$ is attached to the bottom of the cone as shown in the figure and gravity acts in the $-\hat{z}$-direction. Choosing angle θ and distance r defined as shown in the figure as the generalized coordinates, derive the equations of motion of the system using the Euler–Lagrange equations. Compare your results with the equations of motion derived in Exercise 3.9(b) in Chapter 3.

Exercise 6.15   (a) As shown in Figure 6.10(a), two masses m_1 and m_2 are connected by a cable and pulley (both are massless), where a spring of stiffness k with undeformed length l_0 is connected between m_2 and the cable. Gravity g acts downward and assume

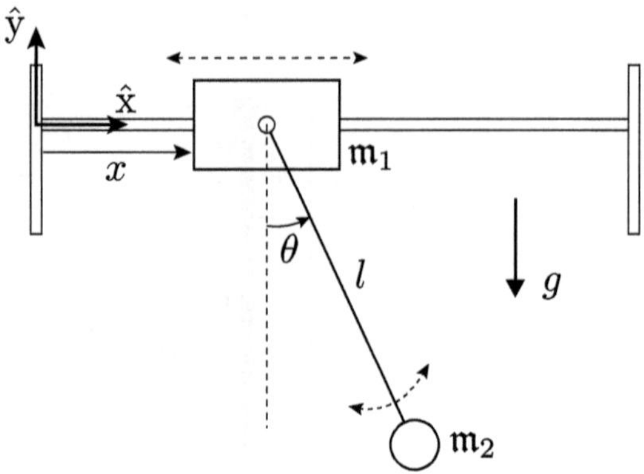

Figure 6.8 Sliding pendulum of Exercise 6.13.

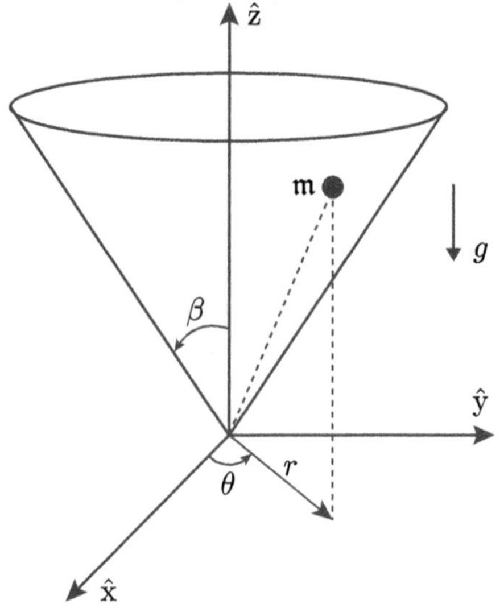

Figure 6.9 Particle sliding inside a cone of Exercise 6.14.

$m_1 = m_2 = m$. Choosing x_1 and x_2 defined as shown in the figure as the generalized coordinates, use the Euler–Lagrange equations to derive the equations of motion of the system. Does $\ddot{x}_2$ depend on x_1?

(b) Now consider the mass-pulley system of Figure 6.10(b), where $m_1 = 2m$, $m_2 = 3m$, and $m_3 = 4m$. Choosing x_1 and x_3 defined as shown in the figure as the generalized coordinates, use the Euler–Lagrange equations to derive the equations of motion of the system.

Exercise 6.16   Figure 6.11 shows a model of a jumping insect. Two masses m and M are connected by a spring of stiffness k and undeformed length L. The spring is initially completely compressed and then suddenly released. The spring length after the release is x, and y is the distance from the ground to mass M. Gravity g acts downward.

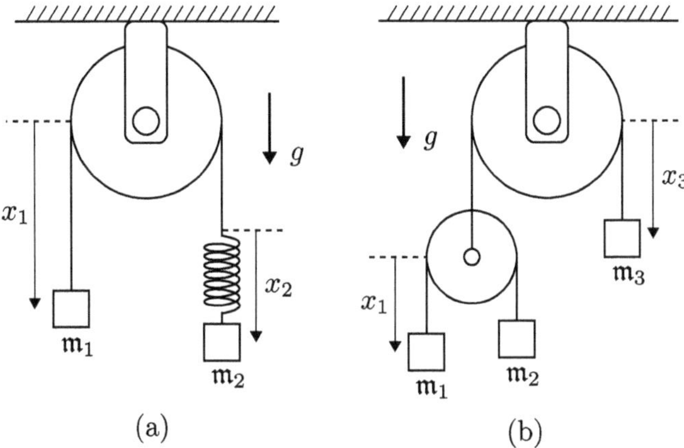

Figure 6.10 Mass-pulley systems of Exercise 6.15.

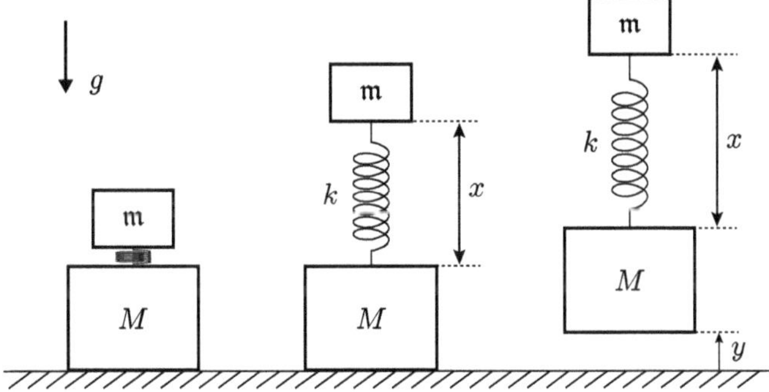

Figure 6.11 Jumping insect model of Exercise 6.16.

(a) Using the Euler–Lagrange equations, derive the equations of motion of the system in terms of x and y (i) before and (ii) after mass M loses contact with the floor. Denote by N the normal force exerted on mass M by the floor before losing contact.

(b) With the equations of motion derived in (a), determine the length x_0 of the spring right before mass M loses contact with the floor. Also, find the inequality constraint that makes it possible for the spring to be extended to x_0.

(c) Now assume $M = 2$ kg, $m = 1$ kg, $k = 10$ N/m, $L = 5$ m and $g = 10$ m/s^2. Suppose mass M loses contact with the floor at $t = 0$, and let $x(0) = x_0$, $\dot{x}(0) = \dot{x}_0$, $y(0) = \dot{y}(0) = 0$. First, find x_0 and $\dot{x}_0$. Then, solve the equations of motion derived in (a) to obtain $x(t)$ and $y(t)$ and draw the graphs of them.

Exercise 6.17 A drum-type washing machine of radius R is modeled as shown in Figure 6.12. Constant torque τ is applied about the drum center, and the clothes being washed are modeled as a point mass m_1 that rotates about the drum center. The mass of the washing machine (with no clothes in) is m_2. A nonlinear spring is placed between the

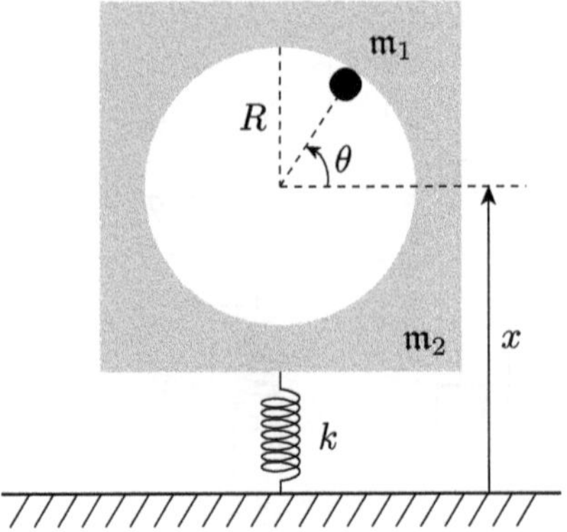

Figure 6.12 Washing machine of Exercise 6.17.

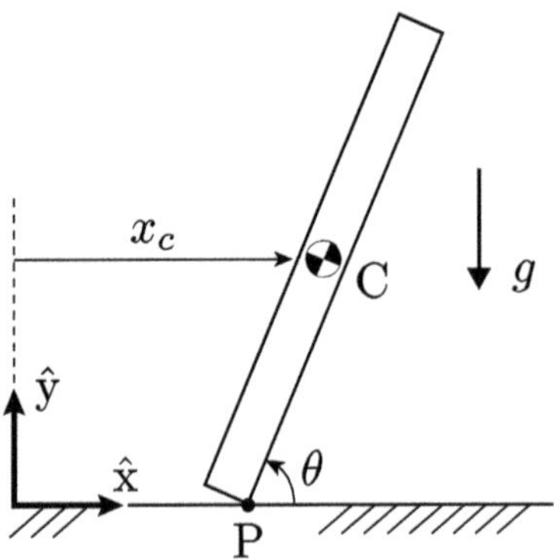

Figure 6.13 Falling stick of Exercise 6.18.

floor and the washing machine; the force generated by the nonlinear spring is $f = kx^2$. Assume the washing machine can only move vertically, and gravity g acts downward. Derive the equations of motion of the washing machine using the Euler–Lagrange equations.

Exercise 6.18   A stick of mass m and length L falls to the ground as shown in Figure 6.13, where gravity g acts downward. The ground is frictionless, causing the contact point P between the stick and the ground to slide. The angle between the stick and the ground is θ and the horizontal distance from the fixed frame $\{\hat{x}, \hat{y}\}$ to the center of mass C is x_c.

(a) Choosing θ and x_c as the generalized coordinates, derive the equations of motion using the Euler–Lagrange equations.

(b) Assuming $\dot{x}_c(0) = 0$, how does $x_c(t)$ change over time?

Exercise 6.19   As shown in Figure 6.14, a planar inverted pendulum is mounted on a cart of mass m_1 moving on a frictionless surface. The rod has mass m_2 and moment of

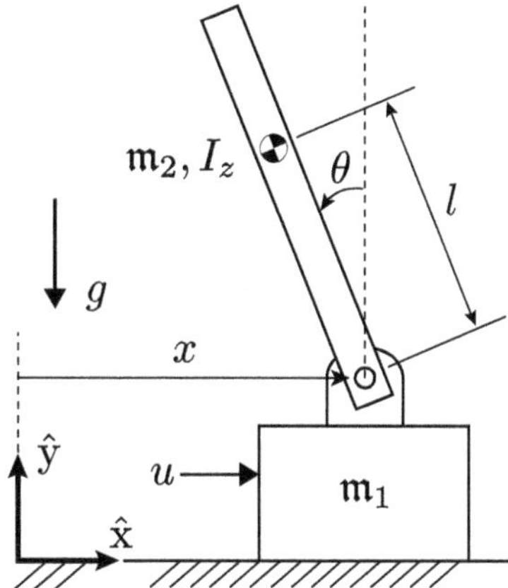

Figure 6.14 Planar inverted pendulum on moving cart of Exercise 6.19.

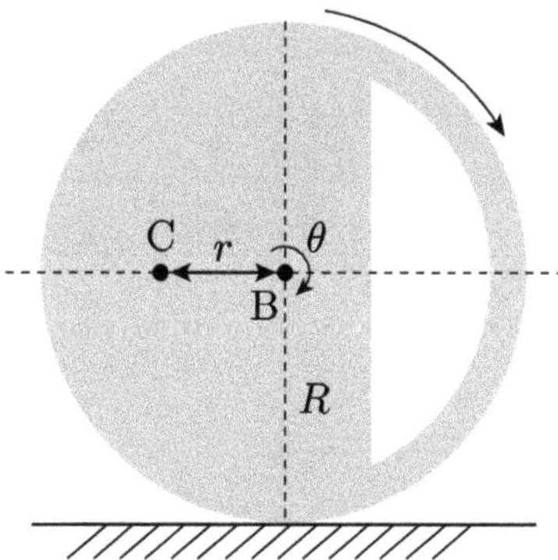

Figure 6.15 Planar wheel with empty space of Exercise 6.20.

inertia I_z about its center of mass, where l is the distance from the joint to the center of mass. A horizontal external force u is applied to the cart and gravity g acts downward. Coordinates x and θ are defined as shown in the figure. Derive the equations of motion of the system using the Euler–Lagrange equations.

Exercise 6.20 As shown in Figure 6.15, a planar wheel with empty space rolls without slipping. Its center of mass C is located at a distance r from its center B. The radius and mass of the wheel are R and m, respectively, and the moment of inertia about center B is I_z. Gravity g acts downward. With angle θ defined as shown in the figure, derive the equations of motion using the Euler–Lagrange equations. The figure shows the orientation of the wheel at $\theta = 0$.

Exercise 6.21 Figure 6.16 shows the one-degree-of-freedom slider-crank mechanism consisting of a rotating wheel (disk) of mass m_w and radius r, a thin uniform rod of mass m_r and length l, and a slider of mass m_s. Angles θ and ϕ are defined as shown in the figure and the distance from the wheel center to the slider is denoted by x_s. Gravity g acts downward.

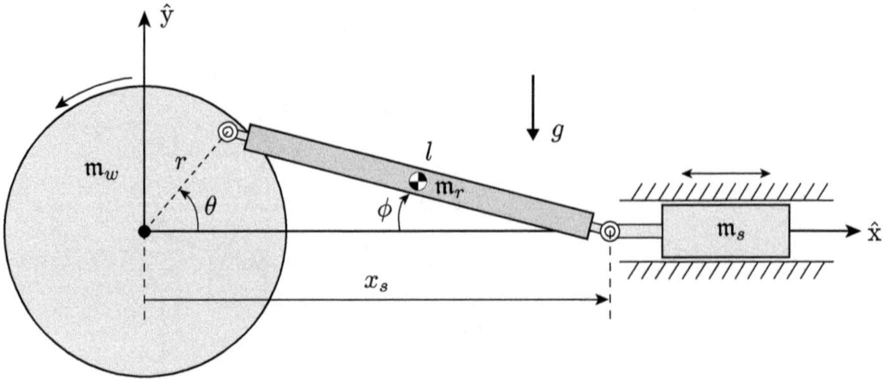

Figure 6.16 Slider-crank mechanism of Exercise 6.21.

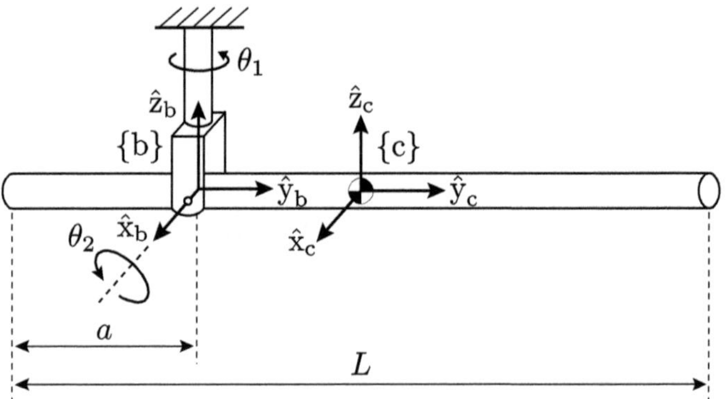

Figure 6.17 Spherical pendulum of Exercise 6.22.

(a) From kinematic constraints, express ϕ and $\dot{\phi}$ in terms of θ and $\dot{\theta}$.

(b) Choosing θ as the generalized coordinate for the system, derive the equations of motion using the Euler–Lagrange equations.

Exercise 6.22   Consider the spherical pendulum of mass m and length L rotating about two orthogonal axes as shown in Figure 6.17. Two frames {b} and {c} are attached to the rod, where {b} is assigned at the intersection of the two axes of rotation and {c} at the rod center of mass. Joint angles θ_1 and θ_2 are defined as shown in the figure. Derive the equations of motion using the Euler–Lagrange equations.

Exercise 6.23   Two identical rods of mass m and length $2a$ are connected as shown in Figure 6.18, forming a rotational inverted pendulum (this is also known as the Furuta pendulum). The fixed frame {0} is assigned at the intersection of joint 1 and rod 1. Moving frames {1} and {2} are attached to rod 1 and rod 2, respectively, as shown in the figure.

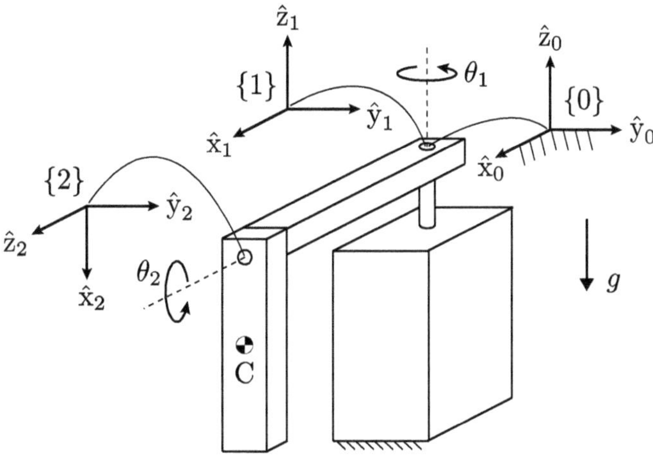

Figure 6.18 Rotational inverted pendulum of Exercise 6.23.

(a) Find ω_1 and ω_2, where ω_i is the coordinates of the angular velocity of rod i expressed in frame $\{i\}$.
(b) Express the velocity of rod 2 center of mass C in frame $\{2\}$ coordinates.
(c) Find the Lagrangian $\mathcal{L} = \mathcal{K} - \mathcal{P}$. Gravity g acts in the $-\hat{z}_0$-direction.
(d) Torques τ_1 and τ_2 are exerted on joints θ_1 and θ_2 respectively. Derive the equations of motion of the system using the Euler–Lagrange equations.

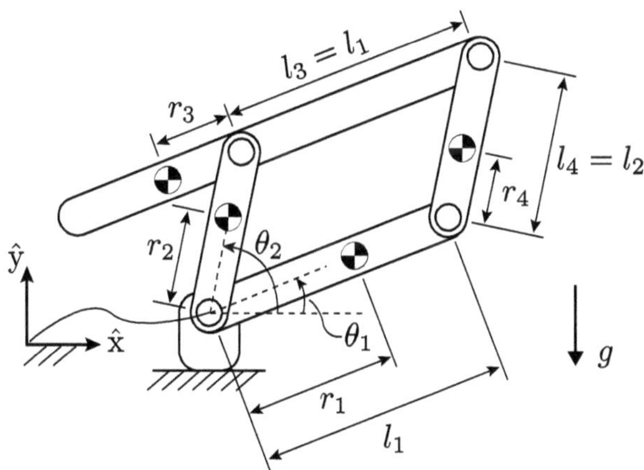

Figure 6.19 Planar five-bar linkage of Exercise 6.24.

Exercise 6.24   The two-degrees-of-freedom planar closed chain mechanism consisting of four links shown in Figure 6.19 is called a five-bar linkage. Suppose $l_1 = l_3$ and $l_2 = l_4$, making the mechanism a parallelogram. Link i has mass m_i and moment of inertia I_i about its center of mass, and the distance from the joint to the center of mass is r_i as shown in the figure. Gravity g acts downward.

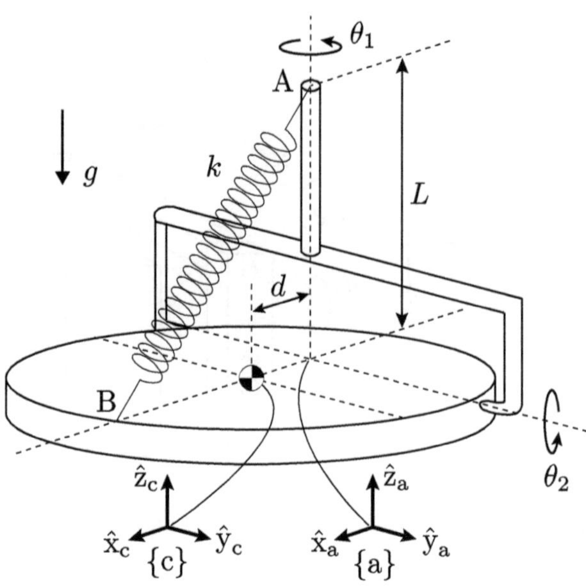

Figure 6.20 Spring-disk system of Exercise 6.25.

(a) Determine the center of mass positions of the four links, where angles θ_1 and θ_2 are the generalized coordinates of the system.

(b) Express the kinetic energy $\mathcal{K}$ of the system as $\mathcal{K} = \frac{1}{2}\dot{\theta}^\top M(\theta)\dot{\theta}$, where $\dot{\theta} = (\dot{\theta}_1, \dot{\theta}_2)^\top$ and $M(\theta) \in \mathbb{R}^{2\times2}$ is a symmetric matrix. Also, express the potential energy $\mathcal{P}$ of the system as $\mathcal{P} = g_1 \sin\theta_1 + g_2 \sin\theta_2$, where g_1 and g_2 are constants. Find the condition that, for arbitrary θ_1 and θ_2, $\mathcal{K}$ becomes a function of $\dot{\theta}_1$ and $\dot{\theta}_2$ only (i.e., $\mathcal{K} = \mathcal{K}(\dot{\theta}_1, \dot{\theta}_2)$; independent of θ_1 and θ_2).

(c) Suppose the condition of (b) is satisfied and torques τ_1 and τ_2 are exerted on joints θ_1 and θ_2 respectively. Denoting the elements of $M(\theta)$ by M_{ij}, $i, j = 1, 2$, derive the equations of motion of the system. You will further be able to express them as $\tau = M(\theta)\ddot{\theta} + g(\theta)$, where $\tau = (\tau_1, \tau_2)^\top$ and $g(\theta) \in \mathbb{R}^2$. This is a very simple example of the robot dynamics equation introduced in Chapter 1.

Exercise 6.25   Referring to Figure 6.20, a thin disk of mass m and radius R is connected to the ceiling by a universal joint. The joint angles are denoted $\theta = (\theta_1, \theta_2)^\top$. The figure shows the spring-disk system's initial configuration $\theta = \dot{\theta} = 0$, where the spring is undeformed. Two moving frames {a} and {c} are attached to the disk in the same orientation: {c} is attached to the center of mass and {a} to the intersection of axis 1 and the disk, a distance d from {c}. The distance between point A and the origin of {a} is L. Gravity g acts downward and no external forces or moments are applied to the system. To simplify the calculation, assume $m = 2$ kg, $R = 2$ m, $d = 1$ m, $k = 12$ N/m, $L = 4$ m, $g = 10$ m/s^2.

(a) For arbitrary θ and $\dot{\theta}$, find the angular velocity of the disk in frame {a} coordinates.

(b) For arbitrary θ and $\dot{\theta}$, find the kinetic energy of the disk. The inertia matrix with respect to frame $\{c\}$ is given by $\text{diag}(\frac{1}{4}mR^2, \frac{1}{4}mR^2, \frac{1}{2}mR^2)$.

(c) Use the conservation of energy to derive a scalar differential equation of the form $f(\theta, \dot{\theta}) = 0$. Is it possible for points A, B, and frame $\{a\}$ origin to lie on a common line?

(d) The equations of motion can be expressed in the form $M(\theta)\ddot{\theta} + b(\theta, \dot{\theta}) = 0$, where $M(\theta) \in \mathbb{R}^{2\times2}$ is symmetric and $b(\theta, \dot{\theta}) \in \mathbb{R}^2$. Find $M(\theta)$ and $b(\theta, \dot{\theta})$.

7   Linearized Dynamics

This chapter focuses on the behavior of dynamic systems subjected to small perturbations around a given solution trajectory. As an example, consider a satellite in a stable periodic orbit that experiences a slight disturbance – such as a brief thruster misfire. Will the satellite return to a stable orbit, or will the disturbance cause it to drift into deep space or fall back to Earth? These and other related questions can be answered by examining the satellite's **linearized dynamics**. As the name implies, the linearized dynamics are a set of linear differential equations that capture the local behavior of the system about a given solution, in the same way that a nonlinear function can be locally approximated by a linear function.

In this chapter we show how to obtain the linearized dynamics of a system about some given solution, derive closed-form analytic solutions to **linear differential equations**, and show how the linearized dynamics and its solutions can be used in applications ranging from **stability analysis** to the study of **mechanical vibrations**.

7.1   Mass-Spring-Damper System

To motivate this chapter, we begin with an analysis of the mass-spring-damper system of Figure 7.1. Let x be the displacement of the mass – assume that $x = 0$ at equilibrium – and let k and b be the spring stiffness and damping coefficient, respectively; from

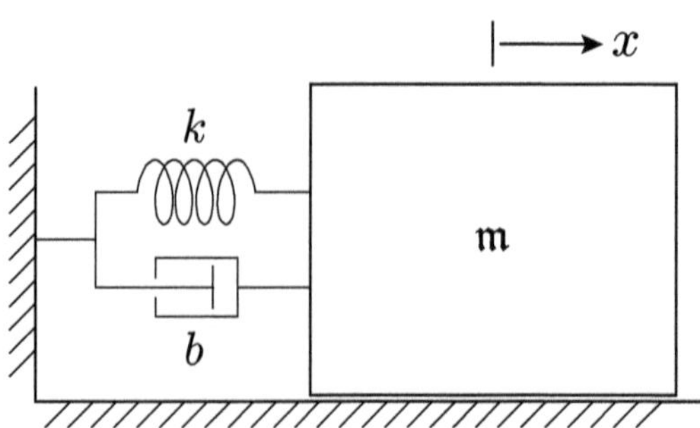

Figure 7.1 Mass-spring-damper system.

physical considerations both k and b can be assumed nonnegative. Assuming no external forces are applied to the mass, the equations of motion are

$$m\ddot{x} + b\dot{x} + kx = 0. \tag{7.1}$$

Before deriving the solution, we define the following two constants: the **natural frequency** ω_n and **damping ratio** β are defined as

$$\omega_n = \sqrt{\frac{k}{m}}, \quad \beta = \frac{b}{2\sqrt{km}}. \tag{7.2}$$

In terms of ω_n and β, the equations of motion can be written

$$\ddot{x} + 2\beta\omega_n\dot{x} + \omega_n^2 x = 0. \tag{7.3}$$

Now, recall that the solution to this second-order, constant-coefficient, linear differential equation is determined by the roots of its characteristic polynomial,

$$s^2 + 2\beta\omega_n s + \omega_n^2 = 0. \tag{7.4}$$

The two roots are

$$s_1 = -\beta\omega_n + \omega_n\sqrt{\beta^2 - 1}, \quad s_2 = -\beta\omega_n - \omega_n\sqrt{\beta^2 - 1}. \tag{7.5}$$

The solution can be divided into three cases:

- **Overdamped**: $\beta > 1$. In this case both roots s_1 and s_2 are real and distinct and the solution to the differential equation is of the form

$$x(t) = c_1 e^{s_1 t} + c_2 e^{s_2 t}, \tag{7.6}$$

  where the constants c_1 and c_2 are determined from the initial conditions. For example, if $x(0) = 1$ and $\dot{x}(0) = 0$, c_1 and c_2 are

$$c_1 = \frac{1}{2} + \frac{\beta}{2\sqrt{\beta^2 - 1}} \quad \text{and} \quad c_2 = \frac{1}{2} - \frac{\beta}{2\sqrt{\beta^2 - 1}}. \tag{7.7}$$

- **Critically damped**: $\beta = 1$. In this case the roots $s_1 = s_2 = -\omega_n$ are equal and real, and the solution is

$$x(t) = (c_1 + c_2 t)e^{-\omega_n t}. \tag{7.8}$$

  For the same initial conditions $x(0) = 1$ and $\dot{x}(0) = 0$,

$$c_1 = 1 \quad \text{and} \quad c_2 = \omega_n. \tag{7.9}$$

- **Underdamped**: $\beta < 1$. In this case the roots are complex conjugates: $s_1 = -\beta\omega_n + i\omega_d$ and $s_2 = -\beta\omega_n - i\omega_d$, where $\omega_d = \omega_n\sqrt{1 - \beta^2}$ is called the **damped natural frequency**. The solution is

$$x(t) = e^{-\beta\omega_n t}(c_1 \cos \omega_d t + c_2 \sin \omega_d t). \tag{7.10}$$

  For the initial conditions $x(0) = 1$ and $\dot{x}(0) = 0$, we get

$$c_1 = 1 \quad \text{and} \quad c_2 = \frac{\beta}{\sqrt{1 - \beta^2}}. \tag{7.11}$$

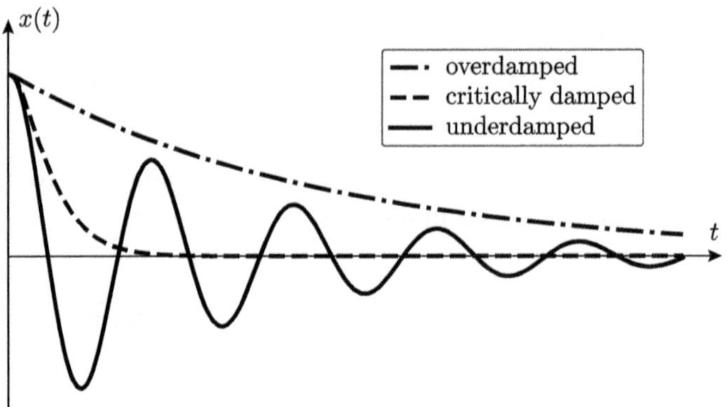

Figure 7.2 Trajectories for the overdamped, critically damped, and underdamped cases.

Trajectories for the three cases are shown for comparison in Figure 7.2. In all three cases the mass asymptotically goes to zero with increasing t, as a consequence of the real part of the roots s_1, s_2 being negative, i.e., $-\beta\omega_n < 0$; in such cases we say that the origin $x = 0$ is a **stable equilibrium**.

The equations of motion (7.1) for the mass-spring-damper system are given by a second-order linear differential equation. We conclude this section by showing how (7.1) can be transformed into a set of first-order linear differential equations. First, define the scalar variables z_1 and z_2 as

$$z_1 = x, \quad z_2 = \dot{x}. \tag{7.12}$$

From the above we can write $\dot{z}_1 = z_2$. Solving (7.1) for $\ddot{x}$, we get $\ddot{x} = -\frac{b}{m}\dot{x} - \frac{k}{m}x$. Writing everything in terms of z_1 and z_2,

$$\begin{bmatrix} \dot{z}_1 \\ \dot{z}_2 \end{bmatrix} = \begin{bmatrix} z_2 \\ -\frac{b}{m}z_2 - \frac{k}{m}z_1 \end{bmatrix} = \begin{bmatrix} 0 & 1 \\ -\frac{k}{m} & -\frac{b}{m} \end{bmatrix} \begin{bmatrix} z_1 \\ z_2 \end{bmatrix}, \tag{7.13}$$

which is a first-order differential equation of the form $\dot{z} = f(z)$. Because $f(z)$ is linear in z, we say that the above is a linear differential equation.

7.2   Linearization

7.2.1   First-Order Differential Equations

To illustrate the notion of linearization of a general differential equation $\dot{x} = f(x)$ about some solution, an analogy with functions is useful. Consider a function $f : \mathbb{R} \to \mathbb{R}$; its linearization about some point $\bar{x}$ is given by the first-order Taylor series approximation

$$f(x) \approx f(\bar{x}) + f'(\bar{x})(x - \bar{x}), \tag{7.14}$$

where $f'(\bar{x})$ denotes the derivative of f at $\bar{x}$. For x near $\bar{x}$, Equation (7.14) approximates $f(x)$ by a linear function. The above can be generalized to a multidimensional mapping

$f: \mathbb{R}^n \rightarrow \mathbb{R}^n$ in the same way:

$$f(x) \approx f(\bar{x}) + \frac{\partial f}{\partial x}(\bar{x})(x - \bar{x}), \qquad (7.15)$$

where $\frac{\partial f}{\partial x}(\bar{x}) \in \mathbb{R}^{n \times n}$ denotes the Jacobian of f at $\bar{x}$:

$$\frac{\partial f}{\partial x}(\bar{x}) = \begin{bmatrix} \frac{\partial f_1}{\partial x_1}(\bar{x}) & \cdots & \frac{\partial f_1}{\partial x_n}(\bar{x}) \\ \vdots & \ddots & \vdots \\ \frac{\partial f_n}{\partial x_1}(\bar{x}) & \cdots & \frac{\partial f_n}{\partial x_n}(\bar{x}) \end{bmatrix}. \qquad (7.16)$$

The linearization of an ordinary differential equation about some solution can be interpreted in a similar fashion. Given a first-order differential equation $\dot{x} = f(x)$, $f: \mathbb{R}^n \rightarrow \mathbb{R}^n$, let $\bar{x}(t)$ be a solution; i.e., $\bar{x}(t)$ satisfies $\dot{\bar{x}}(t) = f(\bar{x}(t))$. The linearization of this differential equation about $\bar{x}(t)$ then proceeds as follows. Let $x(t) = \bar{x}(t) + \eta(t)$, where $\eta(t) \in \mathbb{R}^n$ is some perturbation about $\bar{x}(t)$. Then $\dot{x} = f(x)$ can be written $\dot{\bar{x}} + \dot{\eta} = f(\bar{x} + \eta)$. From the first-order Taylor expansion of f,

$$f(x) \approx f(\bar{x}) + \frac{\partial f}{\partial x}(\bar{x})(x - \bar{x}) = f(\bar{x}) + \frac{\partial f}{\partial x}(\bar{x})\eta, \qquad (7.17)$$

from which the first-order Taylor series approximation of $\dot{x} = f(x)$ about $\bar{x}$ becomes

$$\dot{\bar{x}} + \dot{\eta} = f(\bar{x}) + \frac{\partial f}{\partial x}(\bar{x})\eta,$$

$$\dot{\eta} = \frac{\partial f}{\partial x}(\bar{x})\eta. \qquad (7.18)$$

Equation (7.18) is the linearization of $\dot{x} = f(x)$ about the solution $\bar{x}$. Observe that (7.18) is a linear differential equation of the form $\dot{x} = A(t)x$, where $A(t) = \frac{\partial f}{\partial x}(\bar{x}(t)) \in \mathbb{R}^{n \times n}$ is known.

If the solution $\bar{x}(t)$ varies with time, then the matrix $\frac{\partial f}{\partial x}(\bar{x}(t))$ can be time-varying. If the solution $\bar{x}(t)$ happens to be constant, then $\dot{\bar{x}} = f(\bar{x}) = 0$ and the matrix $\frac{\partial f}{\partial x}(\bar{x})$ will be constant. In this chapter we shall deal exclusively with the latter case: a point $\bar{x} \in \mathbb{R}^n$ at which $f(\bar{x}) = 0$ is said to be an **equilibrium point**, or **stationary point**, of the system $\dot{x} = f(x)$.

7.2.2   Mechanical Systems

So far we have only considered first-order differential equations of the form $\dot{x} = f(x)$, while the equations of motion of mechanical systems involve second derivatives of the form

$$F = M(q)\ddot{q} + b(q, \dot{q}), \qquad (7.19)$$

where $q \in \mathbb{R}^n$ are the generalized coordinates of the system, $F \in \mathbb{R}^n$ is the external generalized force, $M(q) \in \mathbb{R}^{n \times n}$ is an invertible matrix called the **mass matrix**, and $b(q, \dot{q}) \in \mathbb{R}^n$ is a collection of terms representing potential and Coriolis forces. There is in fact a straightforward procedure for converting (7.19) into a first-order differential equation, by setting $x_1 = q$ and $x_2 = \dot{q}$, so that

$$\dot{x}_1 = \dot{q} = x_2, \tag{7.20}$$

$$\dot{x}_2 = \ddot{q} = M(q)^{-1}(F - b(q, \dot{q})) = M(x_1)^{-1}(F - b(x_1, x_2)). \tag{7.21}$$

For unforced systems, i.e., $F = 0$, by defining $x = (x_1, x_2)^\top \in \mathbb{R}^{2n}$, the equations of motion can always be converted into a first-order differential equation of the form $\dot{x} = f(x)$.

Example 7.1   Recall from Example 3.1 in Chapter 3 that the equations of motion for a satellite orbiting the earth is of the form

$$\ddot{r} = r\dot{\theta}^2 - \frac{GM}{r^2}, \tag{7.22}$$

$$\ddot{\theta} = -\frac{2\dot{r}\dot{\theta}}{r}. \tag{7.23}$$

Setting $x_1 = r$, $x_2 = \dot{r}$, $x_3 = \theta$, $x_4 = \dot{\theta}$, we have

$$\begin{bmatrix} \dot{x}_1 \\ \dot{x}_2 \\ \dot{x}_3 \\ \dot{x}_4 \end{bmatrix} = \begin{bmatrix} x_2 \\ x_1 x_4^2 - \frac{GM}{x_1^2} \\ x_4 \\ -\frac{2x_2 x_4}{x_1} \end{bmatrix}. \tag{7.24}$$

The solution corresponding to a circular orbit is given by $r = x_1 = \bar{r}$, $\dot{r} = x_2 = 0$, $\theta = x_3 = \omega t$ (ω constant), $\dot{\theta} = x_4 = \omega$, with $GM = \bar{r}^3 \omega^2$ (the latter equality follows by equating the gravitational force with the required centripetal force for circular motion: $GMm/r^2 = mr\omega^2$). Taking the right-hand side of (7.24) to be $f(x)$, $\frac{\partial f}{\partial x}$ is then

$$\frac{\partial f}{\partial x}(x) = \begin{bmatrix} 0 & 1 & 0 & 0 \\ x_4^2 + \frac{2GM}{x_1^3} & 0 & 0 & 2x_1 x_4 \\ 0 & 0 & 0 & 1 \\ \frac{2x_2 x_4}{x_1^2} & -\frac{2x_4}{x_1} & 0 & -\frac{2x_2}{x_1} \end{bmatrix}. \tag{7.25}$$

Evaluating (7.25) at the circular orbit solution, the linearized equation about the circular orbit is therefore

$$\begin{bmatrix} \dot{x}_1 \\ \dot{x}_2 \\ \dot{x}_3 \\ \dot{x}_4 \end{bmatrix} = \begin{bmatrix} 0 & 1 & 0 & 0 \\ 3\omega^2 & 0 & 0 & 2\omega\bar{r} \\ 0 & 0 & 0 & 1 \\ 0 & -\frac{2\omega}{\bar{r}} & 0 & 0 \end{bmatrix} \begin{bmatrix} x_1 \\ x_2 \\ x_3 \\ x_4 \end{bmatrix}. \tag{7.26}$$

We therefore have a linear differential equation of the form $\dot{x} = Ax$ with A a constant matrix.   □

Suppose the satellite is slightly disturbed from its circular orbit by a small external force – such as a minor impact. What will happen to its motion? Will it settle back into a stable orbit, or will it drift away into space or fall back to Earth? More generally, given a system $\dot{x} = f(x)$ with solution $\bar{x}(t) = \bar{x}$ constant (recall that $\bar{x}$ is called an equilibrium point of f), if $x(0) = x_0$ starts near $\bar{x}$, how will solutions $x(t)$ behave as t goes to infinity?

The answer to this question is provided by the linearization. Loosely speaking, the **linearization theorem** states that the behavior of the system $\dot{x} = f(x)$ near the stationary point $\bar{x}$ is qualitatively identical to that of the linear differential equation $\dot{x} = \frac{\partial f}{\partial x}(\bar{x})x$ for x near the origin. A more precise statement of this theorem as well as its application to mechanical systems will be covered in a later section. Before doing so, we will first derive in the next section closed-form analytic solutions to the linear differential equation $\dot{x} = Ax$, where $A \in \mathbb{R}^{n \times n}$ is constant.

7.3 Linear Differential Equations

7.3.1 Homogeneous Case

In this section we will derive closed-form analytic solutions to the linear differential equation $\dot{x} = Ax$, $A \in \mathbb{R}^{n \times n}$ constant. As a motivational example, recall that the solution to the scalar differential equation $\dot{x} = ax$, where $x \in \mathbb{R}$ and a is a constant scalar, is given by $x(t) = e^{at}x(0)$, where the exponential e^{at} is defined by the series

$$e^{at} = 1 + ta + \frac{t^2}{2!}a^2 + \frac{t^3}{3!}a^3 + \cdots . \tag{7.27}$$

Based on this scalar solution, we conjecture that the solution to the vector linear differential equation $\dot{x} = Ax$ will be

$$x(t) = e^{At}x(0), \tag{7.28}$$

where the matrix exponential e^{At} is defined by the series

$$e^{At} = I + tA + \frac{t^2}{2!}A^2 + \frac{t^3}{3!}A^3 + \cdots . \tag{7.29}$$

To verify whether our conjecture is correct, we can substitute $x(t) = e^{At}x(0)$ into the equation $\dot{x} = Ax$ to see if it is satisfied. Note that

$$\frac{d}{dt}e^{At} = \frac{d}{dt}\left(I + tA + \frac{t^2}{2!}A^2 + \cdots\right)$$

$$= A + tA^2 + \frac{t^2}{2!}A^3 + \cdots$$

$$= A\left(I + tA + \frac{t^2}{2!}A^2 + \cdots\right)$$

$$= Ae^{At} \tag{7.30}$$

$$= e^{At}A. \tag{7.31}$$

The last equality follows from the observation that A could also have been factored to the right rather than the left. From the above,

$$\dot{x}(t) = \frac{d}{dt}e^{At}x(0) = Ae^{At}x(0) = Ax(t), \tag{7.32}$$

which confirms that $e^{At}x(0)$ is indeed a solution to $\dot{x} = Ax$. We also note the following properties of the matrix exponential, whose proofs we omit:

1. e^A is always invertible, with inverse e^{-A}, as can be verified by multiplying the series expansions for e^A and e^{-A}.
2. For any two square matrices A and B, generally $AB \neq BA$, but $Ae^A = e^A A$ holds for any A.
3. For any two square matrices A and B, generally $e^A e^B \neq e^B e^A$, but if $AB = BA$, then it can be shown that $e^A e^B = e^B e^A = e^{A+B}$.

Finally, if the initial condition for $\dot{x} = Ax$ is given at some arbitrary t_0 rather than $t = 0$, the solution $x(t)$ is given by

$$x(t) = e^{A(t-t_0)}x(t_0),\tag{7.33}$$

with the matrix exponential $e^{A(t-t_0)}$ defined by the same series.

7.3.2   Inhomogeneous Case

We now consider the solution to

$$\dot{x} = Ax + f(t)\tag{7.34}$$

for some given forcing term $f(t) \in \mathbb{R}^n$. First, define the vector $z(t)$ via the relation

$$x(t) = e^{A(t-t_0)}z(t),\tag{7.35}$$

or equivalently, $z(t) = e^{-A(t-t_0)}x(t)$. Then

$$\dot{x} = Ae^{A(t-t_0)}z + e^{A(t-t_0)}\dot{z} = Ax + e^{A(t-t_0)}\dot{z},\tag{7.36}$$

from which it follows that we must have $f(t) = e^{A(t-t_0)}\dot{z}$, or equivalently,

$$\dot{z} = e^{-A(t-t_0)}f(t).\tag{7.37}$$

Integrating both sides with respect to t, we get

$$z(t) = z(t_0) + \int_{t_0}^{t} e^{-A(s-t_0)}f(s)\,ds.\tag{7.38}$$

Since $x(t) = e^{A(t-t_0)}z(t)$, we have

$$x(t) = e^{A(t-t_0)}x(t_0) + \int_{t_0}^{t} e^{A(t-s)}f(s)\,ds,\tag{7.39}$$

where we make use of the fact that $z(t_0) = x(t_0)$ and $e^{A(t-t_0)} = e^{At}e^{-At_0}$. Note that the solution (7.39) consists of a homogeneous solution (the first term) and a particular solution (the second term).

7.3.3   Evaluation of the Matrix Exponential

While the solution to the linear differential equation (7.34) can be expressed in closed-form in terms of the matrix exponential e^{At}, there still remains the task of actually calculating e^{At}. First, if A is diagonalizable, i.e., $A = PDP^{-1}$ for some invertible P and diagonal D, then

$$e^{At} = I + tPDP^{-1} + \frac{t^2}{2!}(PDP^{-1})(PDP^{-1}) + \cdots$$

$$= P(I + tD + \frac{t^2}{2!}D^2 + \cdots)P^{-1}$$

$$= Pe^{Dt}P^{-1}. \tag{7.40}$$

Moreover if $D = \mathrm{diag}(d_1, d_2, \ldots, d_n)$, then $e^{Dt} = \mathrm{diag}(e^{d_1 t}, \ldots, e^{d_n t})$.

The **modal solution technique** is a way of determining the exponential of any square matrix. To motivate the technique, recall the procedure for finding the solutions to the scalar second-order linear differential equation $\ddot{x} + 2\beta\omega_n\dot{x} + \omega_n^2 x = 0$ for the mass-spring-damper system. The nature of the solution is determined by the roots of the characteristic polynomial $s^2 + 2\beta\omega_n s + \omega_n^2 = 0$:

- If the roots s_1 and s_2 are both real and $s_1 \neq s_2$, then the solution $x(t)$ is of the form $x(t) = c_1 e^{s_1 t} + c_2 e^{s_2 t}$, with c_1 and c_2 determined from the boundary conditions.
- If $s_1 = s_2 = s$, then $x(t)$ is of the form $x(t) = c_1 e^{st} + c_2 t e^{st}$, with c_1 and c_2 determined from the boundary conditions.
- If s_1 and s_2 are complex conjugates $\mu \pm iv$, then $x(t)$ will be of the form

$$x(t) = e^{\mu t}(c_1 \cos vt + c_2 \sin vt), \tag{7.41}$$

with c_1, c_2 determined from the boundary conditions.

The modal solution technique emulates the above procedure, which we now illustrate with a series of examples.

Example 7.2   For

$$A = \begin{bmatrix} 1 & 1 \\ 0 & 2 \end{bmatrix}, \tag{7.42}$$

the characteristic polynomial is $p(s) = \det(Is - A) = (s - 1)(s - 2)$, and the roots are simply the diagonal elements 1 and 2 (in fact, all triangular matrices have eigenvalues given by their diagonal elements). Since the roots are real and distinct, the matrix exponential is of the form

$$e^{At} = M_1 e^t + M_2 e^{2t}, \tag{7.43}$$

where $M_1, M_2 \in \mathbb{R}^{2\times 2}$ are constant matrices to be determined from an appropriate set of boundary conditions. These are

$$e^{At}\big|_{t=0} = I = M_1 + M_2, \tag{7.44}$$

$$\frac{d}{dt}e^{At}\bigg|_{t=0} = A = M_1 + 2M_2. \tag{7.45}$$

Solving the above linear equations, $M_1 = 2I - A$ and $M_2 = A - I$, and

$$e^{At} = \begin{bmatrix} e^t & e^{2t} - e^t \\ 0 & e^{2t} \end{bmatrix}. \tag{7.46}$$

$\square$

Example 7.3   Given $\omega \in \mathbb{R}^3$, $\|\omega\| = 1$, in this example we compute the matrix exponential of its 3×3 skew-symmetric matrix representation

$$[\omega] = \begin{bmatrix} 0 & -\omega_z & \omega_y \\ \omega_z & 0 & -\omega_x \\ -\omega_y & \omega_x & 0 \end{bmatrix}, \tag{7.47}$$

with $\omega_x^2 + \omega_y^2 + \omega_z^2 = 1$. The characteristic polynomial $p(s) = \det(Is - [\omega]) = s^3 + s$ has roots $s_{1,2} = \pm i$, $s_3 = 0$, and the matrix exponential is of the form

$$e^{[\omega]t} = M_1 \cos t + M_2 \sin t + M_3. \tag{7.48}$$

Three boundary conditions are needed to solve for M_1, M_2, M_3:

$$e^{[\omega]t}\Big|_{t=0} = I = M_1 + M_3, \tag{7.49}$$

$$\frac{d}{dt}e^{[\omega]t}\Big|_{t=0} = [\omega] = M_2, \tag{7.50}$$

$$\frac{d^2}{dt^2}e^{[\omega]t}\Big|_{t=0} = [\omega]^2 = -M_1. \tag{7.51}$$

Solving the three linear equations for M_1, M_2, M_3, the matrix exponential is

$$e^{[\omega]t} = I + \sin t\,[\omega] + (1 - \cos t)\,[\omega]^2. \tag{7.52}$$

$\square$

The matrix exponential for any real $n \times n$ matrix A can be evaluated by generalizing the above procedure: determine the roots of the characteristic polynomial of A, and depending on whether they are real or complex, determine the explicit form of the matrix exponential. For repeated roots, terms of the form e^{st}, te^{st}, $t^2 e^{st}$, and so on should be added to match the number of times the root is repeated.

7.4   Stability Analysis via Linearization

We now state, without proof, the **linearization theorem**.

Theorem 7.1   *Let $f : \mathbb{R}^n \to \mathbb{R}^n$ be differentiable, and let $\bar{x} \in \mathbb{R}^n$ be an equilibrium point of f such that $f(\bar{x}) = 0$. Let $\lambda_1, \dots, \lambda_n$ be the eigenvalues of the $n \times n$ Jacobian matrix*

$$\frac{\partial f}{\partial x}(\bar{x}). \tag{7.53}$$

If $Re(\lambda_i) \neq 0$, $i = 1, \dots, n$ – that is, the real part of the eigenvalues λ_i are all nonzero – then solutions to the differential equation $\dot{x} = f(x)$ in a neighborhood of $\bar{x}$ are qualitatively equivalent to solutions to the linear differential equation

$$\dot{\eta} = \frac{\partial f}{\partial x}(\bar{x})\eta \tag{7.54}$$

for $\eta \in \mathbb{R}^n$ in a neighborhood of the origin.

To better understand the theorem and its implications, we now examine a classical two-dimensional example.

Example 7.4　The Duffing equation $\ddot{q} + \epsilon\dot{q} - q + q^3 = 0$, $\epsilon > 0$, can also be interpreted as the equations of motion for a mass-spring-damper system, but with the spring restoring force governed by the nonlinear relation $q - q^3$ (unlike the linear spring with restoring force kq for some spring stiffness k). The Duffing equation can be converted into the following first-order differential equation in $x \in \mathbb{R}^2$:

$$\dot{x}_1 = x_2, \tag{7.55}$$

$$\dot{x}_2 = x_1 - x_1^3 - \epsilon x_2. \tag{7.56}$$

The equilibrium points at which $f(x) = 0$ are $(0,0)^\top$, $(1,0)^\top$, and $(-1,0)^\top$. The Jacobian of f is

$$\frac{\partial f}{\partial x}(x) = \begin{bmatrix} 0 & 1 \\ 1 - 3x_1^2 & -\epsilon \end{bmatrix}. \tag{7.57}$$

We now examine solutions to the linearized equation at the three equilibria.

1. At $(0,0)^\top$, the Jacobian is

$$A = \begin{bmatrix} 0 & 1 \\ 1 & -\epsilon \end{bmatrix}, \tag{7.58}$$

the characteristic polynomial of A is $s^2 + \epsilon s - 1$, and the eigenvalues are $s_1 = \frac{-\epsilon + \sqrt{\epsilon^2 + 4}}{2}$, $s_2 = \frac{-\epsilon - \sqrt{\epsilon^2 + 4}}{2}$. The eigenvalues are both nonzero and distinct, with $s_1 > 0$ and $s_2 < 0$. The solution to $\dot{\eta} = A\eta$ is of the form

$$\eta(t) = e^{At}\eta(0) = \left(\frac{e^{s_1 t} - e^{s_2 t}}{s_1 - s_2} A + \frac{s_1 e^{s_2 t} - s_2 e^{s_1 t}}{s_1 - s_2} I \right) \eta(0), \tag{7.59}$$

where e^{At} can be determined using any of the methods described in the previous section. Observe that if $\eta(0)$ is an eigenvector of A with corresponding eigenvalue λ – that is, $A\eta(0) = \lambda\eta(0)$ – then

$$e^{At}\eta(0) = \left(I + tA + \frac{t^2}{2!}A^2 + \cdots \right)\eta(0)$$

$$= \eta(0) + tA\eta(0) + \frac{t^2}{2!}A^2\eta(0) + \cdots$$

$$= \left(I + \lambda t + \frac{\lambda^2 t^2}{2!} + \cdots \right)\eta(0)$$

$$= e^{\lambda t}\eta(0). \tag{7.60}$$

That is, $\eta(0)$ is also an eigenvector of e^{At} with corresponding eigenvalue $e^{\lambda t}$. Therefore, for those initial points $\eta(0)$ that lie along an eigenvector direction, the solution $\eta(t)$ is always directed in the same direction as $\eta(0)$:

$$\eta(t) = e^{At}\eta(0) = e^{\lambda t}\eta(0). \tag{7.61}$$

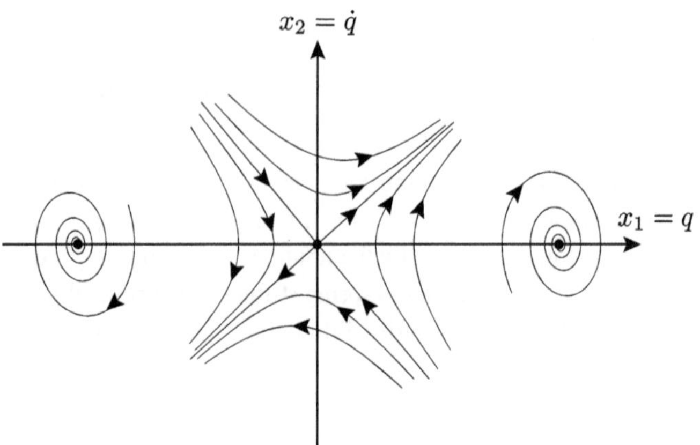

Figure 7.3 Phase portrait for the unforced Duffing equation ($\epsilon^2 < 8$).

For the A given in (7.58), the eigenvectors corresponding to the eigenvalues $\frac{-\epsilon \pm \sqrt{\epsilon^2 + 4}}{2}$ are, respectively,

$$v_1 = \begin{bmatrix} 1 \\ \frac{-\epsilon + \sqrt{\epsilon^2 + 4}}{2} \end{bmatrix}, \quad v_2 = \begin{bmatrix} 1 \\ \frac{-\epsilon - \sqrt{\epsilon^2 + 4}}{2} \end{bmatrix}. \tag{7.62}$$

Solution trajectories in a neighborhood of $(0, 0)$ are shown in Figure 7.3. If $\eta(0)$ lies in the direction of eigenvector v_2, then since the corresponding eigenvalue $s_2 < 0$, the solution trajectory $\eta(t)$ goes to zero as $t \to \infty$. If $\eta(0)$ lies in the direction of eigenvector v_1, then since the corresponding eigenvalue $s_1 > 0$, the opposite occurs, with $\eta(t)$ going to infinity as $t \to \infty$.

2. At the equilibrium points $(1, 0)^\top$ and $(-1, 0)^\top$, the Jacobian is the same:

$$A = \begin{bmatrix} 0 & 1 \\ -2 & -\epsilon \end{bmatrix}. \tag{7.63}$$

The characteristic polynomial of A is $s^2 + \epsilon s + 2$, and the eigenvalues are $s_1 = \frac{-\epsilon + \sqrt{\epsilon^2 - 8}}{2}$, $s_2 = \frac{-\epsilon - \sqrt{\epsilon^2 - 8}}{2}$. Observe that if $\epsilon^2 \geq 8$, then both eigenvalues are real and negative, and

$$\lim_{t \to \infty} \eta(t) = \lim_{t \to \infty} e^{At} \eta(0) = 0. \tag{7.64}$$

If $\epsilon^2 < 8$, then the eigenvalues are in the form of the complex conjugate pair $s_1 = -\mu + i\nu$, $s_2 = -\mu - i\nu$ with μ positive. e^{At} is of the form

$$e^{At} = e^{-\mu t} \left(M_1 \cos \nu t + M_2 \sin \nu t \right) \tag{7.65}$$

for some constant matrices M_1 and M_2. In this case it can also be seen that $\lim_{t \to \infty} \eta(t) = 0$. Solution trajectories in a neighborhood of $(\pm 1, 0)$ are shown in Figure 7.3 for the latter case of $\epsilon^2 < 8$. $\qquad\square$

As the above example illustrates, given an equilibrium point $\bar{x} \in \mathbb{R}^n$ of some differential equation $\dot{x} = f(x)$, i.e., $f(\bar{x}) = 0$, the behavior of trajectories near $\bar{x}$ is governed by the linearized differential equation $\dot{\eta} = \frac{\partial f}{\partial x}(\bar{x})\eta$, provided that none of the eigenvalues of $\frac{\partial f}{\partial x}(\bar{x})$ are pure imaginary. In particular, if all the eigenvalues have negative real parts, then one can conclude that all trajectories approach $\bar{x}$ as t goes to infinity.

Definition 7.1   An equilibrium point $\bar{x} \in \mathbb{R}^n$ of the differential equation $\dot{x} = f(x)$ is said to be a **stable equilibrium** if all the eigenvalues of the Jacobian $\frac{\partial f}{\partial x}(\bar{x})$ have negative real parts.

All solution trajectories that start near a stable equilibrium approach the stable equilibrium in the limit as t goes to infinity.

Example 7.5   Returning to the flyball governor example from Chapter 6, the equations of motion for the system are

$$\ddot{\theta} = \left(-\frac{g}{l} + \omega^2 \cos\theta\right)\sin\theta. \tag{7.66}$$

Setting $x_1 = \theta$ and $x_2 = \dot{\theta}$, Equation (7.66) can be converted into a first-order linear differential equation of the form $\dot{x} = f(x)$, $x = (x_1, x_2)^\top$:

$$\dot{x}_1 = x_2, \tag{7.67}$$
$$\dot{x}_2 = \left(-\frac{g}{l} + \omega^2 \cos x_1\right)\sin x_1. \tag{7.68}$$

The equilibrium points at which $f(x) = 0$ are $(0, 0)^\top$ and $(\cos^{-1}\frac{g}{l\omega^2}, 0)^\top$ (assuming $\omega \geq \sqrt{\frac{g}{l}}$). The Jacobian of f is

$$\frac{\partial f}{\partial x}(x) = \begin{bmatrix} 0 & 1 \\ -\frac{g}{l}\cos x_1 + \omega^2(2\cos^2 x_1 - 1) & 0 \end{bmatrix}. \tag{7.69}$$

We now examine whether the equilibrium point $\bar{x} = (\cos^{-1}\frac{g}{l\omega^2}, 0)^\top$ of the system is a stable equilibrium. At $x = \bar{x}$, substituting $\cos\bar{x}_1 = \frac{g}{l\omega^2}$ into (7.69), the Jacobian is

$$A = \begin{bmatrix} 0 & 1 \\ -\left(\omega^2 - \frac{g^2}{l^2\omega^2}\right) & 0 \end{bmatrix} = \begin{bmatrix} 0 & 1 \\ -a & 0 \end{bmatrix}, \tag{7.70}$$

where $a > 0$ from the condition $\omega \geq \sqrt{\frac{g}{l}}$. The characteristic polynomial of A is $s^2 + a$, whose roots are pure imaginary; the eigenvalues have zero real part, and therefore the linearization theorem is not applicable.

Suppose now that a torsional viscous damper is added to the flyball governor's rotating shaft, generating a resistive torque $\tau = -b_\tau\dot{\theta}$. The equations of motion now become

$$\ddot{\theta} = \left(-\frac{g}{l} + \omega^2 \cos\theta\right)\sin\theta - b\dot{\theta}, \tag{7.71}$$

where $b = \frac{b_\tau}{2Ml^2} > 0$. Converting the above to an equivalent first-order system of the form $\dot{x} = f(x)$, we get

$$\dot{x}_1 = x_2, \tag{7.72}$$

$$\dot{x}_2 = \left(-\frac{g}{l} + \omega^2 \cos x_1\right) \sin x_1 - b x_2. \tag{7.73}$$

The equilibrium points at which $f(x) = 0$ are still $(0,0)^\top$ and $(\cos^{-1}\frac{g}{l\omega^2}, 0)^\top$ (again assuming $\omega \geq \sqrt{\frac{g}{l}}$ is satisfied). At the equilibrium point $\bar{x} = (\cos^{-1}\frac{g}{l\omega^2}, 0)^\top$ the Jacobian of $f(x)$ is

$$A = \begin{bmatrix} 0 & 1 \\ -\left(\omega^2 - \frac{g^2}{l^2\omega^2}\right) & -b \end{bmatrix} = \begin{bmatrix} 0 & 1 \\ -a & -b \end{bmatrix}. \tag{7.74}$$

The characteristic polynomial of A is $s^2 + bs + a$, whose roots are $s_1 = \frac{-b+\sqrt{b^2-4a}}{2}$ and $s_2 = \frac{-b-\sqrt{b^2-4a}}{2}$. Clearly both eigenvalues have negative real parts, so that the linearization theorem is now applicable. The equilibrium point $\bar{x} = (\cos^{-1}\frac{g}{l\omega^2}, 0)^\top$ is therefore a stable equilibrium, and $\lim_{t\to\infty}\eta(t) = 0$. With the addition of the damper, if the flyball governor is slightly perturbed from its equilibrium $\bar{x}$, the system will eventually return to $\bar{x}$. $\qquad\square$

7.5　Mechanical Systems with Compliant and Dissipative Elements

Recall that in the introductory chapter we characterized a mechanical system loosely as a collection of interconnected rigid bodies, connected by devices such as mechanical joints, gears, and cams, and also elements such as springs and dampers. Actuators such as electric motors that can generate forces and torques are also elements of a mechanical system. In this section we revisit the notion of a mechanical system in terms of its basic components.

1. **Inertia elements** can store kinetic energy, and include anything with mass, such as rigid bodies;
2. **Energy sources** deliver energy to the system in the form of forces and moments, and include devices such as electric motors and pneumatic actuators;
3. **Compliant elements** can store potential energy, and include devices such as springs (both linear and torsional), flexible beams and shafts, and various elastic devices (the term compliance is the opposite notion of stiffness; a linear spring with stiffness k is said to have compliance $c = 1/k$);
4. **Dissipative elements** dissipate energy (typically in the form of heat), and include devices such as dampers and other frictional elements.

In this section we will focus on the Lagrangian formulation of the equations of motion for mechanical systems containing compliant and dissipative elements.

7.5.1 Systems with Compliant Elements

In this section we consider systems consisting of masses and one-dimensional springs. We will only consider springs in which the restoring force is linear – that is, $f = -kx$ in the case of a translational spring with x the linear displacement, or $m = -k\theta$ in the case of a torsional spring, where m is the moment and θ is the angular displacement. Recall that spring forces are conservative: in the case of a translational spring, the corresponding spring potential energy is $\mathcal{P}(x) = \frac{1}{2}kx^2$, so that $f = -\frac{d}{dx}\mathcal{P}(x) = -kx$ as desired.

Consider a network of N linear translational springs connected in parallel to a mass m as shown in Figure 7.4(a). Let k_i be the stiffness of spring i, and x be the displacement of the mass; because the springs are connected in parallel to the mass, they all have the same displacement x. The net force applied to m is then

$$f = k_1 x + k_2 x + \cdots + k_N x = \left(\sum_{i=1}^{N} k_i\right) x = k_{eq}x, \tag{7.75}$$

where $k_{eq} = k_1 + \cdots + k_N$ is the equivalent spring stiffness. The N parallel springs can therefore be replaced by a single linear spring of stiffness k_{eq}.

Now consider the case of N linear springs connected serially as shown in Figure 7.4(b). The forces at the two ends of each spring are exactly equal and opposite at all times – this is true of all springs in general (as well as dampers) – and therefore the force experienced by each spring is the same. Denoting this spring force by f, it then follows that $f = k_i x_i$, where x_i is the displacement of spring i. The total displacement of the spring x is then the sum of the spring displacements x_i:

$$x = x_1 + \cdots + x_N = \frac{f}{k_1} + \cdots + \frac{f}{k_N} = \left(\frac{1}{k_1} + \cdots + \frac{1}{k_N}\right) f. \tag{7.76}$$

The equivalent spring stiffness k_{eq} is then

$$k_{eq} = \left(\frac{1}{k_1} + \cdots + \frac{1}{k_N}\right)^{-1}. \tag{7.77}$$

The above can be used to replace more complex spring networks by a smaller number of springs with equivalent spring stiffnesses.

Example 7.6 The mechanical system of Figure 7.5 consists of several springs connected in serial and parallel to a mass m. Using the corresponding reduction techniques introduced above, the network can be reduced to a single mass-spring system with equivalent stiffness

$$k_{eq} = \left(\frac{1}{k_1} + \frac{1}{k_2 + k_3} + \frac{1}{k_4}\right)^{-1} + k_5. \tag{7.78}$$

Note that in the final reduction step, the two springs on either side of the mass act in the same direction (one pushes while the other pulls, both in the same direction); the two springs effectively behave as if they are connected in parallel, so that the equivalent spring stiffness is the sum of the two spring stiffnesses. Deriving the equations of

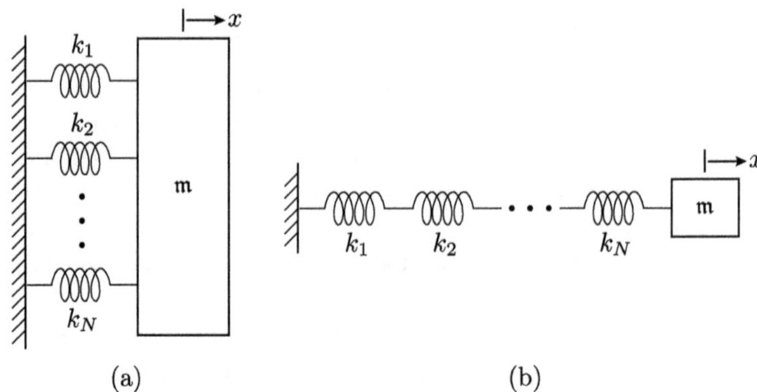

Figure 7.4 (a) Springs connected in parallel. (b) Springs connected serially.

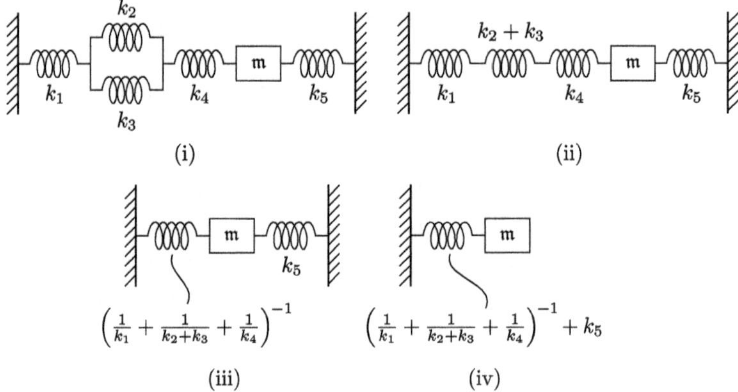

(i)

(ii)

$$\left(\tfrac{1}{k_1} + \tfrac{1}{k_2+k_3} + \tfrac{1}{k_4}\right)^{-1}$$

(iii)

$$\left(\tfrac{1}{k_1} + \tfrac{1}{k_2+k_3} + \tfrac{1}{k_4}\right)^{-1} + k_5$$

(iv)

Figure 7.5 A mass-spring system and its equivalent reduced system.

motion for the mass is now straightforward: choosing the mass displacement $x \in \mathbb{R}$ to be the generalized coordinate for the system, the kinetic energy is $\mathcal{K} = \frac{1}{2}m\dot{x}^2$, the potential energy is $\mathcal{P} = \frac{1}{2}k_{eq}x^2$, and the Lagrangian is $\mathcal{L} = \frac{1}{2}m\dot{x}^2 - \frac{1}{2}k_{eq}x^2$. The Euler–Lagrange equations for $\mathcal{L}$ then lead to

$$m\ddot{x} + k_{eq}x = f, \tag{7.79}$$

where f is the external force applied to the mass. □

Example 7.7 One way to construct a torsional spring from a pair of linear springs is shown in Figure 7.6. With the spring displacements x_1 and x_2 defined as shown, observe that $x_2 = R\theta$ and $x_1 = r\theta$, or $rx_2 = Rx_1$. The total potential energy of the system is

$$\mathcal{P} = \frac{1}{2}k_1 x_1^2 + \frac{1}{2}k_2 x_2^2 = \frac{1}{2}\left(k_1 + \left(\frac{R}{r}\right)^2 k_2\right)x_1^2. \tag{7.80}$$

Substituting $x_1 = r\theta$, the potential energy can be written in terms of the angle θ as

$$\mathcal{P} = \frac{1}{2}\left(k_1 + \left(\frac{R}{r}\right)^2 k_2\right) r^2 \theta^2, \tag{7.81}$$

which can be viewed as the potential for a torsional spring with torsional stiffness

$$k_\tau = \left(k_1 + \left(\frac{R}{r}\right)^2 k_2\right) r^2. \tag{7.82}$$

To derive the equations of motion, observe that the kinetic energy $\mathcal{K}$ and potential energy $\mathcal{P}$ of the system are given by

$$\mathcal{K} = \frac{1}{2}m_1\dot{x}_1^2 + \frac{1}{2}m_2\dot{x}_2^2 = \frac{1}{2}\left(m_1 + \left(\frac{R}{r}\right)^2 m_2\right)\dot{x}_1^2, \tag{7.83}$$

$$\mathcal{P} = \frac{1}{2}k_1 x_1^2 + \frac{1}{2}k_2 x_2^2 = \frac{1}{2}\left(k_1 + \left(\frac{R}{r}\right)^2 k_2\right) x_1^2. \tag{7.84}$$

Defining m_{eq} and k_{eq} by

$$m_{eq} = m_1 + \left(\frac{R}{r}\right)^2 m_2, \tag{7.85}$$

$$k_{eq} = k_1 + \left(\frac{R}{r}\right)^2 k_2. \tag{7.86}$$

The Lagrangian is $\mathcal{L} = \frac{1}{2}m_{eq}\dot{x}_1^2 - \frac{1}{2}k_{eq}x_1^2$, and the equations of motion are given by

$$m_{eq}\ddot{x}_1 + k_{eq}x_1 = f, \tag{7.87}$$

where f is the external force applied to m_1.

Alternatively, using $\theta = x_1/r$ as the generalized coordinate, the kinetic and potential energy terms become

$$\mathcal{K} = \frac{1}{2}\mathcal{I}_{eq}\dot{\theta}^2, \quad \mathcal{I}_{eq} = m_{eq}r^2, \tag{7.88}$$

$$\mathcal{P} = \frac{1}{2}k_\tau\theta^2. \tag{7.89}$$

The Lagrangian is $\mathcal{L} = \frac{1}{2}\mathcal{I}_{eq}\dot{\theta}^2 - \frac{1}{2}k_\tau\theta^2$, and the equations of motion are

$$\mathcal{I}_{eq}\ddot{\theta} + k_\tau\theta = m, \tag{7.90}$$

where m is the external moment applied at joint θ. □

7.5.2　Dissipative Elements

Dissipative elements such as dampers dissipate energy into heat. The ideal translational viscous damper generates a resistive force $f = -b\dot{x}$ that is directly proportional to the velocity $\dot{x}$ (or more accurately, the relative velocity of the two ends of the damper).

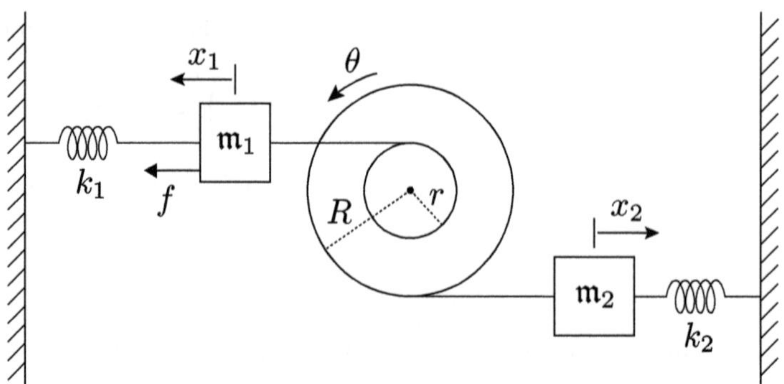

Figure 7.6 A torsional spring constructed from two linear springs.

Similarly, the torsional viscous damper generates a resistive moment $m_\tau = -b_\tau \dot{\theta}$ that is directly proportional to the angular velocity $\dot{\theta}$. Note that these forces and moments are **not conservative**, since they cannot be expressed as the gradient of a potential function.

The equations of motion for mechanical systems with dampers can be obtained in the usual way, by drawing free-body diagrams for each mass, and using the appropriate force or moment relation for the dampers. For viscous dampers, it is possible to extend the Lagrangian method by introducing a dissipative function, called the **Rayleigh dissipation function**, equal to one-half of the total power dissipated in all viscous dampers.

If $q \in \mathbb{R}^n$ are the generalized coordinates for the system, define

$$\mathcal{R} = \frac{1}{2}\sum_{i=1}^{n}\sum_{j=1}^{n} b_{ij}(q)\dot{q}_i \dot{q}_j = \frac{1}{2}\dot{q}^\top B(q)\dot{q}. \tag{7.91}$$

The matrix B is referred to as the **damping matrix**. The dissipative forces F_d due to the dampers are then given by

$$\frac{\partial \mathcal{R}}{\partial \dot{q}} = F_d^\top, \tag{7.92}$$

and the corresponding equations of motion for the system can be written

$$\frac{d}{dt}\frac{\partial \mathcal{L}}{\partial \dot{q}} - \frac{\partial \mathcal{L}}{\partial q} + \frac{\partial \mathcal{R}}{\partial \dot{q}} = F^\top, \tag{7.93}$$

where $F \in \mathbb{R}^n$ denotes all generalized forces not reflected in the left-hand side of (7.93).

We now illustrate the application of (7.93) with an example (the next section contains further examples).

Example 7.8   Figure 7.7 shows a very simple wheel suspension model based on a mass-spring-damper system. The displacement of the mass $\mathfrak{m}$ is denoted x, while y denotes the height of the road surface. For this example, imagine the wheel only moves vertically, while the surface moves with respect to the wheel with varying height $y(t)$;

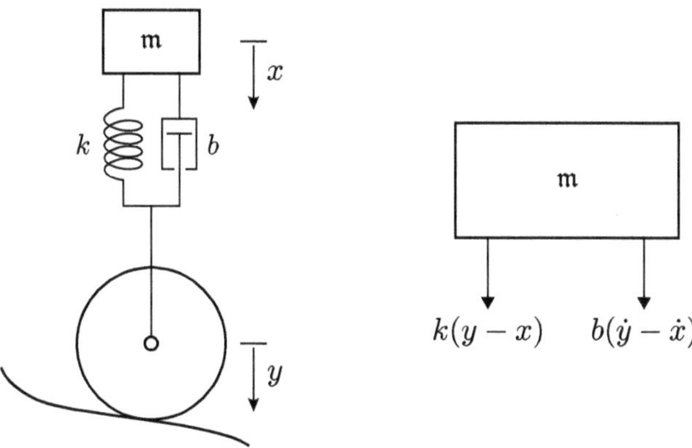

Figure 7.7 A simple model of a wheel suspension system.

the system is therefore a one-degree-of-freedom system. From the free-body diagram in Figure 7.7, it can be seen that the equations of motion are

$$m\ddot{x} + b\dot{x} + kx = b\dot{y} + ky. \tag{7.94}$$

Alternatively, using the Lagrangian formulation with Rayleigh dissipation function, choose x to be the generalized coordinate. In that case,

$$\mathcal{L} = \frac{1}{2}m\dot{x}^2 - \frac{1}{2}k(x-y)^2 \tag{7.95}$$

and

$$\mathcal{R} = \frac{1}{2}b(\dot{x}-\dot{y})^2. \tag{7.96}$$

Substituting $\mathcal{L}$ and $\mathcal{R}$ into (7.93) then leads to the identical equations of motion (7.94). $\qquad\square$

7.6   Mechanical Vibration Analysis

The mechanical systems considered in this section are multi-degrees-of-freedom mass-spring-damper systems, which exhibit a similar range of behavior to the standard one-degree-of-freedom mass-spring-damper system. Our primary focus will be on deriving the equations of motion for such systems using the Lagrangian approach with Rayleigh dissipation functions. We first begin with some examples.

Example 7.9 The three-degrees-of-freedom system of Figure 7.8 consists of three masses connected serially by springs and dampers. Referring to the free-body diagram drawn for each mass, the equations of motion for the system are given by

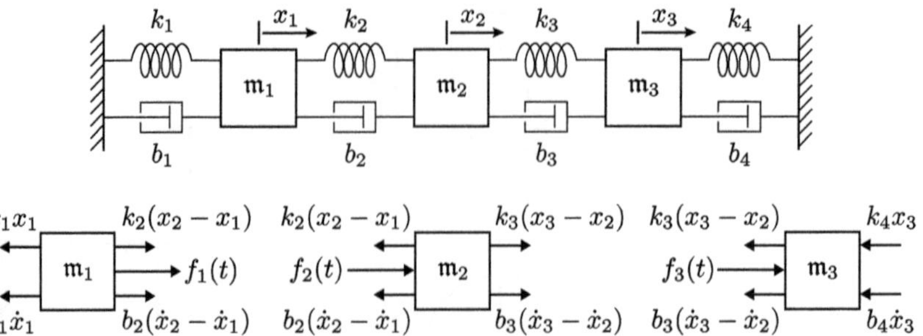

Figure 7.8 A three-degrees-of-freedom mass-spring-damper mechanical system.

$$m_1\ddot{x}_1 = -k_1 x_1 + k_2(x_2 - x_1) - b_1\dot{x}_1 + b_2(\dot{x}_2 - \dot{x}_1) + f_1(t), \tag{7.97}$$

$$m_2\ddot{x}_2 = -k_2(x_2 - x_1) + k_3(x_3 - x_2) - b_2(\dot{x}_2 - \dot{x}_1)$$
$$+ b_3(\dot{x}_3 - \dot{x}_2) + f_2(t), \tag{7.98}$$

$$m_3\ddot{x}_3 = -k_3(x_3 - x_2) - k_4 x_3 - b_3(\dot{x}_3 - \dot{x}_2) - b_4\dot{x}_3 + f_3(t), \tag{7.99}$$

where $f_i(t)$ is an external force applied to mass i. The above can be written equivalently in matrix-vector form as

$$\begin{bmatrix} m_1 & 0 & 0 \\ 0 & m_2 & 0 \\ 0 & 0 & m_3 \end{bmatrix} \begin{bmatrix} \ddot{x}_1 \\ \ddot{x}_2 \\ \ddot{x}_3 \end{bmatrix} + \begin{bmatrix} b_1 + b_2 & -b_2 & 0 \\ -b_2 & b_2 + b_3 & -b_3 \\ 0 & -b_3 & b_3 + b_4 \end{bmatrix} \begin{bmatrix} \dot{x}_1 \\ \dot{x}_2 \\ \dot{x}_3 \end{bmatrix}$$
$$+ \begin{bmatrix} k_1 + k_2 & -k_2 & 0 \\ -k_2 & k_2 + k_3 & -k_3 \\ 0 & -k_3 & k_3 + k_4 \end{bmatrix} \begin{bmatrix} x_1 \\ x_2 \\ x_3 \end{bmatrix} = \begin{bmatrix} f_1(t) \\ f_2(t) \\ f_3(t) \end{bmatrix}, \tag{7.100}$$

or $M\ddot{x} + B\dot{x} + Kx = F(t)$, where M, B, and K are the mass matrix, damping matrix, and stiffness matrix, respectively.

Alternatively, using the Lagrangian approach with generalized coordinates $x \in \mathbb{R}^3$, the kinetic and potential energies of the system are

$$\mathcal{K} = \frac{1}{2}\left(m_1\dot{x}_1^2 + m_2\dot{x}_2^2 + m_3\dot{x}_3^2\right), \tag{7.101}$$

$$\mathcal{P} = \frac{1}{2}\left(k_1 x_1^2 + k_2(x_2 - x_1)^2 + k_3(x_3 - x_2)^2 + k_4 x_3^2\right), \tag{7.102}$$

and the Lagrangian is $\mathcal{L} = \mathcal{K} - \mathcal{P}$. The Rayleigh dissipation function is

$$\mathcal{R} = \frac{1}{2}\left(b_1\dot{x}_1^2 + b_2(\dot{x}_2 - \dot{x}_1)^2 + b_3(\dot{x}_3 - \dot{x}_2)^2 + b_4\dot{x}_3^2\right). \tag{7.103}$$

Substituting $\mathcal{L}$ and $\mathcal{R}$ into (7.93) then leads to the same equations of motion obtained earlier. □

Example 7.10   (Kelly 2011) A car traveling at constant speed can be modeled by a mass-spring-damper network as shown in Figure 7.9. The vehicle's main body has mass m_c and inertia I_c with dimensions as indicated. The front and rear wheel axles have masses m_f and m_r, respectively, and are connected to the car body by a simple suspension system with springs k_s and dampers b_s as shown. The tires are modeled as springs k_t and dampers b_t connecting the axles to ground. As the car travels at constant speed v_c, the front wheel experiences a vertical displacement $h_f(t)$ at time t, while the real wheel experiences the same vertical displacement at time $t + (L/v_c)$, where L is the distance between the front and rear wheels; equivalently, the rear wheel vertical displacement function is $h_r(t) = h_f(t - \frac{L}{v_c})$. The angle θ_c between the car body and ground can be assumed small, so that $\sin \theta_c \approx \theta_c$.

The car system has four degrees of freedom, with generalized coordinates θ_c, x_c, x_f, x_r as indicated (assume $\theta_c = x_c = x_f = x_r = 0$ when the car is traveling over flat ground without any disturbances). Since the car is traveling horizontally at constant speed, it is sufficient to consider only vertical motions of the car when deriving the equations of motion. The kinetic energy $\mathcal{K}$ and potential energy $\mathcal{P}$ of the system are given by

$$\mathcal{K} = \frac{1}{2}m_c\dot{x}_c^2 + \frac{1}{2}I_c\dot{\theta}_c^2 + \frac{1}{2}m_f\dot{x}_f^2 + \frac{1}{2}m_r\dot{x}_r^2, \tag{7.104}$$

$$\mathcal{P} = \frac{1}{2}\left(k_s(x_f - (x_c - a\theta_c))^2 + k_s(x_r - (x_c + (L-a)\theta_c))^2\right)$$
$$+ \frac{1}{2}\left(k_t(h_f - x_f)^2 + k_t(h_r - x_r)^2\right). \tag{7.105}$$

The Rayleigh dissipation function is

$$\mathcal{R} = \frac{1}{2}\left(b_s(\dot{x}_f - (\dot{x}_c - a\dot{\theta}_c))^2 + b_s(\dot{x}_r - (\dot{x}_c + (L-a)\dot{\theta}_c))^2\right)$$
$$+ \frac{1}{2}\left(b_t(\dot{h}_f - \dot{x}_f)^2 + b_t(\dot{h}_r - \dot{x}_r)^2\right). \tag{7.106}$$

Substituting $\mathcal{L}$ and $\mathcal{R}$ into the Euler–Lagrange equations (7.93), the equations of motion are of the form

$$M\ddot{q} + B\dot{q} + Kq = F(t), \tag{7.107}$$

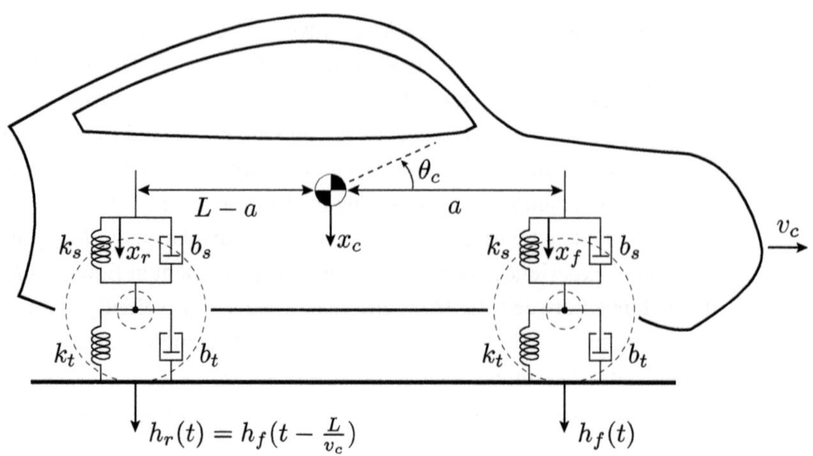

Figure 7.9 A four-degrees-of-freedom vehicle suspension model.

where $q = (\theta_c, x_c, x_f, x_r)^\top \in \mathbb{R}^4$, and

$$
M = \begin{bmatrix} I_c & 0 & 0 & 0 \\ 0 & m_c & 0 & 0 \\ 0 & 0 & m_f & 0 \\ 0 & 0 & 0 & m_r \end{bmatrix},
\tag{7.108}
$$

$$
B = \begin{bmatrix} b_s(a^2 + (L-a)^2) & b_s(L-2a) & b_s a & -b_s(L-a) \\ b_s(L-2a) & 2b_s & -b_s & -b_s \\ b_s a & -b_s & b_s + b_t & 0 \\ -b_s(L-a) & -b_s & 0 & b_s + b_t \end{bmatrix},
\tag{7.109}
$$

$$
K = \begin{bmatrix} k_s(a^2 + (L-a)^2) & k_s(L-2a) & k_s a & -k_s(L-a) \\ k_s(L-2a) & 2k_s & -k_s & -k_s \\ k_s a & -k_s & k_s + k_t & 0 \\ -k_s(L-a) & -k_s & 0 & k_s + k_t \end{bmatrix},
\tag{7.110}
$$

$$
F(t) = \begin{bmatrix} 0 \\ 0 \\ b_t \dot{h}_f + k_t h_f \\ b_t \dot{h}_r + k_t h_r \end{bmatrix}.
\tag{7.111}
$$

$\square$

In both examples the equations of motion are of the form $M\ddot{x} + B\dot{x} + Kx = F(t)$, where $M \in \mathbb{R}^{n \times n}$ is invertible. Assuming $F(t) = 0$, the equation can be converted into a first-order system by setting $q_1 = x$ and $q_2 = \dot{x}$:

$$
\begin{bmatrix} \dot{q}_1 \\ \dot{q}_2 \end{bmatrix} = \begin{bmatrix} 0 & I \\ -M^{-1}K & -M^{-1}B \end{bmatrix} \begin{bmatrix} q_1 \\ q_2 \end{bmatrix}.
\tag{7.112}
$$

The nature of the solution is therefore dependent on the eigenvalues of the matrix

$$A = \begin{bmatrix} 0 & I \\ -M^{-1}K & -M^{-1}B \end{bmatrix}. \tag{7.113}$$

In particular, any complex eigenvalues of A of the form $\mu \pm i\nu$ will result in oscillatory behavior with frequency ν. The subject of mechanical vibrations is concerned with the analysis of such oscillatory behavior in mechanical systems. We do not address this topic in further detail. Rather, we conclude by mentioning that even for mechanical systems whose equations of motion are nonlinear, their linearized dynamics about some equilibrium solution will lead to equations of the form (7.113), which can then be analyzed for oscillatory behavior.

7.7 Summary

- The classical **mass-spring-damper** system is modeled by the second-order linear differential equation $m\ddot{x} + b\dot{x} + kx = f(t)$, where $x(t) \in \mathbb{R}$ is the displacement, $f(t)$ is the external forcing function, and m, b, k denote the mass, damping coefficient, and spring stiffness, respectively. For the case $f(t) = 0$, solutions can be characterized into **underdamped**, **overdamped**, and **critically damped** cases. Qualitatively, the equations of motion for general mechanical systems have similar corresponding terms and behavioral properties with the classical mass-spring-damper system.
- Let $f: \mathbb{R}^n \to \mathbb{R}^n$ be differentiable and consider the differential equation $\dot{x} = f(x)$. Let $\bar{x}(t)$ be a solution to $\dot{x} = f(x)$ – that is, $\bar{x}(t)$ satisfies $\dot{\bar{x}}(t) = f(\bar{x}(t))$. The linearization of $\dot{x} = f(x)$ about the solution $\bar{x}(t)$ is then given by

$$\dot{\eta} = \frac{\partial f}{\partial x}(\bar{x})\eta, \tag{7.114}$$

where $\eta(t) \in \mathbb{R}^n$ is some perturbation about $\bar{x}(t)$ satisfying $x(t) = \bar{x}(t) + \eta(t)$, and

$$\frac{\partial f}{\partial x}(\bar{x}) = \begin{bmatrix} \frac{\partial f_1}{\partial x_1}(\bar{x}) & \cdots & \frac{\partial f_1}{\partial x_n}(\bar{x}) \\ \vdots & \ddots & \vdots \\ \frac{\partial f_n}{\partial x_1}(\bar{x}) & \cdots & \frac{\partial f_n}{\partial x_n}(\bar{x}) \end{bmatrix}. \tag{7.115}$$

The above is a linear differential equation of the form $\dot{x} = Ax$, where $A = \frac{\partial f}{\partial x}(\bar{x}) \in \mathbb{R}^{n \times n}$ is known. If $\bar{x}(t)$ is time-varying then $A = A(t)$ varies with time, whereas if $\bar{x}(t) = \bar{x}$ is constant then A is also constant.
- The general form of the equations of motion for a mechanical system is

$$F = M(q)\ddot{q} + b(q, \dot{q}), \tag{7.116}$$

where $q \in \mathbb{R}^n$ are the generalized coordinates of the system, $F \in \mathbb{R}^n$ is the external generalized force, $M(q) \in \mathbb{R}^{n \times n}$ is the **mass matrix**, and $b(q, \dot{q}) \in \mathbb{R}^n$ is a collection of terms representing potential and Coriolis forces. By setting

$x_1 = q$ and $x_2 = \dot{q}$, the second-order equations of motion can be converted to the first-order system

$$\dot{x}_1 = \dot{q} = x_2, \tag{7.117}$$

$$\dot{x}_2 = \ddot{q} = M(q)^{-1}(F - b(q, \dot{q})) = M(x_1)^{-1}(F - b(x_1, x_2)). \tag{7.118}$$

For unforced systems ($F = 0$) by defining $x = (x_1, x_2)^\top \in \mathbb{R}^{2n}$, the equations of motion can always be converted into a first-order differential equation of the form $\dot{x} = f(x)$.

- The **homogeneous linear differential equation** $\dot{x} = Ax$, $A \in \mathbb{R}^{n \times n}$ constant, has solution $x(t) = e^{At}x(0)$, where

$$e^{At} = I + tA + \frac{t^2}{2!}A^2 + \frac{t^3}{3!}A^3 + \cdots \tag{7.119}$$

  is the **matrix exponential**. If the initial condition for $\dot{x} = Ax$ is given at some arbitrary t_0 rather than $t = 0$, the solution $x(t)$ is given by

$$x(t) = e^{A(t-t_0)}x(t_0), \tag{7.120}$$

  with the matrix exponential $e^{A(t-t_0)}$ defined accordingly.
- The matrix exponential e^A for $A \in \mathbb{R}^{n \times n}$ is always invertible, with inverse e^{-A}. Generally $e^A e^B \neq e^B e^A$ for arbitrary square matrices A, B, but if $AB = BA$, then $e^A e^B = e^B e^A = e^{A+B}$. Also $Ae^A = e^A A$ holds for any A.
- The **inhomogeneous linear differential equation** $\dot{x} = Ax + f(t)$ with the forcing $f(t) \in \mathbb{R}^n$ given has general solution

$$x(t) = e^{A(t-t_0)}x(t_0) + \int_{t_0}^t e^{A(t-s)} f(s)\, ds. \tag{7.121}$$

  The first term is the homogeneous solution while the second term constitutes a particular solution.
- If A is diagonalizable as $A = PDP^{-1}$ for some invertible P and diagonal $D = \mathrm{diag}(d_1, \ldots, d_n)$, then $e^{At} = Pe^{Dt}P^{-1}$, $e^{Dt} = \mathrm{diag}(e^{d_1 t}, \ldots, e^{d_n t})$.
- The **modal solution technique** is a systematic means of finding closed-form expressions for the matrix exponential e^A based on the nature of the eigenvalues of A.
- Let $f : \mathbb{R}^n \to \mathbb{R}^n$ be differentiable and consider the differential equation $\dot{x} = f(x)$. Let $x(t) = \bar{x} \in \mathbb{R}^n$ be a constant solution that is also an **equilibrium point** of f – that is, $f(\bar{x}) = 0$. Let $\lambda_1, \ldots, \lambda_n$ be the eigenvalues of the $n \times n$ Jacobian matrix $\frac{\partial f}{\partial x}(\bar{x})$. As a consequence of the **linearization theorem**, if the real parts of the λ_i are all nonzero, then solutions to $\dot{x} = f(x)$ in a neighborhood of $\bar{x}$ are qualitatively equivalent to solutions to the linearized equation $\dot{\eta} = \frac{\partial f}{\partial x}(\bar{x})\eta$ in a neighborhood of the origin $\eta = 0$. An equilibrium point $\bar{x} \in \mathbb{R}^n$ is said to be a **stable equilibrium** if all the eigenvalues of the Jacobian have negative real parts.
- The basic components of a mechanical system can be classified into four classes: **inertia elements** like rigid bodies can store kinetic energy; **energy sources** deliver energy to the system in the form of forces and moments; **compliant elements**

like springs can store potential energy, and **dissipative elements** like dampers dissipate energy.

- Networks of linear springs connected serially and in parallel can be replaced by equivalent springs. N linear springs of stiffness $k_1, \ldots, k_N$ connected in parallel is equivalent to a spring of stiffness $k_{eq} = k_1 + \cdots + k_N$. N linear springs of stiffness $k_1, \ldots, k_N$ connected serially is equivalent to a spring of stiffness $k_{eq} = ((1/k_1) + \cdots + (1/k_N))^{-1}$. More complex linear spring networks can be reduced to equivalent networks using these simplification rules.
- Forces generated by dissipative elements are not conservative, as they cannot be expressed as the gradient of some potential function. Instead, dissipative forces can be expressed as the gradient of a **Rayleigh dissipation function** that can be used in a Lagrangian dynamics setting.

7.8   Exercises

Exercise 7.1   Suppose a constant matrix $A \in \mathbb{R}^{n \times n}$ has eigenvalues of $\lambda_1, \ldots, \lambda_n \in \mathbb{R}$ and corresponding eigenvectors of $v_1, \ldots, v_n \in \mathbb{R}^n$ (i.e., $Av_i = \lambda_i v_i$, $i = 1, \ldots, n$). Show that if the eigenvectors of A are linearly independent, then A can be expressed as $A = PDP^{-1}$, where

$$P = \begin{bmatrix} v_1 & v_2 & \cdots & v_n \end{bmatrix} \in \mathbb{R}^{n \times n}, \quad D = \operatorname{diag}(\lambda_1, \lambda_2, \ldots, \lambda_n) \in \mathbb{R}^{n \times n}.$$

Exercise 7.2   Compute the matrix exponentials e^{At} of the following constant matrices A.

(a)

$$A = \begin{bmatrix} 2 & 1 \\ 2 & 3 \end{bmatrix}.$$

(b)

$$A = \begin{bmatrix} -1 & -1 \\ 1 & -1 \end{bmatrix}.$$

(c)

$$A = \begin{bmatrix} -4 & -1 \\ 1 & -2 \end{bmatrix}.$$

Exercise 7.3   (a) Compute the matrix exponential e^{At} of the following constant matrix A by using the modal solution technique:

$$A = \begin{bmatrix} 2 & 1 & 0 \\ 0 & 2 & 1 \\ 0 & 0 & 1 \end{bmatrix}.$$

(b) Compute the matrix exponential e^{At} of the following constant matrix A by expressing A as $A = PDP^{-1}$ for some invertible P and diagonal D:

$$A = \begin{bmatrix} 2 & -2 & 1 \\ -1 & 3 & -1 \\ 2 & -4 & 3 \end{bmatrix}.$$

Exercise 7.4   We want to compute the matrix exponential of

$$[\omega] = \begin{bmatrix} 0 & -\omega_z & \omega_y \\ \omega_z & 0 & -\omega_x \\ -\omega_y & \omega_x & 0 \end{bmatrix},$$

where $\omega_x^2 + \omega_y^2 + \omega_z^2 = 1$, this time without finding the eigenvalues of $[\omega]$.
(a) Show that $[\omega]^3 = -[\omega]$.
(b) Find the matrix exponential $e^{[\omega]t}$ by using the definition of matrix exponential (7.29).
Compare your answer with Example 7.3.

Exercise 7.5   A constant matrix $A \in \mathbb{R}^{4\times4}$ is given as

$$A = \begin{bmatrix} 0 & 1 & 0 & 0 \\ a_1 & 0 & a_2 & 0 \\ 0 & 0 & 0 & 1 \\ a_3 & 0 & a_4 & 0 \end{bmatrix}.$$

(a) Find the characteristic polynomial of A.
(b) Express the eigenvalues of A in terms of a_1, a_2, a_3, a_4.
(c) Suppose $a_1 = a_3 = 4$, $a_2 = a_4 = 0$. For $x(t) \in \mathbb{R}^4$, find the solution of the linear differential equation $\dot{x} = Ax$, where $x(0) = (\delta, 0, 0, 0)^\top$ with $\delta > 0$. What happens to $x(t)$ as t goes to infinity?
(d) Now suppose $a_1 = -2$, $a_2 = a_3 = 3$, $a_4 = -2$. Find the solution $x(t)$ of the linear differential equation $\dot{x} = Ax$ for arbitrary $x(0)$.

Exercise 7.6   Find the solution $x(t) \in \mathbb{R}^3$ of the inhomogeneous linear differential equation $\dot{x} = Ax + f(t)$, where

$$A = \begin{bmatrix} 3 & 0 & 1 \\ 0 & 1 & 0 \\ 0 & 0 & 2 \end{bmatrix}, \quad f(t) = \begin{bmatrix} 1 \\ 1 \\ -1 \end{bmatrix}, \quad x(0) = \begin{bmatrix} 1 \\ 1 \\ -1 \end{bmatrix}.$$

Exercise 7.7   (a) Consider the following anti-diagonal constant matrix $B \in \mathbb{R}^{2\times2}$:

$$B = \begin{bmatrix} 0 & b_1 \\ b_2 & 0 \end{bmatrix},$$

where $b_1 b_2 < 0$. Setting $\omega = \sqrt{-b_1 b_2}$, show that the matrix exponential e^{Bt} is computed as

$$e^{Bt} = \begin{bmatrix} \cos \omega t & \frac{b_1}{\omega} \sin \omega t \\ \frac{b_2}{\omega} \sin \omega t & \cos \omega t \end{bmatrix}.$$

(b) Consider the following constant matrix $A \in \mathbb{R}^{2\times2}$ of the form

$$A = \begin{bmatrix} a & b_1 \\ b_2 & a \end{bmatrix},$$

where $b_1 b_2 < 0$. It is known that for any two square matrices P and Q, if $PQ = QP$, then $e^P e^Q = e^Q e^P = e^{P+Q}$. Using this property, show that

$$e^{At} = e^{at} \begin{bmatrix} \cos \omega t & \frac{b_1}{\omega} \sin \omega t \\ \frac{b_2}{\omega} \sin \omega t & \cos \omega t \end{bmatrix},$$

where $\omega = \sqrt{-b_1 b_2}$.

(c) Suppose $x(t) \in \mathbb{R}^2$ satisfies $\dot{x} = Ax$ and $z(t) \in \mathbb{R}^2$ is given as $z(t) = e^{Bt} x(t)$, where

$$A = \begin{bmatrix} -4 & 0 \\ 0 & 1 \end{bmatrix}, \quad B = \begin{bmatrix} 0 & -3 \\ 3 & 0 \end{bmatrix}, \quad x(0) = \begin{bmatrix} 1 \\ 1 \end{bmatrix}.$$

What happens to $\|x(t)\|$ and $\|z(t)\|$ as $t \to \infty$?

(d) First, show that $z(t)$ in (c) is a solution of the linear time-varying differential equation $\dot{z} = C(t)z$, where

$$C(t) = e^{Bt} A e^{-Bt} + B.$$

Then, show that the eigenvalues of $C(t)$ are the same as those of $A + B$, which are independent of time. Explain how this example shows that even if all the eigenvalues of a time-varying matrix $M(t)$ have negative real parts, it is incorrect to conclude that the solution $x(t)$ of $\dot{x} = M(t)x$ satisfies $\lim_{t \to \infty} \|x(t)\| = 0$.

Exercise 7.8   For the mass-spring-damper system $m\ddot{x} + b\dot{x} + kx = 0$, convert the equations of motion into a first-order linear differential equation of the form $\dot{z} = Az$, where $z = (x, \dot{x})^\top$. Then by solving $\dot{z} = Az$, show that for any $x(0)$, $\dot{x}(0)$, m, b, and k, $x(t)$ and $\dot{x}(t)$ go to zero as t goes to infinity.

Exercise 7.9   We want to design a simple feedback control force f that drives the mass of a mass-spring-damper system to a desired position $x = x_d$ from the initial state $x(0) = \dot{x}(0) = 0$. The equations of motion are given by

$$m\ddot{x} + b\dot{x} + kx = f.$$

(a) Define the error between x_d and the current position $x(t)$ as $e(t) = x_d - x(t)$. Rewrite the equations of motion in terms of $e(t)$.
(b) Suppose a force proportional to the error is applied, i.e., $f(t) = K_p e(t)$ for some positive constant K_p. Find the final error $e_f = \lim_{t \to \infty} e(t)$.
(c) To make the final error e_f fairly small, what condition should K_p satisfy? Assuming the condition is satisfied, how does the mass move until it stops near x_d? If it is hard to answer, try drawing $x(t)$ for various values of K_p.
(d) If done correctly, you would have noticed that $x(t)$ of (c) shows an "overshooting" under the condition. To prevent this, we now add a term proportional to the velocity error to $f(t)$, i.e., $f(t) = K_p e(t) + K_d \dot{e}(t)$ with $K_d > 0$. First, determine whether K_d affects the final error e_f. Then, design a K_d that prevents overshooting and exhibits fast convergence of $x(t)$.

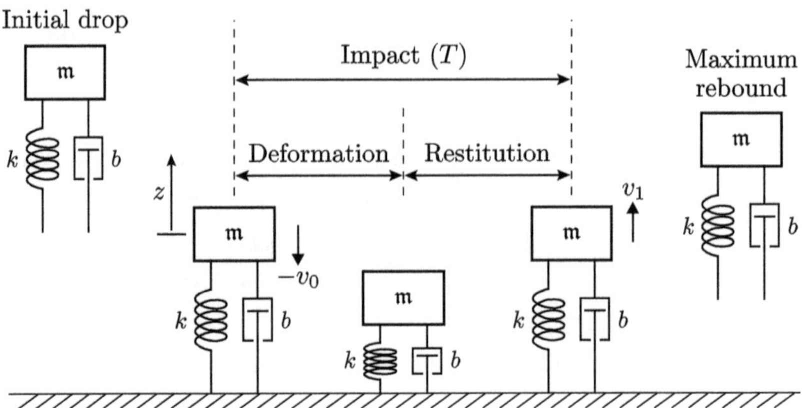

Figure 7.10 Bouncing ball model of Exercise 7.10.

Exercise 7.10   As shown in Figure 7.10, a bouncing ball can be modeled as a mass-spring-damper system with mass m, damping coefficient b, and spring stiffness k. The ball is dropped from rest and strikes the ground at $t = 0$, where the initial condition of the impact is $z(0) = 0$ and $\dot{z}(0) = -v_0$ ($v_0 > 0$). For this exercise, the following formulas can be helpful:

$$\int_0^t e^{-\omega r} \sin \omega r \, dr = \frac{1}{2\omega}\left(1 - e^{-\omega t}(\sin \omega t + \cos \omega t)\right),$$

$$\int_0^t e^{-\omega r} \cos \omega r \, dr = \frac{1}{2\omega}\left(1 + e^{-\omega t}(\sin \omega t - \cos \omega t)\right).$$

(a) Find the solution of the linear differential equation $\dot{y} = My + d$, where

$$M = \begin{bmatrix} 0 & 1 \\ -2 & -2 \end{bmatrix}, \quad d = \begin{bmatrix} 0 \\ -10 \end{bmatrix}, \quad y(0) = \begin{bmatrix} 0 \\ -v_0 \end{bmatrix}.$$

(b) The equations of motion during impact are given by

$$m\ddot{z} + b\dot{z} + kz = -mg,$$

where $m = 1$, $b = 2$, $k = 2$, $g = 10$. What condition should m, b, and k satisfy for the ball to rebound? Do the given values satisfy this condition?

(c) Convert the equations of motion into a first-order linear differential equation of the form $\dot{x} = Ax + c$, where $x = (z, \dot{z})^{\mathsf{T}}$. Then by solving the differential equation, find $z(t)$ and $\dot{z}(t)$. Compare your results with those obtained by solving the equations of motion directly.

(d) Assuming $v_0 = 1000$ and the spring is long enough, find the impact time T (contact duration). Also, find the velocity v_1 of the ball when it loses contact with the ground, i.e., $\dot{z}(T)$. You can use software to solve nonlinear equations. (*Hint*: What should the value of $z(T)$ be?)

Exercise 7.11   Consider the mass-pulley system of Exercise 6.15(a). Convert the equations of motion of the system into first-order differential equations of the form

$\dot{x} = f(x)$ in $x \in \mathbb{R}^4$, where $x = (x_1, \dot{x}_1, x_2, \dot{x}_2)^{\top}$. Then, find a set of equilibrium points of the system.

Exercise 7.12   Consider the spherical pendulum of Exercise 6.11.
(a) Show that $\phi = \omega t$ (ω constant) and $\theta = \cos^{-1}\frac{g}{l\omega^2}$, i.e., a circular orbit, is a solution of the equations of motion of the system $\left(\text{assume } \omega > \sqrt{\frac{g}{l}}\right)$.
(b) Convert the equations of motion into first-order differential equations of the form $\dot{x} = f(x)$ in $x \in \mathbb{R}^4$, where $x_1 = \theta$, $x_2 = \dot{\theta}$, $x_3 = \phi$, and $x_4 = \dot{\phi}$.
(c) Find the Jacobian of f about the circular orbit of (a).

Exercise 7.13   A one-dimensional system with generalized coordinate $z \in \mathbb{R}$ has equations of motion of the form

$$\ddot{z} + \dot{z} - 2z - z^3 = 0.$$

(a) Express the above equations of motion as a pair of first-order differential equations of the form $\dot{x} = f(x)$ in $x \in \mathbb{R}^2$, where $x_1 = z$ and $x_2 = \dot{z}$.
(b) Find the equilibrium point $\bar{x}$ of $\dot{x} = f(x)$ and the Jacobian of f at $\bar{x}$.
(c) For a small perturbation of $\eta(0) = (\delta, -2\delta)^{\top}$ ($\delta > 0$) near $\bar{x}$, what happens to $\eta(t)$ as t goes to infinity?
(d) Then, for another small perturbation of $\eta(0) = (1.1\delta, -2\delta)^{\top}$ near $\bar{x}$, what happens to $\eta(t)$ as t goes to infinity? Explain your answer with a sketch of the solution trajectories in a neighborhood of $\bar{x}$.

Exercise 7.14   In Example 7.5, we verified that $(\theta, \dot{\theta}) = (\cos^{-1}\frac{g}{l\omega^2}, 0)^{\top}$ is a stable equilibrium of the system when $\omega \geq \sqrt{\frac{g}{l}}$. We further analyze additional equilibrium points for various conditions.
(a) Again, suppose $\omega \geq \sqrt{\frac{g}{l}}$. Is $(0, 0)^{\top}$ a stable equilibrium?
(b) Now suppose $\omega < \sqrt{\frac{g}{l}}$. For this condition, is $(0, 0)^{\top}$ a stable equilibrium?
(c) For each condition, sketch the solution trajectories in neighborhoods of the equilibrium points. Draw the phase portrait of the system $\dot{x} = f(x)$ using software and compare it with your sketch (try using the following settings: $l = 1$, $M = 1$, $m = 0.01$, $g = 10$, $b_\tau = 10$, and $\omega = 1$ or 6).

Exercise 7.15   Consider the torus-particle system of Exercise 6.10.
(a) Convert the equations of motion of the system into first-order differential equations of the form $\dot{x} = f(x)$ in $x \in \mathbb{R}^2$, where $x_1 = \theta$ and $x_2 = \dot{\theta}$.
(b) Find the Jacobian of f at $x = (0, 0)^{\top}$.
(c) Determine whether $x = (0, 0)^{\top}$ is a stable equilibrium.

Exercise 7.16   Consider the planar wheel with empty space of Exercise 6.20. Figure 6.15 shows the orientation of the wheel at $\theta = 0$.
(a) Convert the equations of motion of the wheel into first-order differential equations of the form $\dot{x} = f(x)$ in $x \in \mathbb{R}^2$, where $x_1 = \theta$ and $x_2 = \dot{\theta}$.

(b) Show that $\bar{x} = (-\frac{\pi}{2}, 0)^{\top}$ is an equilibrium of the system $\dot{x} = f(x)$. Draw the orientation of the wheel at the equilibrium.

(c) Suppose the system is slightly perturbed by $\eta(0) = (0, \delta)^{\top}$ from $\bar{x}$. Describe the wheel's localized motion by computing $\eta(t)$. Note that $r \le \frac{4R}{3\pi}$, where $r = \frac{4R}{3\pi}$ when the empty space takes up half of the wheel.

(d) Now suppose an air resistance torque $\tau = -b\dot{\theta}$ is applied to the wheel. Derive the equations of motion using the Lagrangian formulation with Rayleigh dissipation function. For this case, how does $\eta(t)$ change as t goes to infinity?

Exercise 7.17   Consider the sliding pendulum of Exercise 6.13. Assume $m_1 = m_2 = 1$, $l = 1$, and $\theta \in [-\frac{\pi}{2}, \frac{\pi}{2}]$.

(a) Convert the equations of motion of the system into first-order differential equations of the form $\dot{z} = f(z)$ in $z \in \mathbb{R}^4$, where $z = (x, \dot{x}, \theta, \dot{\theta})^{\top}$.

(b) Find an equilibrium point $\bar{z}$ of the system and the Jacobian of f at $\bar{z}$. You can use the following approximations: $\sin\theta \approx \theta$, $\cos\theta \approx 1$ for small θ.

(c) Model an appropriate initial perturbation $\eta(0) \in \mathbb{R}^4$ at $\bar{z}$ and find $\eta(t)$ to explain the localized motions of the slider and ball, respectively, for the following cases: (i) the ball is released from a small positive angle, (ii) a small horizontal external impact causes the ball to move. Does the system stay near the equilibrium as t goes to infinity?

Exercise 7.18   The Mars lander of Figure 7.11 consists of three identical spring-damper legs designed to absorb shocks from landing. Let z be the height of the platform. Using z as the generalized coordinate, the equations of motion of the system are given as

$$m\ddot{z} + b\frac{z^2}{z^2 + 3L^2}\dot{z} + kz - \frac{\sqrt{2}z}{\sqrt{z^2 + L^2}}(kL + mg') + mg' = 0,$$

where m, k, b, L, g' are given physical constants.

(a) Convert the equations of motion into first-order differential equations of the form $\dot{x} = f(x)$ in $x \in \mathbb{R}^2$, where the system's state x is defined as $x = (z, \dot{z})^{\top}$. Then find an equilibrium point $\bar{x}$ of the system and the Jacobian of f at $\bar{x}$.

(b) Suppose $m = 1$, $g' = 1$, $k = 1$, $L = 2$. Using the Jacobian derived in (a), find all damping coefficients b that prevent the platform from oscillating when slightly perturbed from $\bar{x}$.

(c) The damping coefficient is chosen as $b = 4$. A small earthquake instantaneously perturbs the system from $(L, 0)^{\top}$ to $(L + \delta, 0)^{\top}$. Find the state trajectory $x(t)$ after the perturbation.

Exercise 7.19   (a) Figure 7.12(a) shows a network of N translational viscous dampers connected in parallel to mass m. The damping coefficient of damper i is b_i and x is the displacement of the mass. Show that the equivalent damping coefficient b_{eq} is

$$b_{eq} = b_1 + b_2 + \cdots + b_N.$$

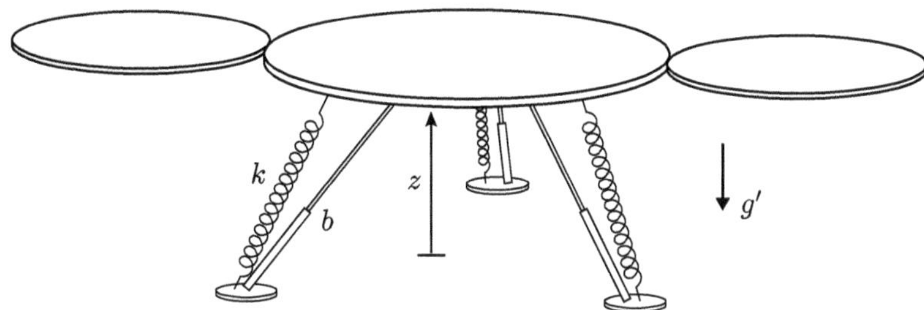

Figure 7.11 Mars lander of Exercise 7.18.

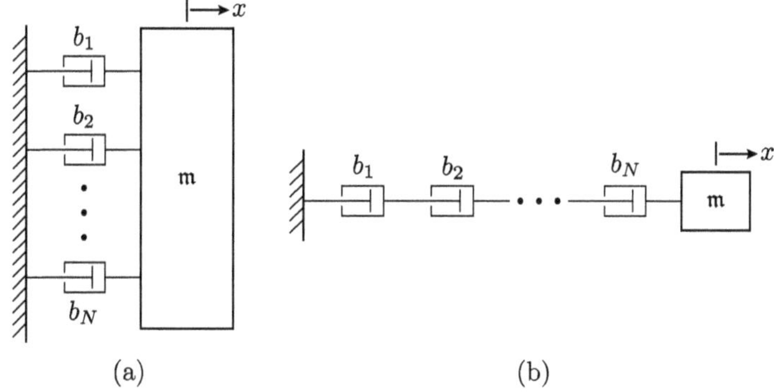

Figure 7.12 Damper networks of Exercise 7.19.

(b) Now the N dampers are connected serially as shown in Figure 7.12(b). Show that the equivalent damping coefficient b_{eq} is

$$b_{eq} = \left(\frac{1}{b_1} + \frac{1}{b_2} + \cdots + \frac{1}{b_N} \right)^{-1}.$$

Exercise 7.20 (a) Suppose the mass-spring-damper system of Figure 7.13(a) is critically damped. Determine whether the mass-spring-damper system of Figure 7.13(b) is overdamped, underdamped, or critically damped.
(b) An external force $f_{ext} = -k_e x - b\dot{x}$ is applied to the mass of the mass-spring-damper system shown in Figure 7.13(b), where x is the displacement of the mass. To make $x(t) \to 0$ as fast as possible, what should k_1 be?

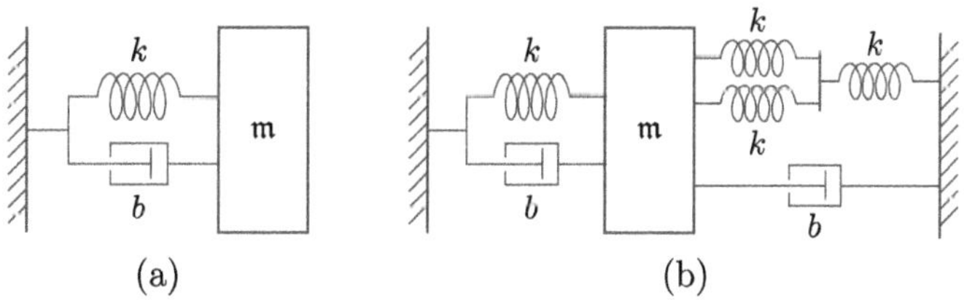

Figure 7.13 Mass-spring-damper systems of Exercise 7.20.

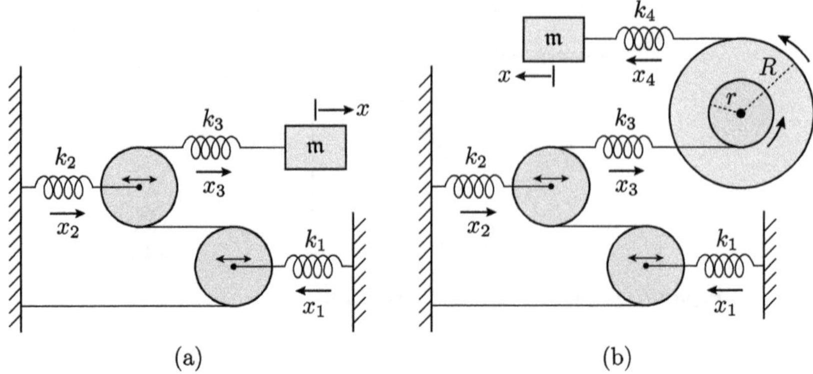

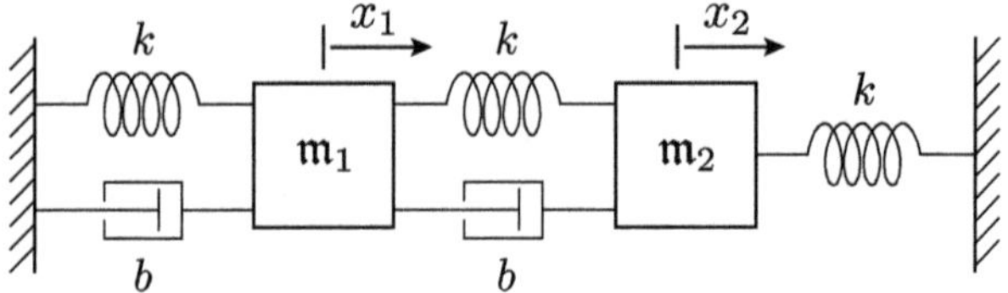

Figure 7.14 Spring-pulley networks of Exercise 7.21.

Figure 7.15 Two-degrees-of-freedom mass-spring-damper system of Exercise 7.22.

Exercise 7.21   (a) Figure 7.14(a) shows a spring-pulley network consisting of two massless pulleys, three linear springs, and mass m. The stiffness and displacement of spring i are k_i and x_i respectively. First, express the displacement x of the mass in terms of x_1, x_2, and x_3 using geometric constraints. Then, find the equivalent spring stiffness k_{eq} of the network.

(b) Now an additional linear spring with stiffness k_4 and a pulley with inner radius r and outer radius R are added to the network as shown in Figure 7.14(b). After expressing x in terms of x_1, x_2, x_3, and x_4, find the equivalent spring stiffness k_{eq} of the network.

Exercise 7.22   Consider the two-degrees-of-freedom mass-spring-damper system shown in Figure 7.15, which consists of masses m_1 and m_2, linear springs with stiffness k, and dampers with damping coefficient b. The displacements of the masses are x_1 and x_2 respectively.

(a) Choosing x_1 and x_2 as the generalized coordinates, derive the equations of motion of the system using the Lagrangian formulation with Rayleigh dissipation function.

(b) Convert the equations of motion into a first-order linear differential equation of the form $\dot{x} = Ax$, where $x = (x_1, \dot{x}_1, x_2, \dot{x}_2)^\top$.

(c) Suppose $k = 2$, $b = 5.037$, $m_1 = 1$, $m_2 = 2$. Mass m_1 and mass m_2 are released from $x_1 = -2$ and $x_2 = 1$ respectively. Draw the trajectories of $x_1(t)$, $\dot{x}_1(t)$, $x_2(t)$, and $\dot{x}_2(t)$ using software and explain the motions of the masses.

Appendix Linear Algebra and Matrix Theory Basics

This appendix reviews some basic results from linear algebra and matrix theory. We first begin with some results on the existence and uniqueness of solutions to the linear equation $Ax = b$, where $A \in \mathbb{R}^{m \times n}$ and $b \in \mathbb{R}^m$ are assumed given and $x \in \mathbb{R}^n$ is to be determined. For square matrices, some formulas for the determinant and inverse of block matrices are given, and basic properties of eigenvalues and eigenvectors of matrices are reviewed. Finally, the Fundamental Theorem of Linear Algebra is used to prove the existence of Lagrange multipliers in the first-order necessary conditions for equality-constrained minimization.

Basic Definitions

1. A set of vectors $\{v_1, \ldots, v_k\}$, $v_i \in \mathbb{R}^n$, $i = 1, \ldots, k$, is said to be **linearly independent** if the equation

$$c_1 v_1 + c_2 v_2 + \cdots + c_k v_k = 0,$$

   where the c_i are scalars, has the unique solution $c_1 = \cdots = c_k = 0$. Otherwise, the set of vectors is linearly dependent. Said another way, if any of the v_i can be written as a linear combination of the remaining vectors, then the set is linearly dependent.

2. The **span** of a set of vectors $\{v_1, \ldots, v_k\}$, each $v_i \in \mathbb{R}^n$, is the set of all possible linear combinations of the v_i:

$$\mathrm{span}\{v_1, \ldots, v_k\} = \left\{ v \in \mathbb{R}^n \mid v = \sum_{i=1}^{k} c_i v_i, \ \text{each } c_i \in \mathbb{R} \right\}.$$

3. A **basis** for $\mathbb{R}^n$ is any linearly independent set of n vectors $\{v_1, \ldots, v_n\}$, each $v_i \in \mathbb{R}^n$. Given a basis, for any $v \in \mathbb{R}^n$ there exists a unique set of scalar constants $c_i, i = 1, \ldots, n$, such that $v = c_1 v_1 + \cdots + c_n v_n$.

4. $V \subseteq \mathbb{R}^n$ is a **subspace** of $\mathbb{R}^n$ if, for any $v_1, v_2 \in V$, we have $c_1 v_1 + c_2 v_2 \in V$ for all possible scalars $c_1, c_2 \in \mathbb{R}$. Any subspace of $\mathbb{R}^n$ itself forms a vector space.

5. Given a matrix $A \in \mathbb{R}^{n \times m}$, its **column space**, also denoted $\mathcal{R}(A)$ and called the **range** of A, is the subspace spanned by the columns of A.

6. Given a matrix $A \in \mathbb{R}^{n \times m}$, its **row space** is the subspace spanned by the rows of A, or equivalently, $\mathcal{R}(A^\top)$.

7. Given a matrix $A \in \mathbb{R}^{n \times m}$, its **rank** is the number of linearly independent columns of A, or equivalently the dimension of the subspace $\mathcal{R}(A)$.

8. Given a matrix $A \in \mathbb{R}^{n \times m}$, its **null space** $\mathcal{N}(A)$, also referred to as the **kernel** of A, is the set of all vectors $x \in \mathbb{R}^m$ that satisfy $Ax = 0$:

$$\mathcal{N}(A) = \{x \in \mathbb{R}^m \mid Ax = 0\}.$$

9. Two vectors $u, v \in \mathbb{R}^n$ are said to be **orthogonal** if $u^\top v = v^\top u = 0$. The notation $u \perp v$ is also used to denote that the two vectors are orthogonal.

10. Given a subspace $V \subseteq \mathbb{R}^n$, its **orthogonal subspace**, denoted $V^\perp$, is given by

$$V^\perp = \{z \in \mathbb{R}^n \mid z^\top x = 0 \ \forall x \in V\}.$$

Linear Equation $Ax = b$

Given $A \in \mathbb{R}^{n \times m}$ and $b \in \mathbb{R}^n$, the equation $Ax = b$ consists of n equations in the m unknowns represented by $x = (x_1, \ldots, x_m)^\top$. Because the equations are linear in the elements x_i, $Ax = b$ is said to be a **linear equation** in x.

1. Given $A \in \mathbb{R}^{n \times m}$ and $b \in \mathbb{R}^n$, $Ax = b$ has a solution $x \in \mathbb{R}^m$ if and only if $b \in \mathcal{R}(A)$ (that is, b is in the column space of A). If the columns of A span $\mathbb{R}^n$ ($\mathcal{R}(A) = \mathbb{R}^n$), then $Ax = b$ has a solution for all $b \in \mathbb{R}^n$.

2. If $n > m$, then there are more equations than unknowns, and $Ax = b$ is said to be an *overdetermined system*. For an overdetermined system, typically (but not always) there may not exist a solution.

3. If $n < m$, then there are more unknowns than equations, and $Ax = b$ is said to be an *undetermined system*. If $\mathrm{rank}(A) = n$, then $Ax = b$ has an $(m - n)$-parameter family of solutions for every $b \in \mathbb{R}^n$. For an undetermined system, typically (but not always) there will be an infinity of solutions x.

Matrices: Basic Properties and Formulas

1. A square matrix $A \in \mathbb{R}^{n \times n}$ has an inverse A^{-1} if and only if it is of full rank n. The inverse is unique and satisfies $AA^{-1} = A^{-1}A = I$. A square invertible matrix is said to be **nonsingular**. If $A, B \in \mathbb{R}^{n \times n}$ are both of full rank n, then

$$(AB)^{-1} = B^{-1}A^{-1}.$$

2. The **parallelepiped** $\mathcal{P}$ determined by a set of n vectors $\{v_1, \ldots, v_n\}$, $v_i \in \mathbb{R}^n$ is the set

$$\mathcal{P} = \{c_1 v_1 + \cdots + c_n v_n \mid 0 \le c_1, \ldots, c_n \le 1\}.$$

3. The **determinant** of $A \in \mathbb{R}^{n \times n}$ is the signed volume of the parallelepiped formed by the columns of A:

$$\text{volume}(\mathcal{P}) = |\det A|.$$

A has nonzero determinant if and only if $\mathrm{rank}(A) = n$. If $A \in \mathbb{R}^{3\times3}$, then an explicit formula for $\det A$ is given by

$$\det A = v_1^\top (v_2 \times v_3) = v_3^\top (v_1 \times v_2) = v_2^\top (v_3 \times v_1),$$

where $v_1, v_2, v_3 \in \mathbb{R}^3$ are the columns of A.

4. Given $A, B \in \mathbb{R}^{n\times n}$, $\det(AB) = \det A \cdot \det B$.

5. If A and B are both square but not necessarily of the same size,

$$\det \begin{bmatrix} A & 0 \\ C & B \end{bmatrix} = \det A \cdot \det B.$$

If additionally A is nonsingular,

$$\det \begin{bmatrix} A & D \\ C & B \end{bmatrix} = \det A \cdot \det(B - CA^{-1}D).$$

6. If $A \in \mathbb{R}^{n\times n}$ and $C \in \mathbb{R}^{m\times m}$ are both nonsingular,

$$(A + BCD)^{-1} = A^{-1} - A^{-1}B\left(DA^{-1}B + C^{-1}\right)^{-1} DA^{-1}.$$

7. If $A \in \mathbb{R}^{n\times n}$ and $B \in \mathbb{R}^{m\times m}$ are both nonsingular,

$$\begin{bmatrix} A & 0 \\ C & B \end{bmatrix}^{-1} = \begin{bmatrix} A^{-1} & 0 \\ -B^{-1}CA^{-1} & B^{-1} \end{bmatrix},$$

$$\begin{bmatrix} A & D \\ 0 & B \end{bmatrix}^{-1} = \begin{bmatrix} A^{-1} & -A^{-1}DB^{-1} \\ 0 & B^{-1} \end{bmatrix}.$$

8. If $A \in \mathbb{R}^{n\times n}$ is nonsingular,

$$\begin{bmatrix} A & D \\ C & B \end{bmatrix}^{-1} = \begin{bmatrix} A^{-1} + E\Gamma^{-1}F & -E\Gamma^{-1} \\ -\Gamma^{-1}F & \Gamma^{-1} \end{bmatrix},$$

where $\Gamma = B - CA^{-1}D$, $E = A^{-1}D$, and $F = CA^{-1}$. If additionally B^{-1} exists, then the $(1,1)$ block becomes $A^{-1} + E\Gamma^{-1}F = (A - DB^{-1}C)^{-1}$.

9. For any $A \in \mathbb{R}^{n\times n}$, its **characteristic polynomial** $p(s)$ is

$$p(s) = \det(Is - A) = s^n + a_{n-1}s^{n-1} + \cdots + a_1 s + a_0.$$

The n roots of $p(s)$, denoted $\lambda_1, \ldots, \lambda_n$, are the **eigenvalues** of A. In terms of the eigenvalues, $p(s)$ can be written

$$p(s) = (s - \lambda_1)(s - \lambda_2) \cdots (s - \lambda_n).$$

10. The eigenvalues of a symmetric matrix are always real.

11. The **trace** $\mathrm{tr}(A)$ of a square matrix A is the sum of its diagonal elements. For any matrices A, B, C for which the product ABC makes sense,

$$\mathrm{tr}(ABC) = \mathrm{tr}(CAB) = \mathrm{tr}(BCA).$$

The trace and determinant of a matrix are related to its eigenvalues via the formulas

$$\operatorname{tr}(A) = \lambda_1 + \lambda_2 + \cdots + \lambda_n,$$
$$\det A = \lambda_1 \lambda_2 \cdots \lambda_n.$$

12. The **Cayley–Hamilton Theorem** states that given a matrix $A \in \mathbb{R}^{n \times n}$ with characteristic polynomial $p(s) = s^n + a_{n-1} s^{n-1} + \cdots + a_1 s + a_0$, then

$$A^n + a_{n-1} A^{n-1} + \cdots + a_1 A + a_0 I = 0.$$

13. Given $A \in \mathbb{R}^{n \times n}$ with eigenvalues $\lambda_1, \ldots, \lambda_n$, any vector $v \in \mathbb{R}^n$ for which $Av = \lambda v$ is said to be an **eigenvector** of A corresponding to the eigenvalue λ. Like eigenvalues, eigenvectors can also be complex depending on the matrix. If v is an eigenvector, then so is αv for any (complex) scalar α. The eigenvectors of a matrix can always be scaled to have unit length.

14. If the eigenvalues of a real symmetric matrix A are all distinct, then the eigenvectors are all orthogonal to each other.

Fundamental Theorem of Linear Algebra

As a motivational example, first consider the following $A \in \mathbb{R}^{3 \times 2}$:

$$A = \begin{bmatrix} 1 & 0 \\ 5 & 4 \\ 2 & 4 \end{bmatrix}.$$

Denote the columns of A by $v_1, v_2 \in \mathbb{R}^3$, and $x = (x_1, x_2)^\top$. Given $b \in \mathbb{R}^3$, the equation $Ax = b$ can be written in terms of the columns of A as

$$Ax = x_1 v_1 + x_2 v_2 = b.$$

That is, b is a linear combination of the vectors v_1 and v_2; a solution exists only if b lies in the plane spanned by v_1 and v_2 (see Figure A.1).

As a second example, consider A of the form

$$A = \begin{bmatrix} 1 & 5 & 2 \\ 0 & 4 & 4 \end{bmatrix}.$$

Again denoting the columns of this A by $v_1, v_2, v_3 \in \mathbb{R}^2$, these three vectors together span $\mathbb{R}^2$. A solution x to $Ax = b$ therefore exists for all $b \in \mathbb{R}^2$. As long as the $\{v_1, v_2, v_3\}$ are not parallel to each other, a solution (in fact, an infinite number of solutions) exists.

With the above as background we now establish some basic results about subspaces. First, it can be seen that if V is a subspace, then so is its orthogonal subspace $V^\perp$, and that $(V^\perp)^\perp = V$. It follows that any vector $x \in \mathbb{R}^n$ can be uniquely decomposed as $x = v + w$, where $v \in V$ and $w \in V^\perp$. We can conclude from this argument that $\dim(V) + \dim(V^\perp) = n$.

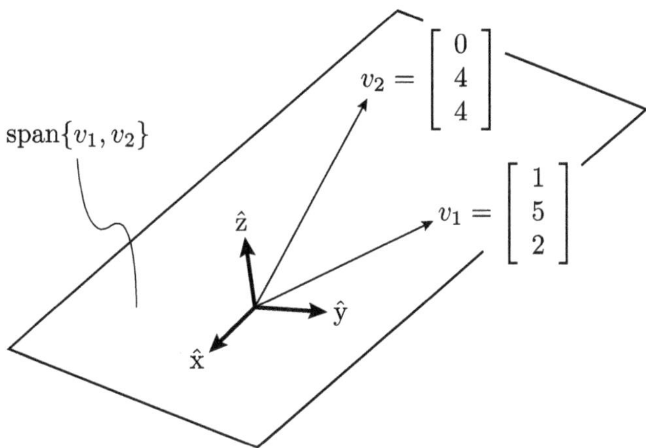

Figure A.1 Geometric depiction of $Ax = b$ for given example.

We next consider the four basic subspaces associated with a matrix $A \in \mathbb{R}^{n \times m}$. Observe that

$$\|Ax\|^2 = x^\top A^\top A x \geq 0,$$

since by definition the norm of a vector can never be negative. Now,

- If $Ax = 0$, this implies that $A^\top A x = 0$, so that $\mathcal{N}(A) \subseteq \mathcal{N}(A^\top A)$.
- If we instead assume $A^\top A x = 0$, this implies that $x^\top A^\top A x = \|Ax\|^2 = 0$, or $Ax = 0$. Thus, we have established that $\mathcal{N}(A^\top A) \subseteq \mathcal{N}(A)$.

What the above two observations establish is that $\mathcal{N}(A) = \mathcal{N}(A^\top A)$.

We next consider some $x \in \mathcal{N}(A)$, i.e., $Ax = 0$. For any $y \in \mathcal{R}(A^\top)$, we have by definition $y = A^\top c$ for some $c \in \mathbb{R}^n$. Then

$$x^\top y = x^\top A^\top c = (Ax)^\top c = 0.$$

What the above implies is that $x \perp \mathcal{R}(A^\top)$ for general $x \in \mathcal{N}(A)$. As a statement about subspaces, we have shown that $\mathcal{N}(A) \subseteq \mathcal{R}(A^\top)^\perp$.

Conversely, for any $z \in \mathcal{R}(A^\top)^\perp$, we have

$$z^\top y = z^\top A^\top c = (Az)^\top c = 0$$

must be true for all c (and therefore true for the particular case of $c = Az$). Using this latter fact,

$$(Az)^\top (Az) = 0 \rightarrow Az = 0 \rightarrow z \in \mathcal{N}(A) \rightarrow \mathcal{R}(A^\top)^\perp \subset \mathcal{N}(A).$$

What the above results establish is that $\mathcal{N}(A) = \mathcal{R}(A^\top)^\perp$. This result is often referred to as the **Fundamental Theorem of Linear Algebra**, and can be summarized by the four following statements:

$$N(A) = \mathcal{R}(A^\top)^\perp,$$
$$N(A)^\perp = \mathcal{R}(A^\top),$$
$$N(A^\top) = \mathcal{R}(A)^\perp,$$
$$N(A^\top)^\perp = \mathcal{R}(A).$$

Finally, since $N(A) = N(A^\top A)$, it follows that $N(A)^\perp = N(A^\top A)^\perp$, or $\mathcal{R}(A^\top) = \mathcal{R}(A^\top A)$.

Example A.1   Given the matrix

$$A = \begin{bmatrix} 1 & 2 & 0 & 1 \\ 0 & 1 & 1 & 0 \\ 1 & 2 & 0 & 1 \end{bmatrix},$$

both its column space and row space are of dimension two. $N(A)$ can be pictured as a two-dimensional plane in $\mathbb{R}^4$, while $N(A^\top)$ can be pictured as a (one-dimensional) line in $\mathbb{R}^3$.   □

Lagrange Multipliers

We now justify the use of Lagrange multipliers appearing in the first-order necessary conditions for minimizing a function $f(x)$, $f: \mathbb{R}^n \to \mathbb{R}$. Before addressing the general formulation, we consider a few motivational examples (see Figure A.2).

Consider $f(x) = x_1^2 + x_2^2$ subject to the equality constraint $x_1 + x_2 = 1$. The constraint defines a line in $\mathbb{R}^2$, while the objective function measures the (squared) distance of x from the origin. The solution to this problem has a nice geometric characterization: it is the point on the line closest to the origin, and can be viewed as the orthogonal projection (more on this later) of the origin onto the line.

As a second example, consider $f(x) = x_1^2 + x_2^2 + x_3^2$ subject to $x_1^2 + 2x_2^2 + 3x_3^2 = 1$. The equality constraint defines a three-dimensional ellipsoid in $\mathbb{R}^3$, while the objective function is, like before, the (squared) distance of the point x from the origin. Again, geometrically the solution can be characterized as the point(s) on the ellipsoid closest to the origin; it is not too difficult to see that there will be two solutions.

With the above as background, let us now consider the general equality constrained problem

$$\min_{x \in \mathbb{R}^n} f(x) \text{ subject to } g(x) = 0,$$

where both $f: \mathbb{R}^n \to \mathbb{R}$ and $g: \mathbb{R}^n \to \mathbb{R}^m$, $n \geq m$, are differentiable. We shall call the set of all points x satisfying $g(x) = 0$ the **constraint surface**, and denote it by $\mathcal{S}$. Suppose x^* is a local minimizer, and for reasons to be made clear later, that

$$\text{rank}\left(\frac{\partial g}{\partial x}(x^*)\right) = m.$$

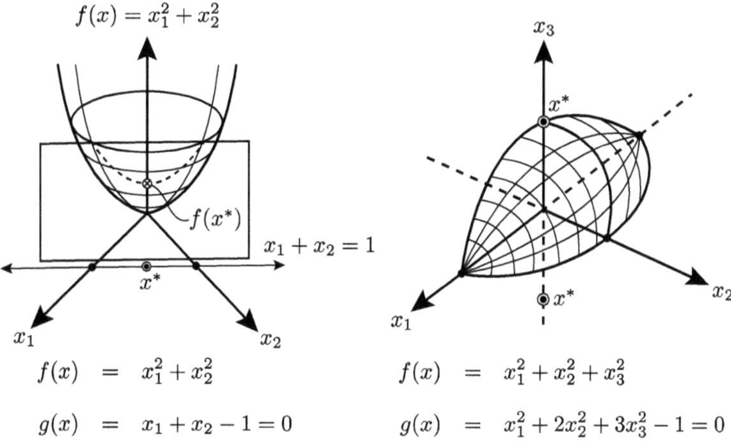

Figure A.2 Examples of equality constrained problems.

Let $x(t)$ be a differentiable curve on $\mathcal{S}$ such that $x(0) = x^*$. If we look at $f(x(t)) = \phi(t)$ as a function of t, and draw the graph of $\phi(t)$ with respect to t, then it is clear that $\phi(t)$ has a local minimizer at $t = 0$, and that $\dot{\phi}(0) = 0$:

$$\frac{d\phi}{dt}(0) = \frac{\partial f}{\partial x}(x^*)\dot{x}(0) = 0$$

(see Figure A.3). Also, since $g(x(t)) = 0$ for all t, it follows that by differentiating both sides we have

$$\frac{\partial g}{\partial x}(x)\dot{x}(t) = 0$$

for all t. The above must hold true for $t = 0$, from which it follows that

$$\frac{\partial g}{\partial x}(x^*)\dot{x}(0) = 0.$$

The above implies that $\dot{x}(0)$ belongs to the nullspace of $\frac{\partial g}{\partial x}(x^*)$. At the same time, the earlier equality $\frac{\partial f}{\partial x}(x^*)\dot{x}(0) = 0$ implies that $\frac{\partial f}{\partial x}(x^*)^\top = \nabla f(x^*)$ is orthogonal to $\dot{x}(0)$. Remembering that $x(t)$ is any arbitrary differentiable curve on $\mathcal{S}$ that satisfies $x(0) = x^*$, it follows that $\dot{x}(0)$ is any arbitrary vector belonging to the nullspace of $\frac{\partial g}{\partial x}(x^*)$. It can therefore be concluded that

$$\nabla f(x^*) \perp \mathcal{N}\left(\frac{\partial g}{\partial x}(x^*)\right).$$

At this point we invoke the Fundamental Theorem of Linear Algebra, specifically,

$$\mathcal{N}\left(\frac{\partial g}{\partial x}(x^*)\right)^\perp = \mathcal{R}\left(\frac{\partial g}{\partial x}(x^*)^\top\right).$$

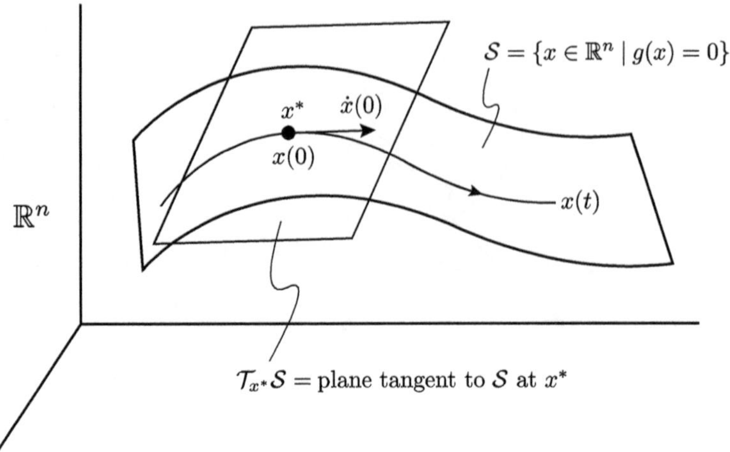

Figure A.3 General setting for the equality constrained problem.

We have shown that

$$\nabla f(x^*) \in \mathcal{R}\left(\frac{\partial g}{\partial x}(x^*)^\top\right),$$

or equivalently, that there exists some $c \in \mathbb{R}^m$ such that

$$\nabla f(x^*) = \frac{\partial f}{\partial x}(x^*)^\top = \frac{\partial g}{\partial x}(x^*)^\top c.$$

For convenience let us introduce a new variable $\lambda \in \mathbb{R}^m$ such that $\lambda = -c$; the above can then be rewritten as

$$\frac{\partial f}{\partial x}(x^*) + \lambda^\top \frac{\partial g}{\partial x}(x^*) = 0.$$

This is the main result; namely, that if x^* is a local minimizer of the equality constrained problem, then there exists some $\lambda \in \mathbb{R}^m$ such that the above equality holds. The first-order necessary conditions for a local minimizer are therefore

$$\frac{\partial f}{\partial x}(x^*) + \lambda^\top \frac{\partial g}{\partial x}(x^*) = 0,$$
$$f(x^*) = 0.$$

λ is called the **Lagrange multiplier**.

Another way to express the above is to define the function

$$H(x, \lambda) = f(x) + \lambda^\top g(x).$$

The first-order necessary conditions then can be equivalently expressed as

$$\frac{\partial H}{\partial x} = 0,$$
$$\frac{\partial H}{\partial \lambda} = 0.$$

We now provide an explanation of why the maximal rank assumption

$$\operatorname{rank}\left(\frac{\partial g}{\partial x}(x^*)\right) = m$$

is needed. The general relationship $\frac{\partial g}{\partial x}(x^*)\dot{x}(0) = 0$ implies that the m rows of $\frac{\partial g}{\partial x}(x^*)$ are orthogonal to $\dot{x}(0)$, where $\dot{x}(0)$ is any arbitrary vector tangent to the constraint surface S at the point x^*. The rank condition therefore implies that the m rows are all linearly independent; in fact, they form a basis for the m-dimensional subspace of $\mathbb{R}^n$ that is orthogonal to $N\left(\frac{\partial g}{\partial x}(x^*)\right)$.

Now, if $\frac{\partial g}{\partial x}(x^*)$ were not of maximal rank m, this would then imply that the normal subspace was not dimension m. This represents a pathological case that arises, for example, in surface self-intersections; the tangent space to the surface no longer becomes well defined at such points.

Index

For EU product safety concerns, contact us at Calle de José Abascal, 56–1°,
28003 Madrid, Spain or eugpsr@cambridge.org.

www.ingramcontent.com/pod-product-compliance
Ingram Content Group UK Ltd.
Pitfield, Milton Keynes, MK11 3LW, UK
UKHW050005161225
465991UK00004BA/25